KV-638-860

Alien Worlds

STEVE
NICHOLLS

Alien Worlds

HOW INSECTS CONQUERED THE EARTH

and why their fate will determine our future

HEAD of ZEUS

First published in the UK in 2023 by Head of Zeus,
part of Bloomsbury Publishing Plc

1 3 5 7 9 10 8 6 4 2

A CIP catalogue record for this book is available from
the British Library.

ISBN [HB] 9781838934767
ISBN [E] 9781838934781

Designed by Isambard Thomas at corvo-uk.com
Colour separation by Dawkins Colour
Printed in Slovenia by DZS Grafik

Head of Zeus Ltd
5–8 Hardwick Street
London EC1R 4RG
www.headofzeus.com

PREVIOUS PAGES
Stag beetle, *Cyclomattus metallifer*

OVERLEAF
African mantis, *Sphodromantis lineola*

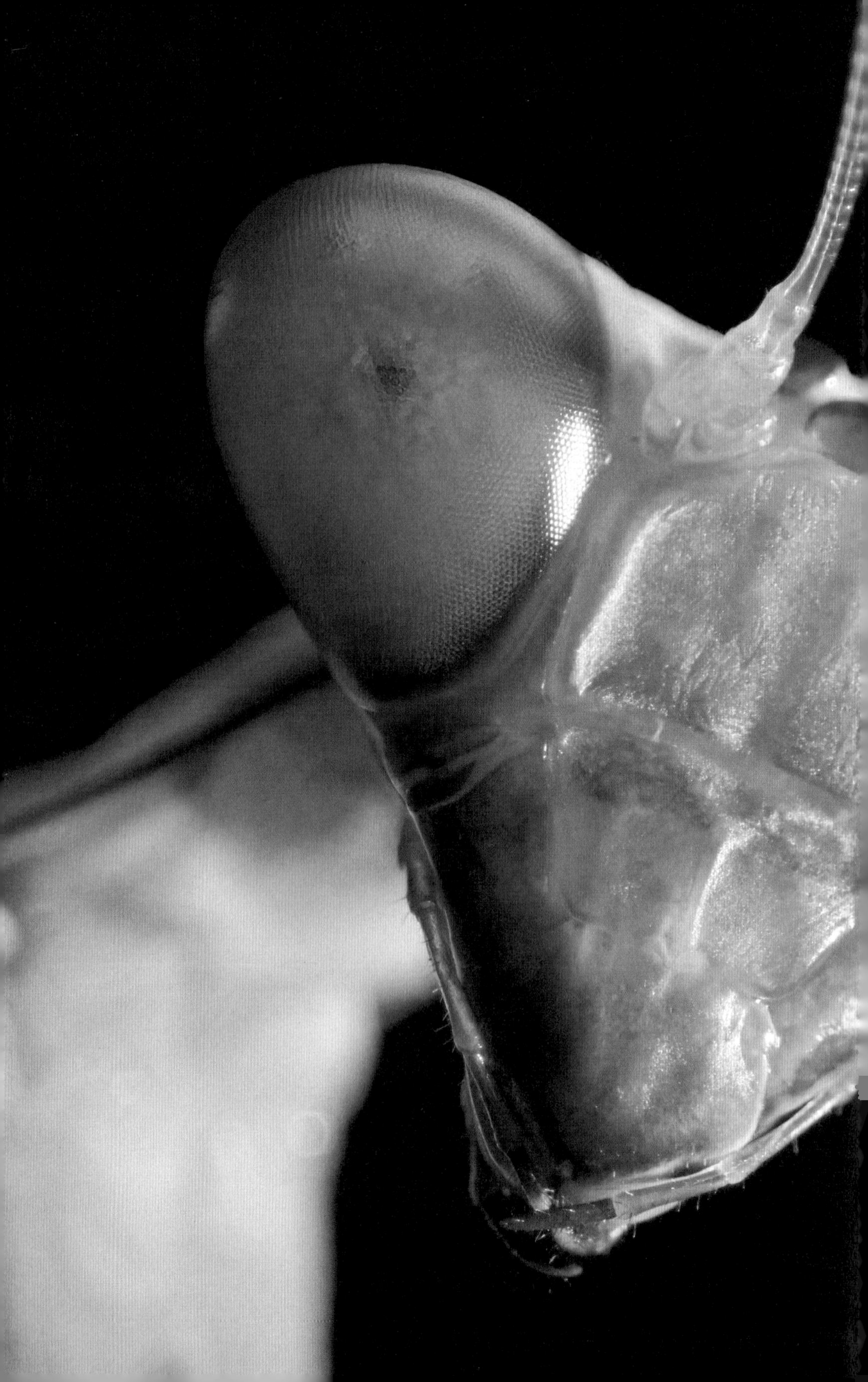

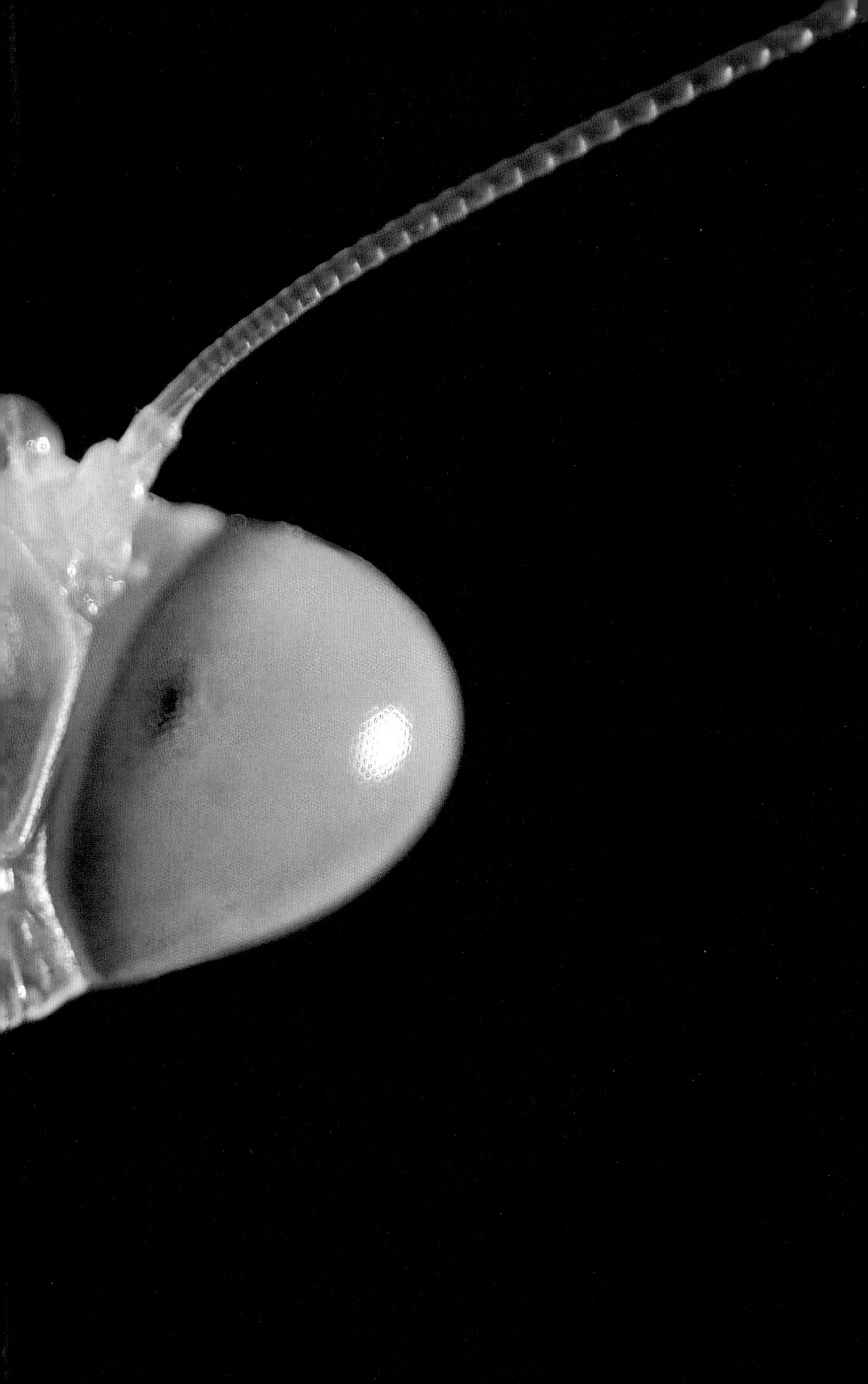

Orders of Insects

Archaeognatha
jumping bristletails

Zygentoma
silverfish and firebrats

Ephemeroptera
mayflies

Odonata
dragonflies and damselflies

Orthoptera
grasshoppers, crickets and katydids

Phasmida
stick insects

Embioptera
web-spinners

Notoptera
ice crawlers and heelwalkers

Plecoptera
stoneflies

Dermaptera
earwigs

Zoraptera
angel insects

Mantodea
mantids

Blattodea
cockroaches and termites

Psocodea
booklice, bark lice and parasitic lice

Thysanoptera
thrips

Hemiptera
true bugs

Hymenoptera
ants, bees and wasps

Strepsiptera
stylops

Coleoptera
beetles

Neuroptera
lacewings

Megaloptera
alderflies, dobsonflies and fishflies

Raphidioptera
snakeflies

Trichoptera
caddisflies

Lepidoptera
butterflies and moths

Diptera
true flies

Siphonaptera
fleas

Mecoptera
scorpionflies

Family Tree of Arthropods and Insects

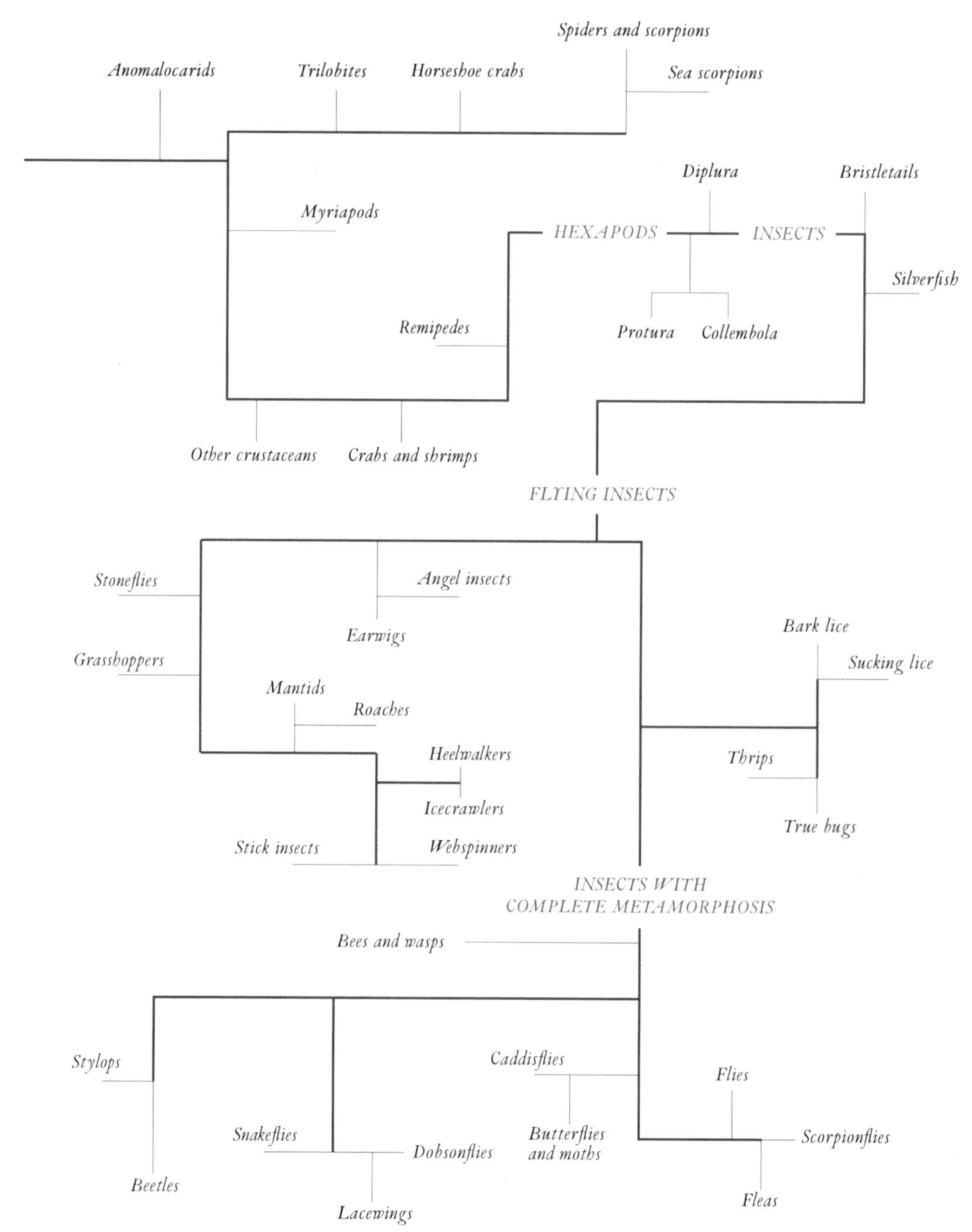

Geological Periods and the Origin of Different Insects

Period	*Millions of years ago*	*Insect events*
Ediacaran	635	*First Arthropods*
Cambrian	538	
Ordovician	485	*First insects*
Silurian	443	
Devonian	419	*First winged insects?* *Hemiptera: True bugs* *Hexapod Gap*
Carboniferous	358	*Hexapod Gap* *Endopterygotes*
Permian	299	*Hymenoptera: Sawflies, wasps, bees and ants* *Coleoptera: Beetles*
Triassic	252	*Orthoptera: Grasshoppers and crickets* *Odonata: Dragonflies*
Jurassic	201	*Lepidoptera: Butterflies and moths* *Diptera: Flies* *Blattodea: Cockroaches and termites*
Cretaceous	145	
Paleogene	66	
Neogene	23	
Quaternary	2.6	

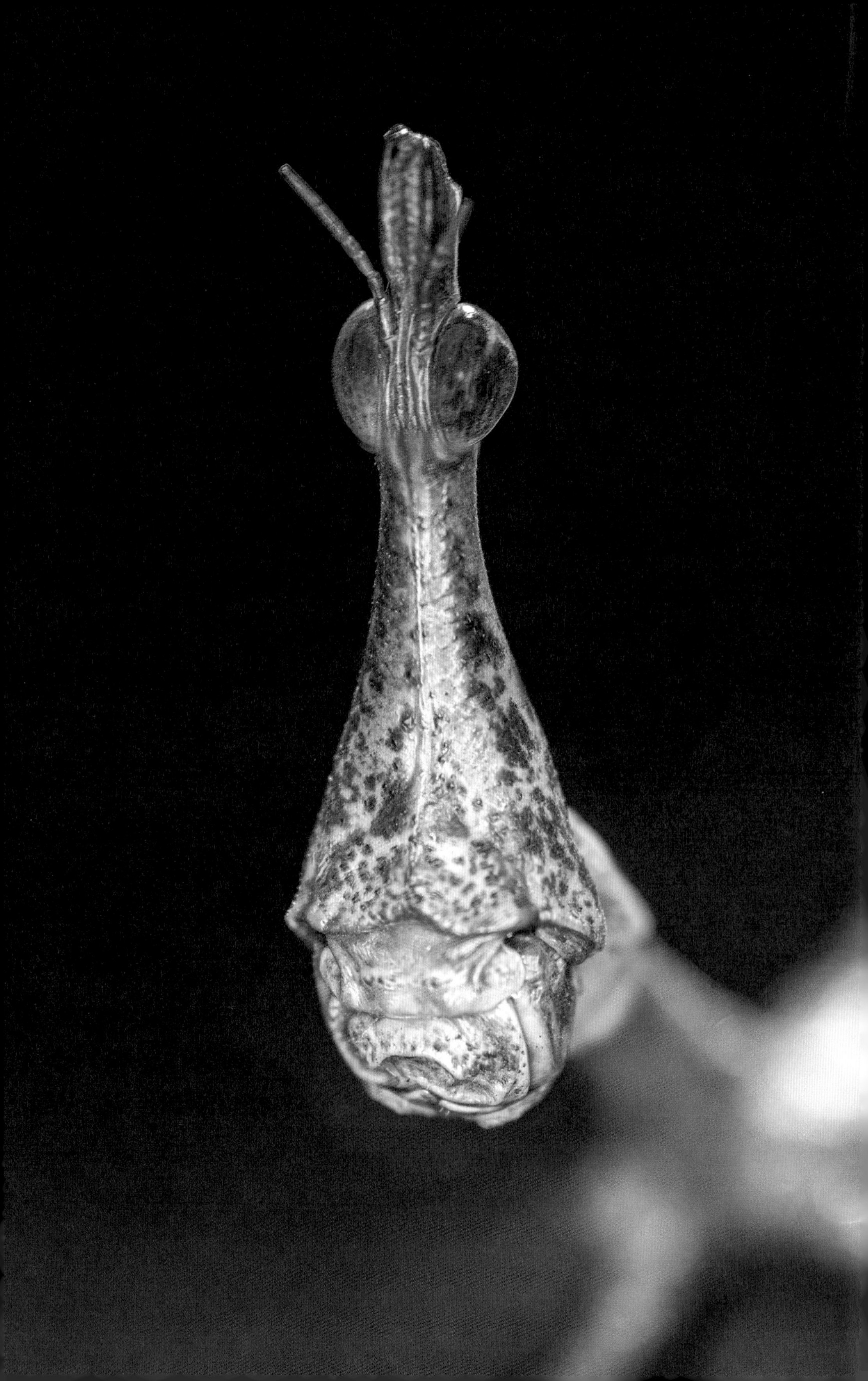

Introduction
Insects 101

I am dying by inches from not having anybody
to talk to about insects...
Charles Darwin, 1828[1]

Every kid has a bug period.
I never grew out of mine.
E. O. Wilson, 2013[2]

Judged by numbers of species (around a million so far described by science – perhaps another 5 million out there… somewhere) or by numbers of individuals (one estimate suggests 10 quintillion of them), insects are the most successful group of animals ever to have lived. One in every four animals on the planet is a beetle, one in every ten is a butterfly or moth. A couple of years ago, thirty entirely new species of flies were discovered just within the city of Los Angeles.[3]

I've been photographing, studying or reading about insects for close on five decades and yet every time I pick up a new research paper or book, or even turn over a log in my local woodland, I come across something new and engrossing to follow up. In my work as a wildlife filmmaker, I've also been privileged to travel the planet and meet with scientists working on many aspects of insect life who have been unfailingly gracious in sharing their work and insights. And, even after all this time and all this global exploration, I still come across whole areas of insect natural history and biology that I had no idea existed. They provide, quite literally, endless fascination – a fascination I hope to share with you over the following pages.

If a book that takes a journey through a world of a million species and perhaps 10 quintillion individuals can be said to have a unifying theme, then it's a search for the reasons behind this phenomenal success of insects. Their recipe for the conquest of the planet (or at least of dry land and fresh water) has many ingredients, some based on the adaptability of the insects' unique ground plan, others on their impressively diverse life cycles and behavioural repertoires.[4] Each chapter is centred on one or more of these key traits of insect life as we explore the reasons behind the triumph of insect-kind – and how they've held dominion over terrestrial and freshwater environments for so long.

However, before we plunge headlong into one of the greatest stories in the natural world, let's get a few essentials tied down with my user-friendly guide to some of the background information that will make our journey easier. In the following pages I'll explain some frequently used terms and answer some frequently asked questions. I'll also outline how the rest of the book is laid out.

Our story unfolds across the whole planet but also through deep time. Insects are an ancient group of animals, and to fully appreciate their achievements we need to understand their long history as well as their current occupation of Planet Earth. The spans of geological time that encompass insect evolution are beyond human imagining, but they can at least be represented graphically, which provides a framework for our

(previous page) The horsehead grasshopper, *Pseudoproscopia scabra*, is just one of a vast range of strange and bizarre insects found in tropical South America.

(right) A keeled skimmer, *Orthetrum coerulescens*. Insects that resembled modern dragonflies flew in the swamp forests of the Carboniferous period, 300 million years ago.

chronological story. In the slice of Earth's history that interests us there are twelve geological periods. The most widely known, thanks to Steven Spielberg, is the Jurassic, but I've included a geological timescale, listing the other periods, beginning 635 million years ago, for quick reference (see p. xi).

Historically, this timescale was assembled piecemeal and, in its early incarnations, it was often based on work in Britain, with what looks like a distinct bias towards the south-west. *Cambrian* refers to Wales and *Devonian* to Devon, while *Silurian* and *Ordovician* are both named after Welsh Celtic tribes. Others are named for the types of rock that predominate (*Cretaceous* – chalk, or *Carboniferous* – coal), or for other regions of the world. The *Permian* is named for the region of Perm in Russia and *Jurassic* for the Jura Mountains on the Swiss–French border. All we need to know, though, is their chronological order and their age, since this is the canvas on which we'll paint the evolutionary history of insects.

Let's begin with an introduction to entomology – not to be confused (as it frequently is) with etymology, the study of words. My own introduction to entomology came in 1973 when I arrived at the University of Bristol as an undergraduate to study zoology. I brought with me a passionate interest in birds and reptiles that dated as far back into my childhood as I can remember, but that soon changed.

At the time, the Zoology Department was headed up by Professor Howard Everest Hinton, an eminent and, at times, controversial entomologist. We students often referred to him by his initials, HEH, and the similarity to HRH – His Royal Highness – seemed to us eminently appropriate. He was often in lively, if not acrimonious, debate with fellow entomologists over a whole range of fundamental concepts in entomology. His lectures were part invective against those who disagreed with his views and part mind-boggling introduction to an alien world. He had an infectious curiosity about all aspects of the insect world. No detail was too small to be ignored, since such minutiae often led to far deeper revelations. Many entomologists might focus on narrow areas of their subject or study insects amenable to life in a laboratory, perhaps a

A membracid bug, *Gigantorhobdus enderleine*, from Malaysia.

sensible approach to such a vast group of animals, but Hinton ranged with enthusiastic abandon across most of insect-kind and questioned everything he saw. He was as much at home in the field as in the lab, and had a wonderful collection of the strange and the bizarre that he would happily show off to anyone who exhibited anything more than bored apathy.

His lectures opened my eyes to an extraordinary world, so I was delighted when he offered me a grant to study for my doctorate in entomology. But, we should begin our journey into the world of insects where Howard Hinton did in the first of his lectures that I attended. We need to ask, *what is an insect?* A great many people are confused about this, including some naturalists. Even scientists have, on occasions, also changed their minds as to which creatures are insects and which aren't – so some degree of confusion is perfectly understandable. However, we perpetuate that confusion daily because of the way these little creatures are categorized in the popular mind.

Many people attach the term 'bug' to any 'creepy-crawly', from spiders and scorpions, to centipedes and millipedes, as well as to all the beetles, flies, butterflies, crickets and similar creatures. Since 'bug' is frequently used interchangeably with 'insect', this widespread view lumps all these creatures together as 'insects', which spiders, scorpions, millipedes, centipedes and the like are decidedly not. However, those who commit this entomological heresy find themselves in good company. Even Aristotle made this same mistake. Along with true insects he included spiders, centipedes and millipedes in his class *Entoma*, which – as we will see shortly – downplays a rich and intriguing history, although Aristotle did at least give us a name for the study of true insects.

Incidentally, 'bug' has a very specific meaning for an entomologist – the term refers to insects belonging to the order Hemiptera (for example, shield bugs or flower bugs). I know entomologists who come close to responding with violence against those who use the word 'bug' to refer to anything that's not a hemipteran. I would never have dared utter such a profanity in Howard Hinton's presence. So, what exactly is an insect, and how do you avoid being decked by an angry entomologist? It turns out to be surprisingly hard to give a simple answer. Yet, exploring this deceptively straightforward question is the very beginning of our journey through the world of insects.

To understand what is (and isn't) an insect, it helps to look at the position insects occupy on the great tree of life. The branches on this tree all have names (of course they do – biologists love classifying and labelling things), so to define an insect in this way, we'll need to delve briefly into the system of hierarchical categories that scientists have invented to impose some kind of order on the exuberant diversity of life. Although life often disregards our attempts to force it into such neatly stacked boxes, this system is nevertheless a convenient way to get a handle on the bewildering variety of living things, and it is certainly useful in making sense of the vast diversity of insects.

First, insects are arthropods – that is, they belong to the *phylum Arthropoda*. Those other 'bugs' – spiders, centipedes and the others – are also arthropods; so, although not insects, they are all related, if only distantly. Unfortunately, there's no scientifically correct way to refer to all these creatures as a whole, so I've no doubt that 'bugs' will retain some currency for the moment as a handy collective term for all these terrestrial arthropods, even among some entomologists. For example, the UK's leading insect conservation charity is called *Buglife*, even if that does raise the blood pressure of more traditional entomologists.

Phyla (singular phylum) are the really broad categories of animals. The term was coined in 1866 by the German zoologist Ernst Haeckel, who derived it from the Greek word *phylon*, meaning a race, tribe or clan. Other phyla include Annelida (various kinds of segmented worm), Cnidaria (anemones, corals and jellyfish), Echinodermata (starfish and sea urchins) and Chordata (all the vertebrates, including ourselves, along with a few obscure creatures that are almost vertebrates). Phyla are divided into further categories, each representing animals that are more closely related to each other, and each of these is then further divided into groups of even closer relatives. There are an awful lot of levels in this hierarchy, and much disagreement over who belongs where, but a few broad categories will help in understanding what an insect is. Phyla are divided into classes, and classes into orders, which are themselves composed of families. Another way to visualize this is as the ever finer branching of the evolutionary tree, from large boughs to tiny twigs.

There are currently thirty-two animal phyla, of which the arthropods are by far the largest and, if judged on that criterion, the most successful.[5] The best way to think of these phyla is that each represents a fundamentally different body plan – for example, the repeated segments of annelids, the five-rayed symmetry displayed by echinoderms, or a body built around a dorsal 'spine' among the chordates. Among other features, arthropods all share a tough outer skeleton – an *exoskeleton* – made largely from a substance called *chitin* but often further reinforced with minerals to make a really tough shell. This is one reason why arthropods seem so

Spiders are not at all closely related to insects,
although frequently thought so in the minds of many people.

alien to our eyes. They are built the opposite way around compared to us. While we drape all our soft bits around the outside of the rigid internal framework of our skeletons, arthropods hang all their vital organs from the inside of a suit of armour. Like a suit of armour, the exoskeleton needs flexible joints to allow arthropods to bend and flex or to swim or walk.

Arthropod literally means 'jointed foot', because another feature common to all arthropods is the possession of such jointed limbs, for walking, swimming or even breathing. This innovation appeared right at the start of arthropod history, and played a significant role in the success of the whole group. A recent dramatic find highlights the flexibility of this fundamental feature of arthropods.

A terrestrial crayfish, *Engaeus* sp., from Tasmania. Crustaceans, like insects, are arthropods and share with them jointed limbs and a hard exoskeleton.

In 2015, an enormous (2-metre-long) and exceptionally well-preserved fossil arthropod was discovered in Morocco. In life, it swam in some ancient ocean around 480 million years ago, during the Ordovician period.[6] This giant was a filter feeder and a kind of anomalocarid, a group of animals that branched off from the very base of the arthropod family tree. Their name means *odd crab*, but that does these magnificent creatures a gross disservice. Many reached large sizes and while some were filter feeders, many more were predators, which has prompted some to call them *killer shrimps*.

It would have been a truly alien experience to have snorkelled in the oceans of the Cambrian and Ordovician periods. While there was little, if any, life on land, the oceans teemed with a bizarre assemblage of creatures, many of which were arthropods. Anomalocarids cruised these ancient seas, propelled by undulating flaps arranged in rows along their bodies. Eurypterids, often called sea scorpions, crawled over the seabed and also grew to enormous sizes. One of their kind, *Jaekelopteris*, reached two and a half metres in length, making it the largest arthropod ever to have lived – at least as far as we know. Trilobites, like giant woodlice, trundled over the ocean floor in great numbers and spectacular diversity. At the same time, our own most distant ancestor was a worm-like creature just a few centimetres long, called *Pikaia*, which snuffled through the ocean sediments, a quintessential bottom feeder.

At 2 metres in length, the Moroccan anomalocarid *Aegirocassis benmoulae* rivalled the largest eurypterids. It was so well preserved that scientists could examine its swimming flaps in unprecedented detail. Each body segment bore double flaps on each side, which appear to have had the same origins as the limbs of other arthropods. Like the swimming flaps, arthropod legs were originally double structures, each leg consisting of an upper prong and a lower prong. In some groups of arthropods, the upper leg branches became specialized as gills for breathing, while the lower ones were used for either swimming or walking. Insects now retain only the lower 'walking' part of that original arthropod double limb since they no longer need gills, although many crustaceans still sport the original double-pronged (*biramous*) limbs.

Arthropods evolved in the ocean, early in – or perhaps just before – the Cambrian period, 541 to 485 million years ago. Due to their tough exoskeletons and versatile legs, they soon rose to dominance. We'll explore more of the early arthropod story in Chapter 2. For now, the main point to take from this is that the earliest ancestors of insects must therefore also have been marine creatures, although no fossil of such an aquatic 'dawn insect' has ever been found, so we have no idea what these creatures might have actually looked like. However, as we'll discover in Chapter 2, we can now make some inspired guesses, fuelled by the unexpected discovery of some bizarre living fossils* as recently as 1979.

The defining arthropod features helped this phylum achieve its long-lasting success, but insects evolved their own unique take on the arthropod body plan, and this further assisted their march to world domination. More than 80 per cent of arthropods are insects, so they definitely got something right. The distinctive insect permutations on basic arthropod design also help in our quest to define what an insect is.

Insects differ from other arthropods in having three distinct body sections – a head (sensory in function, with eyes, mouthparts and one pair of antennae), a thorax (the engine room, with muscles to power legs and wings) and an abdomen (with all the other organs essential for life, including a *fat body* for food storage and the reproductive organs). Attached to the thorax are three pairs of legs and two pairs of wings, although both wings and legs may be further reduced in some insects.

But it's never quite that simple. There are also two groups of true insects that have never evolved wings. You might even come across one of these wingless insects – silverfish – skittering around at the back of a kitchen cupboard. Silverfish belong to the order Zygentoma, while the similar-looking jumping bristletails are placed in a different order, Archaeognatha. Of the two orders, silverfish are the more closely related to the rest of the flying insects, but both groups give us a glimpse of what insects were like before they evolved wings but after they colonized the land.

Although we don't know what the earliest aquatic ancestors of insects looked like, we do know that they must have invaded land at some point, perhaps early in the Ordovician period (485 to 444 million years ago). At that point they had to begin to adapt their marine arthropod heritage to cope with life above the waves. In pursuing this new life, their exoskeleton proved to be extremely handy. It gave them structural support in the much less dense medium of air. Their jointed legs, with a few modifications, were just as useful for walking on land as they were for walking on the seabed or swimming through the ocean. However,

* Let me say at the outset that terms like 'living fossil' and 'primitive', while useful literary shorthand, can be misleading. There are no 'living' fossils. Both they and creatures referred to as primitive exist in the present day and so have been evolving for as long as the rest of us. Some creatures have, however, retained features that are also present in ancient fossils. I'll try and make contexts clear when I inevitably fall into literary shorthand.

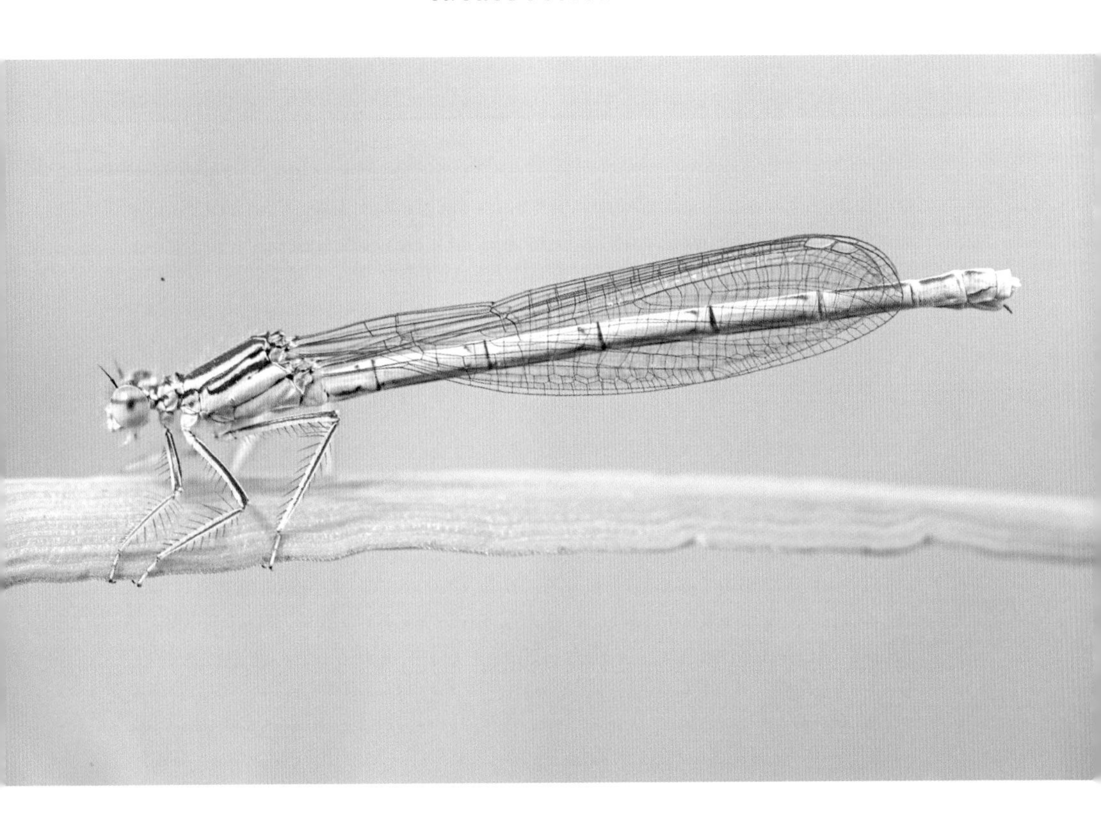

breathing in air presented more of a problem. Aquatic arthropods absorb oxygen dissolved in water through thin, permeable gills. Unfortunately, these structures don't work in air – they just dry out and shrivel up. Insects had to come up with a new solution for breathing on land.

They evolved a *tracheal system*. On each body segment there is a pair of small openings called *spiracles*. Over time, spiracles have been lost from some segments in certain kinds of insect, but the principle of the tracheal system remains the same. The spiracles open into large tubes called *tracheae* (singular: *trachea*), which branch into a network of ever finer tubes, finishing up as tiny *tracheoles*, small enough to penetrate between the insect's cells. Oxygen simply diffuses down the air-filled tubes and out of the thin-walled tracheoles, close to where it is needed to fuel cellular activity. In smaller insects, the passive process of diffusion through the network of tubes is quick and efficient enough to provide ample oxygen, but becomes much slower as insects get larger. So, when they exert themselves, many bigger insects must pump their abdomens vigorously to drive air more quickly along the tracheae. Despite this,

A white-legged damselfly, *Platycnemis pennipes*, clearly shows the basic body plan of insects – a head with eyes and other sense organs, a thorax with attached wings and legs, and an abdomen.

their dependence on the passive diffusion of oxygen along thin tubes imposes a limit on how big insects can grow.

This network of tubes branches throughout the body. Micro CT scans reveal the sheer complexity and extraordinary beauty of this system, but it can also be appreciated without a high-tech laboratory. The branching network of tubes is revealed when the hard exoskeleton is shed during moulting.*

The easiest cast skins in which to admire the insect tracheal system are those of dragonflies. When the aquatic larva is fully grown, it climbs a stem protruding above the water's surface and emerges from the last larval skin as a winged adult. The cast skins, or *exuviae* (singular: *exuvia*), remain perched on pondside plants for a remarkably long time and are not hard to find. The tracheae are easily visible on such exuviae as tufts of white 'hairs'.

The myriapods (centipedes and millipedes) also breathe through tracheae but the latest research suggests that insects and myriapods are not that closely related, so the two groups must have evolved their tracheal systems independently as a good solution for breathing in air. Evolution is never afraid to recycle a good idea.

With their distinctive features, insects are placed in one *class* of arthropods – the division below that of phylum. In the simplest classification, there are five arthropod classes: Chelicerata (spiders and scorpions, along with sea spiders and horseshoe crabs), Crustacea (crabs, shrimps, lobsters and the like), Myriapoda (centipedes and millipedes) and Trilobita (the extinct trilobites) as well as the insects. However, there are no firm rules among taxonomists governing what constitutes a class, so these groups are sometimes given the rank of subphylum and the arthropods as a whole are then divided up into many more 'classes'. I won't delve further into the reasons behind this. It's the proverbial 'can of annelids' and I'll leave you to follow up on that if you're interested – or if you want to see grown taxonomists cry.

For our purposes, insects are either given their own class (nice and easy) or perhaps more accurately placed in a class called Hexapoda – six legs – for obvious reasons. However, Hexapoda is sometimes given the

* The tracheal system evolved as indentations of the insect's outer layer that grew ever deeper to form long tubes that remained connected to the outside. This means that the tubes are lined by the same chitinous exoskeleton that covers the outside of an insect's body. This, in turn, means that the linings of this whole complex network must be shed at every moult. The tracheoles, which may be as small as one thousandth or one ten thousandth of a millimetre in diameter, are not derived from the outer cell layer (as the tracheae are). So their linings are not shed at each moult.

rank of a subphylum or a superclass – as I said, a can of worms. However, in either case the existence of the category Hexapoda creates a slight complication. I know I said that part of the definition of an insect is that it has six legs. And it *is* true that all insects have six legs – but, unfortunately, not all hexapods are insects.

Three obscure groups of 'nearly insects' – springtails (Collembola), two-pronged bristletails (Diplura) and coneheads (Protura) – complicate the simple picture. They are so nearly insects that they used to be classified as true insects, but scientists have recently changed their minds. We'll meet some of these creatures again briefly in later chapters.

So, how are we doing in coming up with an accurate definition of an insect? We've discovered that an insect is an arthropod – but then so are spiders, millipedes and crabs. Insects have six legs, but then so do a few

A flat-backed millipede, *Polydesmus* sp. Insects and millipedes were once considered close relatives, but this no longer seems to be the case, and their shared features, like the tracheal system, evolved independently.

other groups of non-insects, like springtails or coneheads. Insects have two pairs of wings, but some insects, like silverfish, have no wings at all. Insects have their bodies divided into three parts – the head, thorax and abdomen – but then again, so do springtails. However, insects have all descended from one, albeit unknown, common ancestor in the far distant past. They form their own distinct bough on the tree of life, a single bough that has then gone on to branch many times more to create the diversity of insect-kind. This is the most accurate definition of an insect, although, as I said at the start, it's far from a simple one.

To embrace their diversity, insects are divided up into twenty-seven orders, the next major rank below that of class. These represent the broad 'types' of insect, such as beetles (order Coleoptera), flies (Diptera), or butterflies and moths (Lepidoptera). There is also an ever-changing number of insect orders that have long been extinct; it's almost certain that there are still others lurking in rocks we have yet to examine.

A glasswing butterfly, *Greta oto*. Butterflies and moths (Lepidoptera) are one of the most successful groups of insects.

Without living examples, the taxonomy of these extinct orders is, to say the least, fluid – and very complicated. I'll introduce a few of the better-known ones at appropriate points in our story.

Even new orders of living insects occasionally turn up, the most recent in 2001, as we'll see in Chapter 1. The whole family tree that we've been using to help us understand the place of insects in the world is built on shifting taxonomic sands. Classifying the life on our planet is something of a dark art, and it's quite possible that the number of orders may change again with further research. For example, two separate orders, termites (Isoptera) and cockroaches (Blattodea), have recently been merged into one order, still called Blattodea.

So, now we know what an insect is and isn't, and having a map to guide us through the long history of insects, we are ready to embark on an exploration of their extraordinary world. It's generally agreed that there are several key turning points in the insect story, each of which propelled insects to ever greater success, and these form the backbone of the book. Firstly, insects benefitted from features, such as an exoskeleton and jointed legs, which they inherited from their ancient arthropod ancestors (Chapters 2 and 3). But they also came up with several unique evolutionary innovations. It would be hard to overstate the contribution to the insect success story made by the evolution of wings and by the later evolution of mechanisms to fold those wings out of harm's way (Chapter 4). For one thing, wings made it possible for insects to migrate over vast distances and exploit the planet in ways not possible before (Chapter 5).

Insects were presented with new opportunities when flowering plants evolved. Insects and flowers became entangled in a series of complex partnerships, some so intimate and complex that they defy belief. Working together like this, both insects and flowering plants could achieve even greater ecological dominance (Chapter 6).

Insects are very, very good at making more insects. Based on how they do this, they can be divided into two broad groups – those in which the young stages are more or less miniature replicas of the adults (often referred to as *incomplete metamorphosis* or, more technically, as *hemimetaboly*), and those that have a distinct larval stage, like a grub or a caterpillar, which looks nothing like the adult form. This extreme difference necessitated the evolution of a pupal stage during which the larva transforms into the adult. This is called *complete metamorphosis*, or *holometaboly* if you want to impress friends. Some entomologists have suggested that the evolution of complete metamorphosis was the biggest single factor in the success of insects, greater even than the evolution of wings.[7] Certainly, there are more holometabolous insects than any other kind of animal (Chapter 7).

Contributing to the success of future generations, many insects exhibit unexpectedly complex parental care (Chapter 8). But parental care led to another innovation that gave a few groups of insects the key to real ecological dominance – the evolution of social behaviour. A surprising number of insects live social lives (Chapter 9), but among two groups, the termites (Blattodea) and the ants, bees and wasps (Hymenoptera), sociality has developed to its highest degree. All termites and ants live in teeming colonies, but the majority of bees and wasps are solitary creatures or live in small colonies. These familiar insects, therefore, allow us to trace one route from solitary living to life in complex societies (Chapter 10). Between them, ants and termites have achieved the highest levels of social organization, living in societies sometimes millions strong. They build complex nests of such sophistication that they've even inspired human architects, and they shape the ecology of many habitats to a far greater degree than any other insects or any vertebrates (Chapter 11).

We live in a world dominated by insects. They underpin the ecology of most terrestrial and freshwater habitats. Ecologist and entomologist E. O. Wilson of Harvard University, in an address in 1987 to open a new invertebrate exhibition at the National Zoo in Washington DC, described them as 'the little things that run the world'.[8] Without insects, our world would quickly fall apart.

This is their world – and their story.

A caterpillar of the Chinese oak silk moth, *Antheraea pernyi*.

1

Teeming Hordes
The extraordinary diversity and abundance of insects

The total number of species actually named and catalogued is just over one million. Most of these are insects. Indeed, to a good approximation, all species are insects!

Robert May, Baron May of Oxford[1]

Robert May, physicist and mathematician-turned-ecologist, wrote the extraordinary words cited on the previous page in 1986. Since then, legions of taxonomists have added considerably to the number of animal species described. Today we have catalogued around one-and-half-million species, of which two-thirds are insects,[2] so after nearly four decades of intensive research, this startling fact hasn't changed. Most animals on 'our' planet are insects. In my 1996 TV series *Alien Empire*, and in the title of this book, I'm having fun with the idea that there are creatures sharing the planet that are so different from us in every way that they seem like aliens. But the truth is that it's we vertebrates that are the aliens. The standard kit of parts for an Earthling is six legs and an exoskeleton.

Insects, along with crustaceans, spiders, scorpions, millipedes, centipedes and various other less well-known beasts, are arthropods. Throughout their history arthropods have been remarkably successful both in variety of species and numbers of individuals. There are around 40,000 species of spiders and 67,000 species of crustaceans currently known to science. Yet, although there are doubtless large numbers of these creatures still to be discovered, both groups are insignificant compared to the number of insects inhabiting Planet Hexapod – by far the most successful of all the arthropods.

EXTREME INSECTS

One facet of their ascendency is that insects have occupied every corner of *their* planet, at least on dry land and in fresh water, if not always in teeming hordes. Only one species that we know of, a wingless midge, *Belgica antarctica*, is native to the Antarctic continent, although a second species, *Parochlus steinenii*, which has retained the power of flight, survives on the South Shetland Islands, off the northern tip of the Antarctic peninsula.[3] In contrast, all the other continents abound in an amazing variety of insects. One reason for this diversity is that insects can live in such a wide range of habitats – in every habitable spot of land and in quite a few that seem, to us at least, totally uninhabitable. It's these latter 'extreme insects', living life on the edge, that demonstrate the extraordinary adaptability of insect-kind.

That little midge, braving life in the Antarctic, possesses an endurance beyond belief. Its larvae spend most of the year encased in ice and when they finally get a reprieve for a couple of months, during what passes for summer down there, they must face desiccating winds and salt spray, all under a witheringly intense sun. Larvae may freeze and thaw

(previous page) A roost of four-spotted chaser dragonflies, *Libellula quadrimaculata*, warmed by the first rays of the sun.

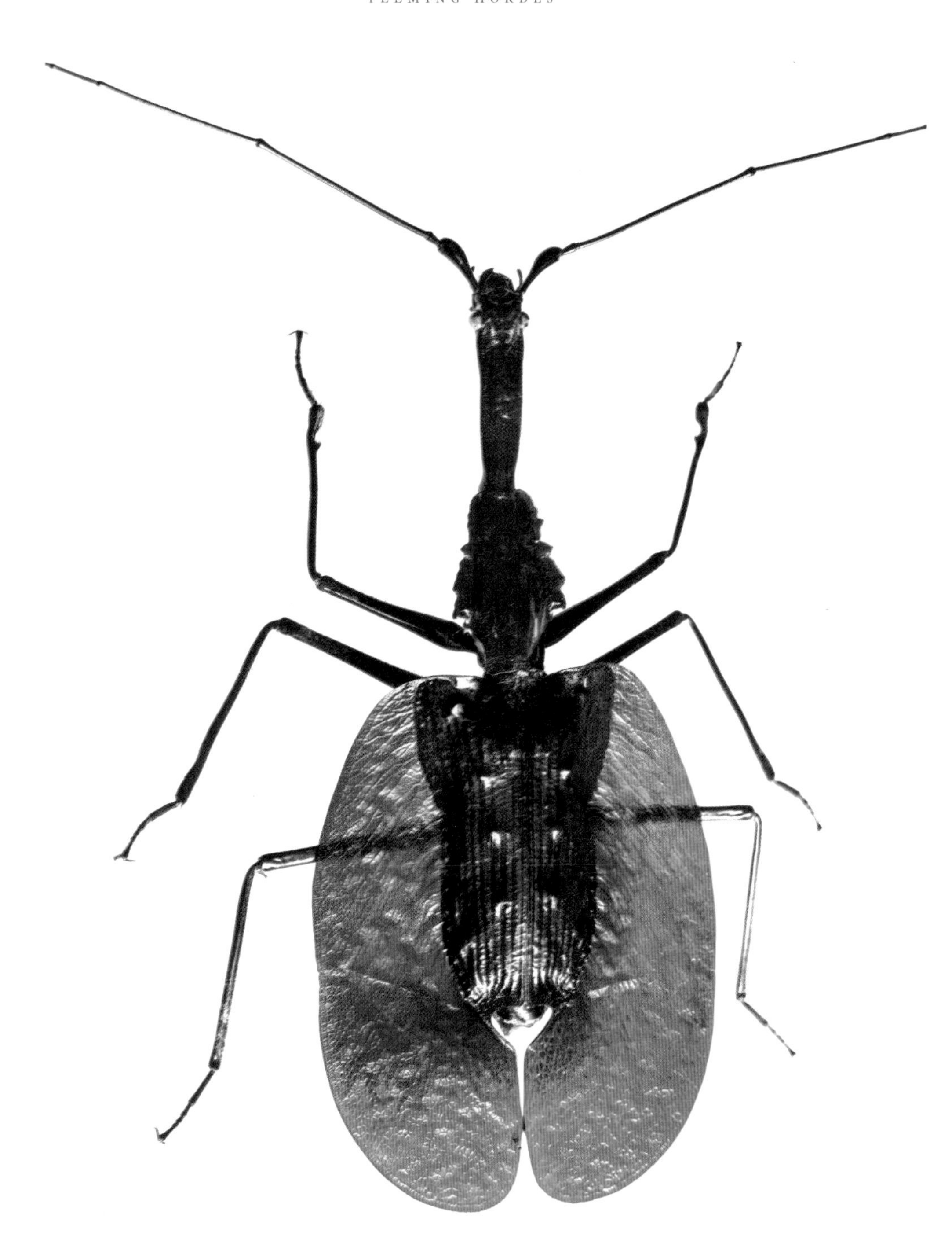

A violin beetle, *Mormolyce phyllodes*, from southeast Asia. Insects exist in a bewildering diversity of shapes.

up to 140 times each year and it takes them two years to feed for long enough to turn into adult midges. To survive being frozen solid, these long-suffering midges must also lose 70 per cent of their body's water and enter a state of suspended animation.[4] That's all for a brief seven-day life as an adult. Still, it's a living.

With such adaptability to extreme conditions, it's surprising that insects haven't conquered the largest living space on the planet, the open ocean. It can't be the physical problems of living in salt water – we'll see shortly that insects thrive in environments that are a lot tougher than seawater. On top of that, the far distant ancestor of insects was a marine creature.

One frequently cited reason for the absence of insects from the ocean is that crustaceans got there first. They diversified before insects evolved and the ocean world has remained relatively stable since that time; so, when insects began to diversify, crustaceans were just too entrenched to dislodge. Crustaceans are now as ubiquitous in the ocean as insects are on land. However, a few insects are perfectly at home on the open ocean, far more so than we humans, as I discovered while trying to find and film these marine insects in the rolling swell off Hawaii's Big Island.

Ocean striders or sea skaters, *Halobates* spp., are relatives of the more familiar pond skaters that skip over the surface film of ponds and slow-moving rivers right around the world. However, tiny, silvery ocean striders ride out storms and hurricanes on the surfaces of tropical oceans worldwide. We could only find these half-centimetre-long creatures in the vastness of the ocean by trailing a *neuston net* behind our boat. Such nets are designed to collect organisms in the top few centimetres of water, including those floating on top. Once we had collected some striders, it was time to transfer to a much smaller boat, from which we could reach the ocean's surface with a special probe lens that allowed us to see these frail-looking insects in their endless ocean home.

Lying over the gunwales with the tip of the lens perilously close to corrosive salt water, I released our small collection of *Halobates*. As I did so, I felt for the cameraman. I've worked with Kevin Flay for more than thirty years, filming in all kinds of situations, all over the world, so I knew he was not a great sailor. Within a few minutes of staring down the distorting view of the probe lens, which magnified every bit of movement in the swell, he was violently sick. Refusing to be put off by such distractions, he continued to work, with frequent pauses to feed the fish, until we had completed the sequence of an insect that few have even heard of, let alone seen.

Playing back the footage, I found it hard to believe that these tiny creatures, skittering over the ocean surface like sparkling water droplets, could ever survive out here amid mountainous waves and wind-whipped foam. The key to this amazing seaworthiness, allowing them to ply the high seas with impunity, has only recently been discovered. It's that silvery coat, made up of closely packed, specialized hairs of several different types. The most numerous are a mere one five-hundredth of a millimetre long but mushroom-shaped, covering the whole body in a dense layer. These water-repellent hairs make ocean striders so completely unwettable that the familiar expression regarding ducks would be far more appropriately phrased as 'like water off an ocean strider's back'.

In addition, this coat serves as an emergency scuba kit. Ocean striders breathe air but if they are submerged, the hairs trap a layer of air that allows the insect to breathe underwater. Should their submergence be prolonged, they use up their oxygen supply, but there's still no need to panic. The hairs hold the bubble so tightly that it can't shrink. Instead, the pressure inside the bubble drops as the insect uses up oxygen, which forces more oxygen to diffuse into the bubble from the ocean, refilling the scuba tanks. The trapped bubble creates a physical or 'plastron' gill,[5] allowing ocean striders to cope with whatever the tropical seas throw at them.

However, this neat trick also suggests another possible reason for the paucity of insects in the open ocean. The swarming hordes of small crustaceans that live in the surface waters of the ocean make an epic migration every day, descending into the dark depths during the daylight hours, and only returning to the surface to feed when night falls. Taking into account the tiny size of these crustaceans and their extraordinary numbers, this daily vertical trek is the largest migration on the planet and is driven by an abundance of ocean predators. The crustaceans must hide in the deep, dark water during the day to avoid falling prey to the hungry mouths of countless fish and other predators that feed close to the surface. However, under cover of darkness, they can feed in the richer surface layers in relative safety.

To compete with crustaceans in the marine world, insects too would need to avoid predation by joining this migration. In freshwater environments, plenty of insects live beneath the surface, where they use plastron gills or more conventional gills with permeable surfaces to extract oxygen from the water. But all these structures connect to the network of tubes (tracheae) that transport oxygen throughout insect bodies. Calculations suggest that a plastron gill would collapse at depths of more than 30 metres, where the pressure would buckle the supporting hairs. Deeper still, beyond a hundred metres, the tracheal system itself would be so compressed that it could no longer function.[6]

So, in the open ocean, an insect's options are limited to the surface film – in part due to its physiological limitations and in part because no other marine invertebrate lives full-time on the surface film, so it's the one oceanic niche that remained free to colonize. They are not entirely safe from predators here, but *Halobates* makes its life on the ocean waves work because it is so tiny, with extremely long legs. Combined with super-fast reactions, its legs allow it to spring from the water's surface almost instantaneously if it senses any threat.[7] Even so, there are only five species of *Halobates* striding over the planet's open oceans, although around forty other species live on the sea surface close to shore, where they are joined by three species of *Asclepios*, their close relatives.[8] However, they are not the only insects to have recolonized the world in which their ancestors originally evolved.

They are joined by three or four species of tiny flightless midge, *Pontomyia* spp. But it is highly likely that there are more species of these midges out there. Finding a one-and-a-half-millimetre-long fly on the vast ocean surface is far from easy, especially since a ripe old age for an adult *Pontomyia* midge is just three hours and most of the midges live for only one or two hours. In this very limited window of opportunity, the male midge must use his shrivelled wings to row over the surface in search of a wingless and legless female.[9] It seems like an extremely precarious lifestyle, but the midges cope pretty well. Occasionally, they reach high enough densities that they become easy to spot as an orange scum on the water's surface; such concentrations must make finding a mate in less than three hours a whole lot easier.[10]

Despite the assumed limitations imposed by their structure and physiology, some insects have even invaded the deep ocean, although not without help. Around the world seals and sea lions are plagued by lice (Phthiraptera). These insects are more or less permanent residents in the fur of seals, so must endure deep dives when their hosts go fishing – those that specialize on elephant seals regularly face depths of nearly 2,000 metres.[11] It seems that they have little option but to shut down bodily functions for the duration of the dive, although seal lice do exhibit a few differences from their landlubber kin that may help them survive prolonged submergence. They are covered in short, spiky scales that Howard Hinton, my professor during my years at the University

Frequent eruptions from Kilauea on Hawaii's Big Island create extensive, barren lava fields on which lava crickets, *Caconemobius* sp., thrive.

of Bristol, suggested might work as a more robust version of a plastron gill, allowing them to breathe while underwater.[12]

However they do it, adult lice can cope with this extreme lifestyle, but eggs and nymphs are soon killed by immersion in seawater, so the lice also must tie their breeding cycle with that of their hosts. When they breed, seals and sea lions spend most of their time out of the water, which gives the lice an opportunity to produce their own offspring. Even in just this brief survey, it seems that there are quite a few insects that haven't heard the widespread idea that there are no marine insects.

From one extreme to another. Gratefully back on the dry land of Hawaii's Big Island, Kevin and I headed up the slopes of Kilauea, one of the most active volcanoes on the planet. By the simmering crater of Halemaumau, we met with Frank Howarth of Honolulu's Bishop Museum, an entomologist who has worked extensively on Hawaii's curious insect fauna and who had agreed to help us find another extreme insect, the lava cricket, *Caconemobius fori*. These crickets weren't described by science until 1978, although long before that they were well known to native Hawaiians, who called them *ūhini nēnē pele*,[13] which means 'chirping grasshopper of the lava'. This ordinary-looking cricket has chosen a place to live every bit as hostile as the open ocean sparkling far below the blackened lava slopes. Kilauea's frequent outpourings have created a landscape streaked by black rivers of lava that often cool in exotic swirls and cascades, like water frozen in mid-flow, and it's this magnificent if barren landscape that lava crickets like to call home.

More recently, a second species, *C. anahulu*, has been discovered inhabiting the lava flows of Hualālai Volcano in the Kona district of the Big Island. Often, lava crickets move in just a few months after an eruption cools, when molten rock is probably still flowing not far beneath the surface. The first living things to colonize this new land, they subsist on other, less hardcore insects that get blown on to the lava flows and perish. Once plants begin to soften the landscape, the crickets move out, to find another sterile flow of lava, although there's still a lot that we don't know about what makes a lava cricket happy. Studies carried out in 2021 showed that in some places the crickets were still living contentedly on flows that were fifty years old, while in others the crickets shunned recently cooled flows entirely.[14] They might survive in one of the harshest environments on the planet, but they are clearly very fickle little creatures.

Were it not for a little local know-how, lava crickets would be as impossible to find on these flows as ocean striders or marine midges are in the vastness of the Pacific. Frank uses pitfall traps set among recently cooled blocks of lava, which he baits with ripe blue cheese. I never did ask him how he discovered the little crickets' fondness for the pungent smell, but his traps catch many hundreds of lava crickets, which goes to show how successful these insects are at coping with such an extreme environment – and just how much they love cheese.

It's hard to imagine more of a contrast between the baking rivers of lava on the Big Island and the frozen rivers of ice that descend from the Patagonian Ice Field, in the Andes of southern Chile and Argentina. A few years after chasing lava crickets and ocean striders in Hawaii, Kevin and I found ourselves walking around the cracked and creviced ice at the snout of the Perito Moreno Glacier in Argentina, one of the fastest-moving glaciers anywhere in the world. We were here to film the spectacular blocks of ice that calve off the face of the glacier, a huge ice cliff towering over the chilly waters of Lago Argentino. Not infrequently, a resounding crack echoed off the surrounding mountainsides as an enormous block of ice detached itself and tumbled into the water. Perched on a narrow beach close to the glacier face, we had to first grab the shot and then grab the kit, to retreat to higher ground as the wave created by the falling ice rushed up the beach.

Our guide here was an old friend, Daniel Gomez, who I first met when he lived in Bristol, in the UK, the city that I still call home. However, he's a native son of Patagonia and an excellent and intrepid naturalist with a vast knowledge of South American wildlife, and he knew we'd be interested in another intriguing aspect of this glacier. He led us into a massive fracture in the blue-white ice and into an unearthly, if slightly frightening world. Pressing my face against the ice I could see deep into the glacier – and deep into the past. Fragments of plants hung suspended in the translucent ice, frozen in both space and time. They may have been incorporated into the glacier hundreds of years ago in the upper reaches of this river of ice and now, close to the snout, they would shortly be freed again in a crashing avalanche. All the time, the glacier creaked and groaned as it moved imperceptibly downslope. Surrounded by ice in a metre-wide fissure, I was only too keenly aware of the unstoppable power of the glacier. High above, condors were circling. It felt as though they were waiting for a sudden surge in the ice to provide them with a frozen meal.

Out on the surface of the glacier itself, small pools of meltwater collect, and these too look as if they have accumulated small fragments of twigs. However, these are no inanimate scraps of plants – they are

insects, living on the open glacier, as desolate a world as the lava flows of Hawaii. They are the larvae of glacier stoneflies, *Andiperla willinki*,[15] somewhat aggrandized by their local name 'Patagonian Dragon'. If they are dragons, then they are ice dragons, which need the protection of their own brand of antifreeze to survive in their bitterly cold world.

It seems that there's no end to what insects can endure. As pools form on the top of the Perito Moreno Glacier, so they also form in shallow depressions on sun-baked rocks in Nigeria and Uganda. These pools are so shallow that they fill and dry several times during the wet season and disappear completely under the baking sun of the dry season. Yet this is the chosen home of another little midge, *Polypedilum vanderplanki*. I first learnt about this extreme insect during my undergraduate years in the early 1970s from Howard Hinton, its discoverer. Hinton called me into his office one morning to show me a jar of dry soil containing the shrivelled remains of something that might once have been an insect. They were, he pointed out with great excitement, the larvae of midges that he had collected in the late 1950s. He had discovered that they could dry out completely and enter suspended animation, only to revive when the pool fills again, a process called *anhydrobiosis*. The midges can dry and reanimate as many times as they need to, until they complete their larval growth.

Snow fleas, *Boreus hyemalis*, are active during the winter when they're happy crawling over ice and snow.

A question-mark roach, *Therea olegrandjeani* – and it's a very good question as to how many kinds of insects there are in the world. We can only guess.

The larvae can survive losing 95 per cent of the water in their bodies (far more even than the larvae of the Antarctic midge). In their dry state, Hinton immersed them in liquid nitrogen and boiling water, and they still revived when wetted, despite having endured temperatures of -270°C and +102°C.[16] Almost all larvae revived after being stored dry for more than three years at room temperature and humidity, and many survived an even longer period when stored over calcium chloride, which absorbed any moisture from the atmosphere. Hinton died from cancer in 1977 at just sixty-four years of age, as I was beginning my PhD studies at Bristol. While sorting through his collections in the university Zoology Department, we came across some remaining jars of dry dust labelled as *P. vanderplanki*, from which we were able to revive a couple of larvae that had been stored for more than a decade and a half. Just add water for instant midge.

This particular midge was long thought to be the only creature capable of this astonishing trick, but recently several other species that are capable of anhydrobiosis have been discovered, in Namibia in southern Africa, and in Malawi.[17] [18] In 2022, yet another species turned up, this time in short-lived rock pools in the Drakensberg Mountains of South Africa.[19] Such creatures demonstrate that a major factor behind the success of insects is their ability to live anywhere and everywhere.

Even so, these extreme insects of lava flows and glaciers, open ocean and baking rock pools represent a vanishingly small fraction of insect-kind. In more benign places there are uncountable numbers of different insects – literally uncountable, or at least so far uncounted.

HOW MANY INSECTS?

You might think that, as the third decade of the twenty-first century dawns, the question of how many different kinds of animal occupy the planet would have a definitive answer. Unfortunately not. In Robert May's 1986 *Nature* paper, from which this chapter's opening quote was taken, he begins by pointing out that we can only give the vaguest of answers – somewhere between 1.5 and 30 million. And here again, the situation hasn't changed much.

We think we've done a pretty good job of enumerating our own vertebrate kind, particularly mammals and birds – both of them popular, conspicuous and therefore well-studied groups. Many would consider that our estimates of around 10,000 birds and 5,500 mammals are close to reality. But even here there may be surprises. Since nature pays scant regard to our compulsive desire to fit things into neat boxes, it's no easy task to ascertain exactly what constitutes a species. Depending on what your criteria are, the true number for birds could be as high as 18,000 to 20,000 species, double the generally accepted figure.[20] That's for a group of animals studied by armies of amateur naturalists as well as professional biologists. It shows the daunting scale of the task facing entomologists – who, at the absolute minimum, are dealing with one hundred times as many species.

Since insects make up such a large proportion of animal species, the question of how many different kinds of animal occupy the planet is, to a large extent, answered by working out how many insects there are. Surprisingly, these apparently fundamental questions about life on Earth didn't occupy the thoughts of many biologists until the 1980s. Before that, a few early naturalists had made a stab at an answer, based mostly on guesswork backed up by a lack of knowledge of the world at large.

Carl Linnaeus, the eighteenth-century Swedish botanist who invented the binomial system of naming species that we still use today,* was

* Every living organism is described by a unique two-part name – genus and species, for example *Homo sapiens*. A genus is a group of closely related species, so there are several kinds of *Homo*, for example *Homo neanderthalensis* or *H. erectus*, but *H. sapiens* is uniquely applied to our own species.

(*previous page*) Rainforest in Kubah National Park, Sarawak. Places like this reveal plenty of new species of insects every time someone looks.

perhaps the first to attempt to catalogue the diversity of life. In his *Species Plantarum* (1753) he listed around 6,000 plant species and, taking a somewhat optimistic view of the thoroughness of his own work, reasoned that the global total was therefore likely to be no more than 10,000 species. The first to try such an estimate for insects was John Ray (1627–1705), one of the earliest in that great English tradition of parson-naturalists. In his book *The Wisdom of God Manifested in the Works of the Creation*, published in 1714 after his death, he suggested that Britain was home to 2,000 different kinds of insects, and so perhaps there might be 20,000 species worldwide.[21] With the benefit of hindsight, it's clear that these early guesses were well wide of the mark.

By using more carefully thought-out assumptions and extrapolations, later scientists came up with what they hoped were more accurate figures. William Kirby, continuing the parson-naturalist tradition, along with co-author William Spence, produced one of the first entomology textbooks in 1826.[22] Kirby and Spence suggested that there might be up to 120,000 species of plant and fungus worldwide and since, in Britain at least, there were roughly six kinds of insects associated with each plant species, they reasoned that there must be 600,000 insect species crawling, hopping and flying over Planet Earth. The number was going up, but it sky-rocketed in the 1980s, when a Smithsonian scientist called Terry Erwin began work in the rainforests of Central America.

Erwin used a similar logic to that of Kirby and Spence, extrapolating from the number of plants species to reach a figure for insects, but he used intensive sampling techniques to add weight to his assumptions. He shrouded nineteen specimens of a common rainforest tree, *Luehea seemannii*, with clouds of insecticide, and gathered all the insects that fell from its branches in collecting funnels arrayed beneath each tree. He amassed around 1,000 beetle species from this one kind of tree alone. He made some informed guesses about how many of these beetles lived only on this one species of tree and how many were creatures that ranged over many kinds of rainforest tree. Since he reckoned that there are likely to be around 50,000 different tropical trees worldwide, he came up with the astounding figure of 8 million species of beetle globally. It is generally assumed that beetles are the richest order of insects (although see below) and make up about 40 per cent of all arthropods – so, making a few other assumptions about the richness of the rainforest canopy compared to the forest floor, Erwin came up with a figure of 30 million species of tropical arthropod,[23] far higher than anyone had ever considered possible.

Because this audacious number was based on so many assumptions,

some of which were no more than inspired guesswork, it soon came in for criticism.[24] Refinements to Erwin's figure based on detailed surveys carried out in the rich rainforests of New Guinea suggested somewhere between 2 and 7.4 million arthropods, most of which are insects.[25] Many scientists today are comfortable with figures in the range of 5–10 million species, which, for me at least, is still an unimaginably large number.[26] However, these figures are based on the widespread assumption that insect diversity is correlated with plant diversity. Rainforests are so rich in insects because such a large variety of plants live there – but this relationship between numbers of insects and diversity of plants may not always hold true.

The mangrove forests that line many tropical coasts are not botanically rich, and therefore have, until recently, been regarded as places of low insect diversity. However, that supposition has now been challenged. Surveys of mangrove swamps in Singapore revealed an unexpected total of 3,000 insect species. Throughout south-east Asia, mangroves proved to be similarly rich in insects. Furthermore, many of these insects were found in relatively small geographical areas, so mangroves in different areas have rich but different populations of insects.[27] Taking all the

A Doris longwing butterfly, *Heliconius doris*, one of several thousand species of butterflies that live in the rainforests of South America.

south-east Asian mangroves together, these studies suggest that this habitat hides an unexpectedly high diversity of insects, which in turn suggests that estimates of insect numbers based on the assumption that the highest diversity is found in tropical rainforests may underestimate the true total. Moreover, mangroves are not the only source of hidden diversity.

HIDDEN DIVERSITY

As taxonomy moves from a reliance on physical features to identify different kinds of insect to maps of their genomes at a molecular level, it's clear that we've substantially underestimated their diversity. A single species can turn out to comprise several different ones – and not just among obscure or little-studied groups of insects.

It's always a special moment to encounter a wood white butterfly, *Leptidea sinapis*. It still flies in a few woodlands close to where I live, although – like many of our insects – it has declined significantly in the last few decades. Its characteristic slow flight makes it look like a scrap of paper wafted on the breeze, but the males are tougher than they look and can stay aloft for most of the day as they patrol glades and rides on the lookout for females. In 2001 this unobtrusive little butterfly created a bit of a stir among lepidopterists.

European butterflies are one of the most comprehensively studied groups of insects anywhere in the world, and British species in particular have had a long history of close scrutiny by both professional and amateur entomologists. Many kinds occurring in the British Isles were described as early as 1634, in a book called *The Theatre of Insects*, by Thomas Moffet,* and subsequent collectors scoured every bit of Britain in search of rare and desirable specimens. Yet in 2001, after nearly 400 years of intensive butterfly hunting, a new species turned up. Careful observation, later confirmed by molecular analyses, showed that wood whites in Ireland are a different, if almost identical species from those in southern England and Wales. It was originally thought that the Irish species was Real's wood white, *L. reali*, which otherwise flies in southern France, Italy and Spain. But wood whites had yet another secret to reveal.

Further molecular studies uncovered a third species hidden under the umbrella of 'wood white' and it was this third species, understandably

* It is sometimes said – without any evidence that I can find – that Thomas's daughter, Patience, was the inspiration for the nursery rhyme 'Little Miss Muffet', who was frightened from her tuffet by a spider. Hardly the behaviour of an entomologist's daughter.

named the cryptic wood white, *L. juvernica*, which flew across Ireland. Well – most of Ireland. The wood whites flying over the Burren in County Clare are all the original species, *L. sinapis.* Yet as far as we know, no cryptic wood whites live in England or Wales, even though they are now known to occur widely across Europe.[28] However, all three of these species are identical to the naked eye, so we probably don't yet know the full picture. The importance of the wood white story to the question of how many insects there are is simply that this is just one case of a single species becoming three.

Even the most familiar insects sometimes turn out to be complexes of multiple cryptic species. You have to know where to look to find wood whites, but I can sit in my garden and see another species that has now become three. The white-tailed bumblebee, *Bombus lucorum,* is one of the most common bees in Europe, known for centuries, but it turns out that it is really three virtually identical species, *B. lucorum* itself, plus the northern bumblebee, *B. magnus*, and the cryptic bumblebee, *B. cryptarum.* Now we know what we are looking at, each of these species does exhibit differences in behaviour, for example in the flight times of the queens, and in habitat choice. The ecology of this group of species has been studied in most detail in western Scotland, where the cryptic bumblebee is the most abundant. In this area, it prefers uplands and cooler climates, whereas the white-tailed bumblebee favours lowland or urban habitats. The northern bumblebee is a heathland specialist, feeding on bell heather, of which there is certainly no shortage in this part of the world.[29]

This phenomenon, sometimes called the 'biodiversity wildcard', seems to be widespread, so we may have greatly underestimated the number of insect species, although there's no consensus yet on the scale of this phenomenon. While some studies suggest that only 1 or 2 per cent of described species hide undiscovered cryptic species, others believe that numbers could be much higher.[30] In one recent study, the authors assumed as many as six cryptic species of arthropod for every one distinguishable by morphology.[31] That would make Terry Erwin's figures a gross underestimation. It also shows that we have no clear idea of how wild the biodiversity wildcard really is.

Even non-cryptic insect species continue to be discovered, often in astounding numbers, especially where a wide range of habitats occurs close together. One such place is the eastern slope of the Andes. I once drove down from the Altiplano, the central plateau around 6,000 metres in elevation that runs between the even higher ranges of the Eastern and Western Cordilleras, to the lowland forests of Bolivia. The journey took

A membracid bug, *Heteronotus* sp., belongs to a family of bugs that flamboyantly illustrate the flexibility of insect design. In this species, the top of its thorax is shaped to look like an ant.

me through a whole range of habitats in a remarkably short distance. Driving east, the route first crosses the Eastern Cordillera, where tough alpine plants form perfect domed cushions against the extreme conditions. It then heads down through brightly coloured fields of quinoa, painted in red and yellow and carved into impossibly steep slopes, and finally descends through woodland that imperceptibly turns into the rainforests of the Amazon Basin. In a similar transect in Peru, running for just 65 kilometres through forests at different elevations, scientists found 2,500 butterfly species and estimated that there could be another 500 species that they missed.[32]

This staggering figure represents 15 per cent of all known butterflies, but it's very unlikely that this one strip of the eastern Andes, as rich as it is, is really home to such a large proportion of the world's butterflies. Instead, this survey suggests that there are still an awful lot of butterflies out there that we haven't yet found – and this, as I've pointed out, is one of the most studied and collected groups of insects.

Far fewer people spend their time seeking out chalcids, tiny parasitic wasps that are only millimetres long. However, obscure groups like these highlight the task ahead in enumerating insects. In one study in Costa Rica, just six hours' collecting yielded 790 species of these wasps. In Guanacaste, a famous and well-studied nature reserve in Costa Rica, another survey in 2014 found 186 new species of *Apanteles*, another kind of parasitic wasp from a different family (Braconidae). Before this,

only 19 species of *Apanteles* had been described for the whole of Central America.[33] This group of wasps is found right around the world, so imagine this same story multiplied many times over.

You may well have seen one of these tiny wasps – or at least its handiwork. *A. glomeratus* (now renamed *Cotesia glomerata*) is commonly found in gardens and allotments in Europe and North America, where its larvae live and grow inside the bodies of the caterpillars of large and small white butterflies, *Pieris brassicae* and *P. rapae.* These caterpillars are only too familiar to vegetable gardeners as insatiable cabbage eaters, so gardeners should appreciate these tiny wasps. When the wasp larvae are fully grown, a dozen or more erupt from the caterpillar's body in a manner reminiscent of that famous scene in Ridley Scott's *Alien*. In fact, it was these tiny wasps that provided the writers of *Alien* with the inspiration for this seminal moment of science-fiction film history.

Because of their sci-fi life cycle, growing inside a living host only to kill it as they burst forth, these real-life aliens are more correctly termed parasitoids rather than parasites. Once on the outside, they spin a mass of silk within which they pupate and turn into adult wasps. In short order, these wasps set off in search of more caterpillars to serve as living larders for the next generation. The stricken caterpillars, anchored to a cabbage leaf by a shroud of yellowish silk, are often found by gardeners.

These various groups of parasitic wasps might represent another biodiversity wildcard. Parasitoids have to be finely tuned to both the life cycle of their host and to its physiology, so they don't trigger any immune responses. Many are specialized to live inside just one or only

Beetles come in a dazzling variety of shapes and colours. This little jewel is a weevil, *Eupholus magnificus*, from New Guinea.

a few species of host. In addition, wasps in some families are so small – like the Mymaridae or fairy flies – that they can parasitize the *eggs* of other insects. This form of egg parasitism has arisen eighteen times independently among the wasps.[34] Thus an individual insect might have to run the gauntlet of parasitoids at every stage from egg to adult. If that's not enough of a nightmare (as we'll see in Chapter 10), some of the larval parasitoids, far from being safely hidden away inside their host's body, themselves fall victim to their own even tinier parasitoids. They are the living embodiment of the lines penned by Jonathan Swift in 1733:

> So, Nat'ralists observe, a Flea
> Hath smaller Fleas that on him prey;
> And these have smaller yet to bite 'em;
> And so proceed ad infinitum.

About half of the twenty-seven insect orders have found various ways to escape egg parasites, for example by hiding their eggs or by diligently guarding them,[35] and some parasitoids attack a broad range of species. Even so, for every free-living insect there are potentially several different kinds of associated parasitoid, dramatically increasing the diversity of insects on the planet.

Such are the uncertainties in our knowledge that it is still impossible to make more than an inspired guess as to the number of insect species alive today, but the realization that there are many potentially cryptic species, and that orders like the Hymenoptera and Lepidoptera are far more diverse than originally thought, is giving some scientists the confidence to push up estimates for the numbers of insects to as high as 20 million species, perhaps even more, slowly edging back towards Erwin's audacious figure from the 1980s.[36] For the sake of biological fairness, I should mention that insects are not the only group whose diversity we have probably grossly underestimated. There are also very large numbers of mites and nematode worms in the world, each group with plenty of both parasitic and free-living species. There could be several mites or nematodes living in or on every insect species, which would vastly inflate the numbers of these two groups. However, nematodes and mites are much less appealing than insects and therefore much less studied, so we are deep into guesswork again. It's probably safest to round off these thoughts by recognizing that there are three mega-diverse groups of animals – mites, nematodes and insects – and time may one day tell us which is top of the league.[37]

WINNERS AND LOSERS

Insect diversity is not equally distributed across all the insect orders. While we know of close on half a million different kinds of beetle,[38] we've only ever found a few dozen kinds of the little-known heelwalkers (Mantophasmatodea) from Africa. Heelwalkers were not discovered until 2001,[39] and the fact that a whole new order of insects was happily going about its business unknown to us until the start of the twenty-first century again illustrates how many surprises still await us as we try to come to terms with the diversity of insects. The heelwalkers were the first new insect order to be discovered for nearly a hundred years, since ice crawlers (Grylloblattodea) crept into our awareness in 1914. As their name suggests, ice crawlers are yet more insects at home in extreme conditions, in this case around glaciers in the mountains of North America, China, Siberia, Korea and Japan, although more than likely these inconspicuous insects will turn up in other high mountain areas if anyone bothers to look. Currently, only a few dozen species have been described.

These two orders are clearly related to each other, and they are now lumped together in a newly created order called Notoptera. Together, they number just over fifty species, but even notopterans are not bottom of the league. Zorapterans seem somewhat overglorified by their common name of angel insects. About 3 millimetres long, they hide away beneath bark and number only about forty species. In fact, the whole order Zoraptera consists of just one family and one genus, the antithesis of the hyper-diverse insect orders.

The vast bulk of the exceptional diversity of insects resides in just a few orders. The big four are the flies (Diptera), the butterflies and moths (Lepidoptera), the ants, bees and wasps (Hymenoptera) and the beetles (Coleoptera). All four of these orders comprise insects that exhibit larval, pupal and adult stages. We'll see in the next chapter that this kind of life cycle – complete metamorphosis – was one of the evolutionary innovations that gave insects a real edge. However, there is one other mega-diverse order, the true bugs (Hemiptera), which is the fifth largest order of insects. Although exhibiting only incomplete metamorphosis, with no pupal stage, the true bugs have clearly come up with their own strategies for success.

There is a tale, probably apocryphal, that the eminent evolutionary biologist John (J. B. S.) Haldane, seated next to the Archbishop of Canterbury at a Cambridge college dinner, was asked by the eminent theologian what his studies of the natural world had taught him about

A small selection of moths from my moth-trap in an urban garden in the UK. *Top row, l–r*: the herald, *Scoliopteryx libatrix*, dusky thorn, *Ennomos fuscantaria*. *2nd row*: orange swift, *Hepialus sylvina*, lunar spotted pinion, *Cosmia pyralina*. *3rd row*: the gothic, *Naenia typica*, buff ermine, *Spilarctia luteum*. *Bottom*: elephant hawkmoth, *Deilephila elpeno*.

the Creator. Haldane's reply was that He had 'an inordinate fondness for beetles'. This sentiment was later shared by the Coleopterists Society. This august society, which may of course have a slightly biased view, tells us that 'we live in an age of beetles'.[40]

To be fair, it has long been a central tenet of entomology that beetles are the most diverse order of insects, indeed the most diverse group of organisms on Earth.[41] This was one of the assumptions made by Terry Erwin as he extrapolated from the beetles he collected in Panama a total number of arthropods in the tropics. However, when scientists fogged the forest canopy in Brunei, on the island of Borneo, as Erwin had done in Central America, they found more hymenopterans than beetles. Likewise, in Britain there are more species of both Hymenoptera and Diptera than of beetles. One study has suggested that for every beetle, there may be two or three associated parasitic wasps, which would push the Hymenoptera right to the top of the league.[42]

Flies are not far behind, however. Gall flies (Cecidomyiidae) are tiny flies that induce plants to make often elaborate structures (galls*) to house and feed their larvae. Genetic studies in Canada suggest that there may be 16,000 species living there and possibly 2 million species worldwide – in just one family among the 188 described fly families.[43] In 2015, thirty new species of fly belonging to a different family (Phoridae) were discovered living within the city boundary of Los Angeles. There are obviously a lot of flies left for us to find, even in the heart of our biggest cities.[44]

The jury is still out on which order of insects holds the most species – we still don't know whether God's inordinate fondness was for beetles or for flies, wasps or moths (or even for mites or nematodes). If we still can't give a definitive answer to the question of how many kinds of insect there are, we do know that the number is still rising, and it wouldn't be at all surprising to see future estimates rise even higher than the largest current assumptions.

This richness has arisen not, as might be expected, by insects evolving species more quickly than other groups. Indeed, their rates of speciation are generally equivalent to those of many other animal groups. Rather, insects seem less prone to extinction than other groups of animals, which means that, over time, their diversity has just grown and grown. For ex-

* A gall is an abnormal growth of plant tissue, often creating a large structure, caused by an insect, fungus, bacteria or even a mechanical injury. A gall often takes on a distinctive shape and texture depending on what caused it, and the gall tissue often serves as food for the creature that caused it.

ample, for reasons not fully understood, a remarkable number of insects survived the famous extinction event at the end of the Cretaceous period (the K-Pg extinction),* which brought about the end of the dinosaurs.[45] That's not to say that some insect groups in some cases haven't experienced increased rates of speciation, but in searching for factors that explain the diversity of insects, we need to assess them on the basis of how they confer resistance to extinction rather than how they promote speciation,[46] and one such factor is size.

BIGGEST AND SMALLEST

Insects are small, which means that large numbers of both species and individuals can be crammed into a limited space. A single acacia tree in Africa can't feed even a single giraffe, but it's more than enough for a whole community of insects. On one such tree in Uganda, thirty-seven different sorts of insects were making a good living. Large local populations make it far less likely that an insect species will become extinct.

Small size also opens more opportunities for insects. I recently filmed a colony of acorn ants, *Temnothorax* sp., which – as their name suggests – house their whole colony inside a single acorn. On a broader scale, a single acre of soil might contain 3–25 million individual wireworms (the larvae of click beetles). As we'll see in later chapters, locusts swarm in their billions, while lake flies (Chironomidae) rising over Lake Victoria in Africa form dense, towering clouds that look like smoke and are visible from miles away.

Melbourne Museum's splendid 'Bugs Alive' exhibit tells its visitors that across the planet there could be 1 quintillion (that's one with *eighteen zeros* – or 10^{18}) insects alive at any one moment. Others have suggested 10 quintillion, but both figures can only be wild guesses, and both are figures that are so large as to be impossible to imagine. Trying to give this some context, this works out to be 150 million insects for every one of us lucky human beings. According to the *New York Times*, that translates to 300 pounds of insects for each and every one of us. These

* The K-Pg extinction marks the boundary between the Cretaceous period (the 'K' comes from the German word *Kreide*, meaning chalk, used to avoid confusion with 'C' for Cambrian) and the Palaeogene period. It is often referred to as the K-T (Cretaceous-Tertiary) extinction, but many geologists favour phasing out the term Tertiary. The extinction event was caused by a massive asteroid impact and wiped out three-quarters of all animals and plants. It spelt the end of all the non-avian dinosaurs, although avian dinosaurs (birds), like many insects, survived.

figures are even more speculative than those for numbers of species, but at least they serve to hammer home how successful insects are – as well as why they should become an integral part of our diet.

If size has been a key part of insects' success, how small can they get? For once we do have an answer to that question – 139 microns, or just over one-tenth of a millimetre. This is the adult size of the world's smallest insect, the males of a mymarid wasp, *Dicopomorpha echmepterygis*, whose name printed here is about two hundred times longer than the insect it describes. These male wasps are smaller than some species of *Paramecium*, which are single-celled protozoans. It's possible that an even smaller species exists among the vast number of undescribed species of parasitic wasps, but it probably won't break the record by much. Studies on these micro-insects show that they are approaching the lower size limits of the insect body plan, and they have already had to come up with some innovative workarounds to cope with being so tiny.[47]

The basic insect body plan, with its regional specialization, has to be abandoned in the very smallest insects. *D. echmepterygis* has only two visible abdominal segments, reduced from the eleven of most insects. A tiny feather-winged beetle, *Mikado* sp. (Ptiliidae), has such a small head that it has had to shift its brain into its thorax.[48] Additionally, due to their tiny size micro-insects can fit far fewer neurons into their brains, perhaps just a few thousand in the smallest examples, which creates problems for controlling complex behaviour. Many feather-wing beetles are so small that they've had to make other significant sacrifices. A male has room for only one testis and a female just one ovary, and in some cases female ptiliids can lay only one egg in each cycle. There is also a size limit below which conventional insect eyes won't work. The tiny males of *D. echmepterygis*, for example, are blind. However, the hallmark of the success of insects is their adaptability, and several insect groups have come up with a new eye design to solve that problem.[49]

A major limit to miniaturization is that an insect needs to be large enough to lay an egg containing sufficient yolk to nourish the growing embryo. However, even with such a seemingly unbreakable barrier, insects have found a way to circumvent the problem – just use someone else's yolk. This is exactly what the tiniest parasitic wasps do. They can afford to lay microscopic eggs with no yolk at all because their eggs are laid directly into the eggs of much larger insects. We've already seen how successful a strategy that is.

A few insects grow to much larger sizes, dwarfing these tiny species. Such a large variation in size is another factor promoting insect diversity,

The giant scarab beetles such as this *Eupatorus birmanicus* from Asia are among the biggest of all insects.

by opening up a wider variety of different habitats and lifestyles. However, it's unlikely that size range is one of the main reasons for the richness of insects. The size range of insects covers only three orders of magnitude, small compared to fish, for example, whose size range covers eight orders of magnitude.

Just as the insect body plan imposes limits on miniaturization, so it also prevents insects from growing to the monster sizes depicted in some sci-fi films. The problem lies in the way they are built and in the way they breathe. Insects inherited an exoskeleton from their arthropod ancestors – a tough suit of armour protecting all their soft tissues on the inside. But to grow, an insect must shed its exoskeleton; then, while the new one is still soft, it must pump up its body with air or water, so that when this new cuticle hardens, it's a size bigger, creating space for the insect to grow into.

If an insect was too big, it would be hard for it to support itself after discarding its old exoskeleton before its new one hardened. It could end up as a shapeless blob. However, this constraint only applies in air. Water is denser than air and therefore provides more support; so crustaceans, which also must moult to grow, can reach enormous sizes. The Japanese spider crab, *Macrocheira kaempferi*, has a leg span approaching 4 metres, making it the world's largest crab, forty times the size of the largest insects (giant wētās from New Zealand, about which more in a moment).

However, the limitations of moulting on dry land can't be the whole story.

Christmas Island in the Indian Ocean is a crab-lover's paradise. Once a year the forest floor comes alive with land crabs, *Gecarcoidea natalis*, tens of millions of them, a great red army all marching downslope to the ocean where they will spawn. Their migration is one of the great natural spectacles – it has appeared in so many natural history films that I've lost count. It's so famous that you can buy soft-toy versions of the Christmas Island red crab bearing greetings from the island, one of which is perched on the corner of my desk as I write this. There is, however, another kind of land crab on Christmas Island that to me is even more impressive. It lurks under cover in the forests but ventures into town at night to raid dustbins. And it is huge. At 4 kilograms in weight and with a leg span of 1 metre, robber crabs, or coconut crabs, *Birgus latro*, are the biggest terrestrial invertebrates in the world. They are more than fifty times heavier than the heaviest recorded insect, and with their massively powerful claws, they demand some serious respect. However, like their insect relatives, these crabs must moult to grow. So simply having an exoskeleton can't be the sole reason for a limit on the size of insects.

One big difference between crustaceans and insects is in the way they supply their tissues with oxygen. Aquatic crabs breathe using gills and land crabs have adapted this system to work on dry land by enclosing the gills in a damp chamber that works a little like our lungs. In addition, crabs have an internal circulatory system to carry oxygen to their tissues. Insects, as we've seen, have adopted an entirely different approach. They breathe through a series of holes or spiracles, no more than two on each segment, which connect to an elaborate network of tubes. These tubes (tracheae) branch throughout the insect's body, getting finer and finer until they end as microscopic tracheoles close to the cells they serve. In tissues such as flight muscles, which have a huge demand for oxygen, the tracheoles actually penetrate individual cells, but in both these cases the insect depends on oxygen simply diffusing along the tubes to reach the cells – and herein lies the problem.

If an insect doubles in size, the rate of diffusion along its larger tracheae also doubles, but the oxygen demand of the insect's tissues increases by four times. This means that, as insects get larger, their tracheal system can't keep up – and the insect body plan based on an exoskeleton doesn't help. As insects get larger, the exoskeleton of their legs must get thicker to support their increased weight. However, this narrows the space through which the tracheae need to pass to supply the leg muscles with oxygen, so making it even harder for insects to evolve into larger sizes.[50]

Today, the heaviest insect is a flightless cricket from New Zealand, a giant wētā, *Deinacrida heteracantha*, weighing 71 grams. But even this mega-bug is an exception. The record is held by a female heavy with an exceptionally large number of eggs, and most giant wētās are a lot smaller. There are five kinds of giant beetle that are generally accepted as the biggest insects, at least by bulk: the long-horned beetle, *Titanus giganteus* (167 mm), the elephant beetles *Megasoma elephas* (137 mm) and *M. actaeon* (135 mm), and the goliath beetles *Goliathus goliatus* and *G. regius* (110 mm). But there are other ways of measuring size.

The longest insect is, unsurprisingly, a stick insect – Chan's mega-stick, *Phobaeticus chani*, from Borneo. The body of a big female (not including its long legs) can reach nearly 40 centimetres.[51] This just beats a related stick insect from Borneo, the previous record holder, *P. kirbyi*. In 2006 another giant stick insect was discovered, this time in the rainforests of Queensland in Australia. Only a few female gargantuan stick insects, *Ctenomorpha gargantua*, have ever been found in the wild. They are much bigger than the males and the first one found, named Lady Gaga, was sent to Melbourne to start a captive breeding colony. The gargantuan stick insect is currently the third largest stick insect, although there are rumours that someone came across a really huge individual, photographed it and re-released it; admirable behaviour but precluding this species from perhaps claiming the world record. However, these mega-insects are dwarfed by some of the giants from the past.

Around 300 million years ago, a monstrous beast called *Meganeura monyi* roamed the skies. It probably resembled a modern dragonfly, although only distantly related, but had a wingspan of about 70 centimetres and body length of 30 centimetres. Assuming similar proportions to a modern dragonfly, it must have weighed more than 200 grams, nearly three times more than the heaviest insect known today.[52] This makes it the heaviest insect ever, at least as far as we know, and it flew rather than crawled like today's giant wētās, which demands a lot more oxygen. Nor was it the only ancient giant.

From the late Carboniferous period and on through the following Permian period, insect evolution produced some real monsters. Most impressive were the Meganisoptera, the group to which *M. monyi* belonged. Most fossils of these creatures, now commonly called griffin-flies, are fragmentary wings, but more complete fossils of one species,

Meganeurites gracilipes, have been found and show it to have had large eyes with acute vision and spiny front legs. These are features also found in modern hawker dragonflies, making it likely that griffinflies were active hunters, grabbing prey in flight much like modern dragonflies.[53] Another extinct order of flying insects, the Palaeodictyoptera, reached wingspans of more than 40 centimetres, as did the mayflies of the time. An entomologist exploring the swamp forests of the Carboniferous period would have needed a very large collecting net.

During this time, other arthropods produced even more impressive giants. Without the constraints imposed by aerodynamics, a millipede called *Arthropleura armata* reached more than 2 metres in length. In 2021, a chance rockfall along the coast of Northumberland in north-eastern England revealed the largest specimen of *Athropleura* yet discovered. This monster was half a metre wide and more than two and a half metres long, making it the largest arthropod in Earth's history – at least that we know of.[54] Giant arachnids also stalked the great forests of the coal swamps. The likes of these giant terrestrial arthropods have never been seen on Earth since – and many might be very grateful for that. The reason is that the late Carboniferous period presented them with a unique opportunity. Extreme gigantism during this period was possible because atmospheric concentrations of oxygen reached 35 per cent, compared to just 20 per cent today. Such high levels of oxygen loosen the constraints on insect body size by making insect tracheal systems much more efficient. Oxygen could diffuse further and faster down longer tubes, allowing much larger species to evolve.[55]

There's evidence from living insects that oxygen concentration really can affect the size of arthropods. The life cycle of fruit flies is so quick that they can squeeze many generations into a short time, making them a favourite laboratory animal for studying evolutionary processes. Colonies reared in high oxygen levels do produce larger individuals over time, as long as the temperature is also high. In the wild, too, there is a correlation between the maximum size reached by aquatic amphipod crustaceans and the oxygen content of the water in which they are living. Such studies suggest that higher oxygen concentrations really do allow arthropods to get bigger.[56]

The giant meganeurids probably averaged around 150 grams in weight and, even in their high oxygen world, insects of this size faced problems. In the late Carboniferous and Permian periods, the climate was much warmer than it is today. In such a hothouse world, the energy needed to power these heavy hunters in pursuit of prey would soon cause them to overheat. None of the ways in which flying insects cool themselves today would have been sufficient to prevent burnout, although calculations show that if the atmosphere was also denser than today's, meganeurids could balance their heat budgets. The existence of these giant flying insects suggests that at the time they flew, the atmospheric pressure must have been about one and half times that of today.[57]

Times have changed. Today's thinner atmosphere with its lower oxygen concentration can't support these real giants. Given time, the ever-adaptable insects may have come up with a way around this, but something else happened that doomed giant flying insects. Vertebrates conquered the air. Vertebrates evolved flight on at least four separate occasions, beginning with the pterosaurs in the late Triassic period, around 220 million years ago. They were followed by birds, then by bats. Recently, some curious little dinosaurs have turned up in China that had winged hands, not unlike bats. There is still much debate about these creatures but it seems that dinosaurs may have evolved the ability to fly twice, one line using feathers to create a wing surface (birds) and another using bat-like membranous wings.[58] All of these creatures no doubt relished large flying insects.

The vertebrate endoskeleton, an internal scaffolding of bone, is more suited to large body sizes, as is the vertebrate lung system, so flying vertebrates easily reached sizes greatly exceeding those of flying insects. One pterosaur, with a wingspan of more than 10 metres, grew to the size of a light aircraft. Birds later achieved similar scales. *Pelagornis sandersi*, described as a seagull on steroids, had a wingspan of more than 7 metres. Even today, birds reach sizes far greater than any of the giant insects of the past, and all the flying vertebrates had no trouble evolving to sizes that made giant flying insects nothing more than bite-sized snacks. The mega-insects didn't stand a chance.

RARE – AND GETTING RARER

Predation still plays a role in cutting insects down to size. Wētās, of which there are a whole variety in New Zealand, can only grow so big because, until humans arrived, their home was free of predatory mammals.[59] Many of the largest insects today are confined to similar isolated and predator-free islands. *Polposipus herculeanus*, a giant darkling beetle (Tenebrionidae), is only found on Frégate Island in the Seychelles. The Lord Howe Island stick insect, *Dryococelus australis*, at 20 centimetres in length, is large enough to have earned the name *tree lobster*. It thrived

in its only home, a tiny volcanic speck 600 kilometres due east of Port Macquarie in New South Wales.

Until 1918, Lord Howe Island was also free from mammalian predators. Unfortunately, in this year a British ship ran aground on the island. Inevitably, it had plenty of stowaways in the shape of black rats, which soon escaped to shore. Two years later, they had eaten their way through the entire world population of Lord Howe Island Stick Insects – and, in 1920, the tree lobster was declared extinct. Then, in 2001, two intrepid scientists made an exciting discovery. Intrigued by tales from climbers of hand-sized dead insects littering the rocks at the base of Ball's Pyramid, a 600-metre remnant volcano close to Lord Howe Island, they explored the place at night and found a small population of living Lord Howe Island stick insects.

We saw earlier that the potentially enormous populations of small insects give them some degree of protection from extinction, but giant insects on tiny islands are the exact opposite. They are much more vulnerable to extinction. The Lord Howe Island stick insect is sometimes cited as the world's rarest insect and if black rats ever make it to Ball's Pyramid it will probably be gone for good this time. Other modern giants, though, have already gone the way of the ancient griffinflies.

The giant earwig, *Labidura herculeana*, a creature that reached 8 centimetres in length, was last seen in 1967. It lived only on St Helena, one of the British Overseas Territories and an island so remote that the British chose it as a place to keep Napoleon Bonaparte out of harm's way after his defeat at Waterloo. It sits in the Atlantic Ocean, a tiny pinprick on the map nearly 2,000 kilometres off the coast of southern Africa and it's not an easy place to get to, or at least it wasn't until recently. In 2016, a somewhat precarious-looking airport was opened to receive flights from Africa, but when I travelled there in the early 1990s, it required a flight with the RAF to Ascension Island some distance to the north-west, followed by a sea journey of 1,300 kilometres on a Royal Mail Ship, the RMS *St Helena*.

The strangest thing about finally arriving on this remote island is how little out of place the main town, Jamestown, would seem if it were relocated to the coast of Dorset or Devon. But beyond this home from home, the island, surrounded by dramatic cliffs, rises to spectacular heights cut by deep valleys. When it was discovered by the Portuguese in 1502, it was a verdant place, covered in trees, many of which were found nowhere else on Earth. Giant earwigs were thought to have lived in forests of endemic gumwood trees. In 1995, some dried remains of

Despite their evolutionary success, many insects are currently suffering catastrophic declines, caused by pesticides and habitat loss among other reasons. Few flower-rich meadows, like this one on the machair of the Isle of Lewis, in the Outer Hebrides, still remain.

this earwig were found in an area that suggested that the species also frequented the vast seabird colonies that once covered large areas of the island before they too were wiped out.

Much of the native vegetation has also now been cleared, either by human hands or by the mouths of introduced grazing mammals, so there weren't many places left for the earwig to live. Living specimens were found under boulders at Horse Point Plain in 1965 and there were other reports from 1967. I had been asked by concerned entomologists to spend some time looking for this charismatic earwig in the places where it was last seen alive, and I was happy to create some space in my filming schedule to go on a giant earwig hunt. An expedition mounted by London Zoo in 1988 had failed to find living earwigs and, sadly, I too failed. Subsequent searches in 2005 also proved fruitless, and in 2014 the giant earwig was officially declared extinct.

Since they evolved around 480 million years ago, insects have gradually increased in diversity to the astounding levels that we have encountered in this chapter. Many aspects of insect life have contributed to this success (and we'll discover more of them in the coming chapters). Many of these factors enhanced insect diversity by making them less prone to extinction. Extinction, however, is as much a part of evolution as speciation and happens all the time, although the history of life on Earth has been punctuated by five mass extinctions, in which extinction rates reached much higher levels. The worst of these, at the end of the Permian period, saw 96 per cent of marine life disappear along with 70 per cent of terrestrial vertebrates. This 'Great Dying', as it has been called, also had a huge impact on insects. More than half of all insect families disappeared, the worst extinction suffered by insects in their long history. Yet in the long run, insects continued to thrive and increase in diversity.

Unfortunately, the fate that befell the giant earwig on St Helena – and that might yet befall the Lord Howe giant stick insect – is now being repeated around the whole planet. The scale of this became headline news in 2017, when a report re-analysed data from an entomological society in the small German town of Krefeld near Düsseldorf. The society's headquarters were stacked with alcohol-filled specimen jars crammed with insects collected locally since the society's formation in 1905. In the past they needed so much alcohol to preserve their specimens that the local narcotics bureau took a serious interest in their activities. But society member Martin Sorg noticed that recently their alcohol bill had dropped dramatically.

The great yellow bumblebee, *Bombus distinguendus*, was once widespread across Britain but is now confined to flower-rich meadows along Scotland's west and north coasts.

The data held by the society is so complete that they allowed Sorg to look at how the abundance of insects on local nature reserves had changed over the last hundred years or so. The results shocked not only members of the society, but scientists and naturalists around the world. On one reserve, insect abundance today was 80 per cent lower than in 1989. That pattern repeated itself across the sixty-three reserves they looked at. Overall, in just the last few decades, insect abundance in this corner of Germany had fallen by three-quarters.[60] The speed and scale of the drop was so startling that the paper rapidly became one of the most widely discussed that year among scientists and naturalists around the world. It received a lot of global press coverage too; though, as is often the case, it was soon replaced by other news.

At the same time, a Danish naturalist had noticed that on a drive through the countryside, his windscreen remained free of bugs. He remembered that the same journey in his youth had resulted in a windscreen so spattered with dead bugs that it necessitated frequent stops to clean it. So, with the support of colleagues, he began to gather quantitative data – by equipping cars with large nets on their roofs. These drew plenty of attention from passing motorists, but also from scientists around the world. As in Germany, insect populations had crashed in the last few decades, and following this study the phenomenon

of insect decline is often called the windscreen effect.

To cut an increasingly long and depressing story short, this catastrophic collapse of insect populations seems to be global in scale and is attracting epithets such as 'Insect Apocalypse' and 'Insect Armageddon'. Data from many places is still sparse and it's already clear that not all insect groups are declining. Some are stable and some have even shown recent increases.[61] For example, a wide-ranging analysis shows that terrestrial insects are declining at a rate of 1 per cent a year, while aquatic insects are increasing at the same rate.[62] Some studies are also contradictory. One study claimed that insects have suffered a catastrophic collapse in Puerto Rico's Luquillo Experimental Forest,[63] while another suggests that insect numbers have been more or less stable.[64]

In 2019, the Entomological Society of America hosted a symposium in St Louis, Missouri, to pull together data from around the world to clarify patterns of decline. The participants outlined the wide range of threats faced by insects, describing it as 'death by a thousand cuts'. Although the picture is far more complicated than some of the earlier reports suggested, the scale of the decline is nevertheless alarming, since insects underpin virtually all terrestrial ecosystems.[65] Many scientists now regard our current impacts on the natural world as a sixth mass extinction. Insects survived all five of the others, but human activity seems to be having a far greater impact on insects than any of the previous five. To continue on the same path and further erode insect numbers and diversity is the height of folly. Three-quarters of all the different crop types that we rely on require insect pollination, and many other large animals are equally dependent on insects for their survival.[66]

Insects are also food for people. Around the world, some 2,000 species of insects are eaten by humans. Some are curious luxury items, such as chocolate-covered ants or mealworms embedded in candy. The late Emperor Hirohito of Japan was apparently very partial to boiled wasps with rice, and tinned grasshoppers (*inago*) are still sold in Japan as a luxury food. However, many more people rely on harvesting insects for at least part of their livelihood.

From our biased Western perspective, we might view such people as struggling to survive, forced into eating insects as a last alternative. In fact, they have the moral high ground. Insects are by far the most ecologically sound way of producing animal protein. Crickets, for example, convert their food into edible protein with more than ten times the efficiency of cows – and in doing so, they need over fifty times less water. Eating insects could, quite literally, save our planet.

For the economically minded, there's already big money in insects as food. Tinned silkworms, to pick just one example, are worth $50 million in exports to Thailand. Because economics rather than ecology makes most world leaders sit up and take notice, scientists have also tried to put a monetary value on the pollination services that insects carry out for us. It's based mostly on inspired guesswork, but the global figure is somewhere between $235 and $577 billion dollars annually. In one sense, though, these are meaningless figures. If we lose these pollination services, no amount of money will buy us out of a diet based solely on wheat, rice and maize, our main wind-pollinated plants. Forget your 'five-a-day'.

Already, in south-west China, pollinators are so scarce that farmers have to pollinate apples and pears by hand, an unbelievably tedious and time-consuming process. Even in Britain, apple crops today are smaller than they could be. It's estimated that the recent drop in pollinator abundance is costing growers here around £6 million each year. In response, engineers are trying to design robotic bees that can take on the duties that insects already provide for free. Is it just me – or is this utterly crazy?

The decline of insects is having a huge impact on economy and ecology alike, enough that you might think that by now this problem should have risen up the political agenda. Sadly not. Part of the reason lies in the fact that most people don't really care about insects. At a 1968 meeting of the International Union for the Conservation of Nature, a Senegalese forester, Baba Dioum, remarked: 'In the end we will conserve only what we love, we will love only what we understand.'

He was absolutely right. We need to learn to understand and love insects if we want a future on this planet. In the following pages, I hope to bring a greater understanding of why insects are so vitally important, by exploring their long history and their diverse and often surprising behaviour and ecology. Perhaps then we'll realize why our future depends on these little creatures. Because they've been so successful over their evolutionary history, these teeming hordes have become the lynchpins of terrestrial ecosystems. The simple fact is that we need insects far more than they need us.

2

Origins
The long history of insects

...what can be more different than the immensely long spiral proboscis of a sphinx-moth, the curious folded one of a bee or bug, and the great jaws of a beetle? Yet all these organs, serving for such widely different purposes, are formed by infinitely numerous modifications of an upper lip, mandibles, and two pairs of maxillæ.

Charles Darwin, *On the Origin of Species*, 1859[1]

Cape May, protecting the entrance to Delaware Bay on America's East Coast, is less than a couple of hours' drive from Philadelphia. Historically, this 'city of brotherly love' served briefly as the nation's capital and today lies at the centre of the Northeast Corridor, a relentlessly bustling and claustrophobic strip of concrete, glass and asphalt anchored at one end by the modern capital of Washington DC, and at the other by the financial capital of New York City. But as dawn broke and the bay slowly revealed itself, I could have been half a planet away from the modern world with all its incessant noise and bustle. A long beach curved away into the pre-dawn glow, backed by low dunes of rustling grasses. From where I stood, a few deserted beach shacks poked above the dune line, loosened shutters clattering in gusts of wind. The braying of laughing gulls as they greeted the new day echoed out of the receding darkness. As more of the landscape was illuminated, my sense of remoteness only grew. Soon I would feel distant from the civilized world not just in space, but in time as well. The event that I had come here to film would transport me back beyond the time the first Europeans stood on these shores, further back even than the first American Indians to see this bay, all the way past the last and first dinosaurs, to a point nearly 450 million years ago.

It began with a few shapes exposed by the backwash of the waves, to be covered again as the next wave rolled in, more and more over the next ten minutes until the shapes started to haul themselves clear of the water. Now, they revealed themselves as strange alien creatures, some up to half a metre in length, including a long spike of a tail that projected from a domed body topped by two frowning eyes. They could have stepped straight from a *Star Wars* film. This was the first time I'd ever seen a horseshoe crab, although in just an hour's time I would have seen tens of thousands of them. On the spring tides in May and June, they emerge from the murky depths of the bay in vast numbers to lay their eggs at the high-water mark. There the eggs will develop, safely buried in the sand, until the next spring tide in a couple of weeks' time washes newly hatched larvae out into the bay.

Horseshoe crabs have been around for a very, very long time. The oldest known fossils, at 445 million years old, date back to the Ordovician period.[2] Although we can't know whether these ancient horseshoe crabs spawned in the same way as their modern descendants, it would make a lot of sense. The Ordovician seas teemed with predators, whereas no animal had yet set foot permanently on dry land. It was by far the safest place to lay their eggs.

(*previous page*) Every spring and early summer, on spring tides, thousands of Atlantic horseshoe crabs, *Limulus polyphemus*, crawl on to the beaches of Delaware Bay on America's east coast to lay their eggs at the high-tide mark.

Horseshoe crabs are not insects – they're not even crabs. They're more closely related to spiders, but like crabs, spiders and insects they *are* arthropods; by crawling ashore, they are reliving a turning point in arthropod history. I've revisited this spectacle on many occasions since, and every time, as I watch the drama of their mass spawning unfold, I can't help but imagine that at some time in the past, the far distant marine ancestors of insects must have done something similar, their first faltering steps on land driven by the need to keep their eggs safe or perhaps to graze on bacterial mats just above the tide line. Over time, generations of these unknown creatures could have ventured ever further from the water until finally they became true land dwellers – ready to spawn the most successful group of animals ever to have evolved, above or below the waves.

THE ARTHROPOD DIASPORA

We have absolutely no idea what the ancient ancestors of the insect dynasty looked like and much of their early story remains a mystery, lost in deep time. Even so, we can now lay out a generally accepted broad timeline for their conquest of the planet, informed both by good old-fashioned fossil digging and by ingenious new molecular and imaging techniques. We know for certain that the ancestors of insects, like those of all other arthropods, began life in the ocean and at some point crawled ashore to colonize dry land, although they weren't the only arthropods to take this bold step.

Arthropods have invaded land on several separate occasions, at least three times in the ancient past, to give rise to three different terrestrial dynasties – the arachnids (spiders and scorpions), the myriapods (centipedes and millipedes) and the insects.[3] In more recent times, some crustaceans have also successfully emerged from the waves. The woodlice in your garden are crustaceans, as are the strange little amphipods called lawn shrimps, originally from Australia although now successfully invading many parts of the world thanks to human assistance. True crabs have also made it a surprisingly long way from the ocean. I've followed hordes of large yellow land crabs, *Johngarthia lagostoma*, as they descended more than 300 metres from the high central mountain on Ascension Island to the jagged lava coast where they spawn, an arduous trek that takes them several days.

However, the most impressive example that I've ever come across is the desert crab, *Austrothelphusa transversa*, which lives in Australia's dry interior, admittedly buried in a small, damp hole deep below ground for most of the year, only emerging when occasional rains form temporary pools into which females can release their young. Vertebrates, including our far-distant ancestors, also successfully invaded land, but only once and much later than the three oldest terrestrial arthropod groups. Clearly, there is something special about arthropods.

Arthropod means 'jointed foot', and such jointed limbs are shared by all these varied creatures, along with their extinct relatives such as the trilobites. Articulated legs allowed for efficient crawling on the seabed or swimming through open water, but would eventually prove equally useful for crawling out onto land. The earliest arthropods had lots of these legs, whereas our own forebears had to make do with just four, derived from the cumbersome, fleshy, lobed fins of a fish-like creature, perhaps resembling a modern lungfish or coelacanth. 'Creepy crawly' might be a term of disparagement that we heap on 'bugs' of all sorts, but those creeping, crawling jointed limbs allowed arthropods to run before we could even walk.

Multiple multi-jointed legs make arthropods seem very alien to us, so much so that these creatures are a frequent inspiration to designers of sci-fi monsters for film and television. However, there's an even bigger reason why extraterrestrials often bear an uncanny resemblance to terrestrial arthropods. They have a hard external skeleton that can be sculpted into a profusion of bizarre spikes or horns, or textured to look like stone, bark, leaves or lichens. In the previous chapter I pointed

out that an exoskeleton limits the size that insects can reach, but any such disadvantages are more than offset by the advantages it gives an arthropod setting off to conquer dry land. Along with their jointed legs, their exoskeleton was a pivotal feature that gave arthropods the edge over our own ancestors in the invasion of dry land. Out of water, the exoskeleton provides support in the drastically less dense world of air, and it can easily be rendered waterproof, simply by a coating of wax, to avoid drying out.

However, these crucial attributes of terrestrial arthropods, their jointed legs and hard exoskeleton, didn't evolve as a solution to life on land. They are *exaptations* – features that became co-opted for uses different from those that drove their original evolution. They are fortuitous 'pre-adaptations' that gave arthropods a distinct advantage when it came to leaving the ocean world behind. So, in our quest for the reasons behind the ultimate success of insects, we must first ask where these key arthropod features came from. How and when did the first arthropods appear on the scene, and what originally drove the evolution of the features that made them arthropods? That search takes me back in time, further even than ancient horseshoe crabs – and, unfortunately, back into the busy Northeast Corridor, where I crawled slowly south through nose-to-tail traffic on Interstate 95 until I hit the Capital Beltway, a 64-mile-long car park encircling Washington DC.

The great cities of America's north-east are not my natural habitat, although I visit Washington frequently because it's the home of the Discovery Channel, the Smithsonian Channel and National Geographic, for whom I've made many documentaries over the years. Washington does have one other major draw for me – the National Mall, running from the US Capitol to the Lincoln Memorial and lined in part by the museums of

(left) Yellow land crabs, *Johngarthia lagostoma*, from Ascension Island are just one of a surprising range of crustaceans that have conquered dry land.

(top) A lantern bug, *Cathedra serrata*, from South America, illustrates just how versatile the insect exoskeleton is. It can be moulded to almost any shape.

the Smithsonian Institution. Halfway along the museum strip stands the National Museum of Natural History (NMNH), a monumental edifice to the natural world within which I'll find a major clue in our pursuit of arthropod origins.

The fossil hall of the NMNH is, inevitably, dominated by dinosaurs, a group of only middling success at best when compared to arthropods, but for some reason universally more popular. Tucked away in one corner, largely ignored, is a display of rocks covered in faint impressions, some like delicate etchings, others like photographic negatives, of creatures far more alien than even horseshoe crabs. These rocks originated in a remote corner of the Canadian Rockies, and they revolutionized our view of early life on Planet Earth.

WHERE DID ARTHROPODS COME FROM?

The fossils of the Burgess Shales date from around 505 million years ago, in the Cambrian period. They are housed here at the Smithsonian because they were originally discovered by Charles Walcott, who was at the time the fourth Secretary of the Smithsonian Institution and an expert on trilobites, a group of arthropods common in the Cambrian period. Towards the end of a busy field season in 1909, studying the Cambrian strata of the Rockies in British Columbia, he stumbled across a previously unknown treasure trove of ancient fossils. The often repeated, although probably apocryphal, story is that his wife's horse (his wife often accompanied him on these long field seasons) stumbled on a rock, which cleaved to reveal a fossil preserved in incredible detail. Whatever the true circumstances were, Walcott immediately recognized the very high quality of preservation, if not the true value of his find. In what later turned out to be a classic understatement, he wrote to a colleague that he had 'found some very interesting things', which – at the time – he identified as phyllopod crustaceans.

It was too late in the season to explore the site further, but when Walcott returned the following year the true importance of his find became apparent. The marine world of the early Cambrian period was laid out in unprecedented detail – a fauna of strange creatures that seemed to defy imagination. After nine more seasons in the field, Walcott had recovered 65,000 specimens of some 120 species, ranging from *Hallucigenia*, a creature that, as its name suggests, would not be out of place in a drug-induced vision, to *Anomalocaris*, a half-metre-long killer 'shrimp'. Since Walcott's day other locations around the world, with similar exquisitely

A tadpole shrimp, *Triops* sp. Only a few species of these primitive crustaceans survive today but their fossil record stretches back nearly 400 million years – a glimpse into the early history of arthropods.

preserved fossils, have been discovered, at Chengjiang in China, Sirius Passet in Greenland and Emu Bay in Australia, all of which pre-date the Burgess Shales. More recently, palaeontologists came across a remarkable site not far from Walcott's original find that gave up even more of the Cambrian period's secrets. In just one season, in 2012, 3,000 specimens belonging to 50 different kinds of animal were extracted from Marble Canyon, in a remote part of Kootenay National Park in British Columbia. Ten of these were entirely new to science.[4]

Together these discoveries paint a picture of the early Cambrian period as a time of rapid proliferation of life, a time when all the modern body plans, including that of the arthropods, first came into existence. These creatures appeared so suddenly that this period of evolution is often referred to as the 'Cambrian Explosion'. Whether this really *was* an explosion or just an artefact of a woefully incomplete fossil record has long been the subject of debate, although most scientists now agree that something special really did happen at the start of the Cambrian.[5] So, were arthropods born in this explosion, and if so, what was the trigger?

One intriguing suggestion, made by Andrew Parker of Oxford University, is that eyes evolved. Predators with the power of sight would have been a devastating new force in the ocean, prompting the rapid evolution of protective measures, such as, for example, the tough exoskeleton developed by the earliest arthropods. An escalating arms race of measure and countermeasure, as predators evolved ever more ways of overcoming their prey's defences, could fuel many such evolutionary innovations and create an explosive increase in diversity.

Others have suggested that the key factor in igniting the Cambrian Explosion was oxygen. For much of Earth's four and a half billion-year history, our atmosphere was devoid of oxygen. It wasn't until a group of bacteria (cyanobacteria), sometime around 2 billion years ago, evolved the neat trick of photosynthesis – of using the power of sunlight to make carbohydrates from water and carbon dioxide – that things began to change. Life arose on Earth about three and a half billion years ago, and for several billion years existed as nothing more than microbes, including many and varied forms of bacteria. This period has been called the 'boring billions', which is a little harsh. Microbes are extraordinary – they are invisible chemical factories that even today still have a massive impact on the way our planet works.

Over their long history, bacteria came up with lots of ways of using the energy in light to make food. One of these, the type of photosynthesis that all modern plants now depend on, is called oxygenic photosynthesis, because its main waste product is oxygen. And it was a global catastrophe – at least at first. For all the organisms that existed at the time oxygen was deadly, so reactive that it could rip their molecules apart. However, in time, life would not only work out how to live with oxygen but turn it to a huge advantage.

It didn't take long for a different group of those little chemical engineers to evolve a way of harnessing oxygen, by allowing it to react in a controlled way with food molecules to release far more energy than was available to any of the other microbes around at the time. This evolutionary innovation changed the world forever, eventually paving the way for more complex cells, then for multi-cellular life – the first animals and plants.

After this *great oxygenation event*, oxygen levels in the atmosphere fluctuated over geological time, driven by biological and geological processes. We've already seen that such changes have had profound effects on insect evolution but, back in the Cambrian period, long before insects appeared on the scene, oxygen may have been the fuel that caused the Cambrian Explosion to burn so bright. Energized by raised oxygen levels, active hunters evolved, for example with jointed legs for crawling or swimming in fast pursuit of prey, or powerful, muscular claws for grabbing and crushing their victims. At the same time, their prey was not entirely helpless. These creatures also had energy to burn and could afford to build elaborate defences in the form of exoskeletons and mineralized shells or faster legs to escape. In this scenario, the controlled combustion of oxygen fuelled a diverse new array of lifestyles.[6]

What does all this mean for our search for the origins of arthropods, and for those arthropod features that insects would later put to such good use? When Walcott began to examine the Burgess Shale fossils, he tried to fit them into existing animal groups. Later, some scientists were so bemused by the weirdness of creatures like *Hallucigenia* that they assigned many of the organisms of the Burgess fauna to entirely new groups of animals – strange experiments in evolution that blossomed and died in this first flowering of complex life. Later still, and with many more sites yielding fossils of a similar age or older, scientists have returned to Walcott's original interpretation. As weird as they are, these early Cambrian creatures show features that tie them to modern groups. And among these creatures are the first arthropods.

The earliest arthropod fossil that we know of dates to 537 million years ago, almost at the start of the Cambrian, which began 541 million years ago. It's a fossil called *Rusophycus*, but it's not an actual animal. It's a trace fossil, formed by an unknown, multi-legged beast as it rested in the mud, contemplating the strange new world in which it found itself.[7] Arthropods, it seems, arose very quickly as the Cambrian Explosion ignited – so quickly that some have their suspicions that they may have had their origins even deeper in time.

The belief that the Cambrian period saw the birth of complex life was so entrenched up until the middle of the twentieth century that we consistently ignored evidence to the contrary. Then, in 1957, a schoolboy, Roger Mason, climbing in Charnwood Forest in Leicestershire, came across some strange markings in the rocks. He took a rubbing of the markings and showed it to scientists, who recognized it as a form of complex life, now called *Charnia*, in rocks that pre-dated the Cambrian period. *Charnia* became the first form of complex life to be described from Precambrian rocks.

The story of our discovery of these first glimmerings of complex life is one of misinterpretations and missed opportunities. Even *Charnia* should have been in the spotlight of geological fame earlier than it was. It turns out that the Charnwood fossil was originally seen by a fifteen-year-old schoolgirl called Tina Negus a year before Mason's climbing adventure. She had a deep interest in geology and knew she was exploring Precambrian strata, so was perplexed at seeing a clear, frond-like impression in the rocks. Of course, no one believed her, and her geography teacher informed her in no uncertain terms that there were no fossils in Precambrian rocks. When she insisted that she had definitely seen a fossil, the teacher's response was that therefore these were not Precambrian rocks.[8]

(overleaf) Black sea nettles, *Chrysaroa achlyos*, belong to a group (Cnidaria – the jellyfish and sea anemones) that probably evolved and thrived in the Ediacaran Period, before the Cambrian explosion, when much simpler body plans predominated.

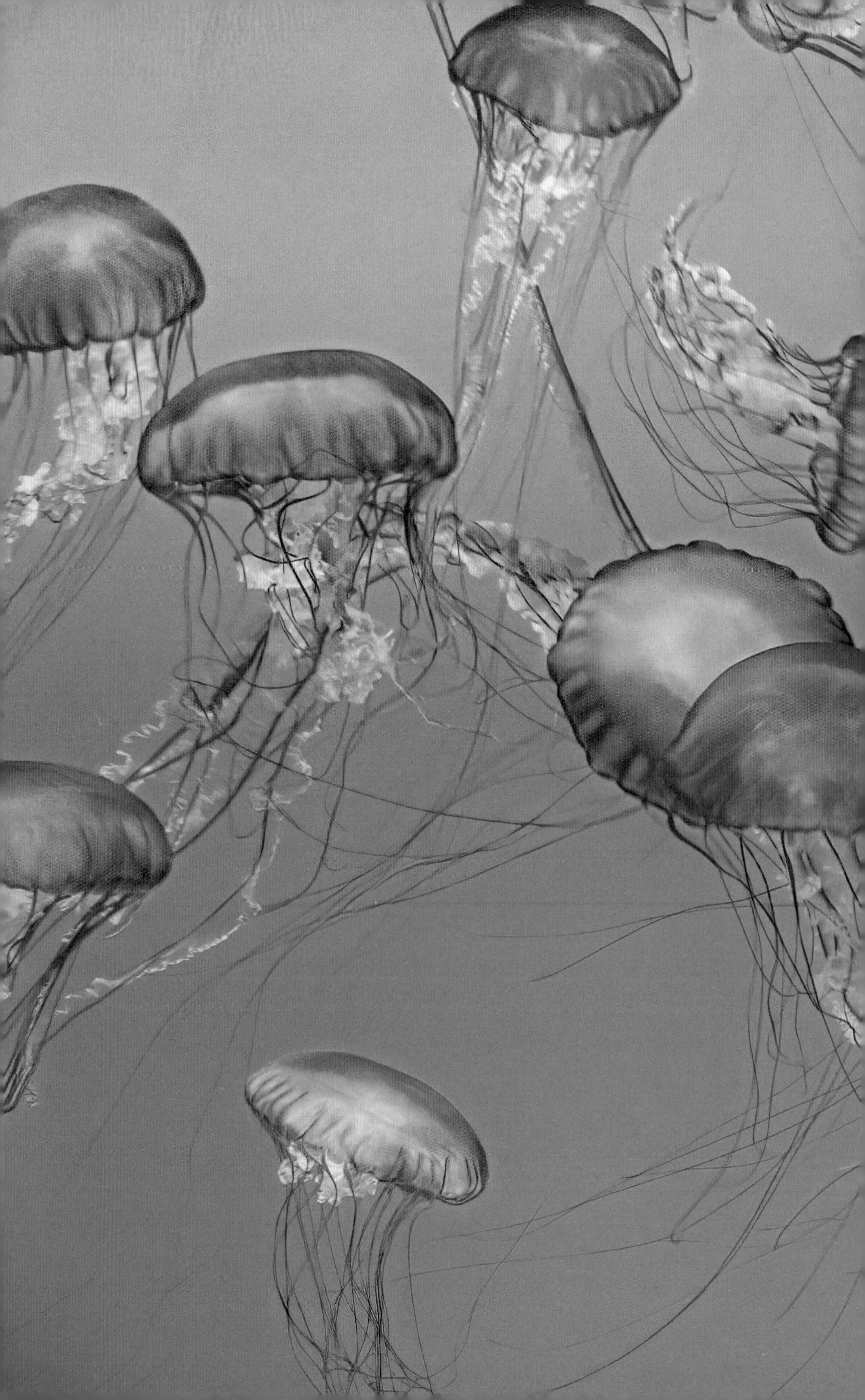

That same logic had been applied to similar fossils, found long before in England, and to others from Newfoundland, Namibia and Australia. In 1946, the Australian geologist Reg Sprigg was working in the Flinders Ranges of South Australia when he came across faint impressions in the ancient rocks in the Ediacara Hills. These were clearly complex multicellular life forms, so by the reverse logic of the time, the rocks were therefore presumed to be of Cambrian age. After the description of *Charnia*, all these errors were corrected, and it became clear that we now needed a new geological period to encompass these discoveries. Eventually, this period, immediately preceding the Cambrian, became known as the Ediacaran, after the Ediacara Hills, which continued to reveal new fossils of this age.

The creatures that lived in this period are, to say the least, enigmatic. They are even harder to place on the tree of life than the Burgess Shale animals. They may represent an early explosion of multicellular life, unrelated to anything that followed. Some have even suggested that they are intermediate life forms between animals and plants. Others think that, among this strange fauna, they have glimpsed the earliest ancestors of the arthropods. This suggestion, based on existing fossil evidence, is hotly disputed, but there are other ways of looking at the problem of arthropod origins.

Evolution proceeds by the slow accumulation of mutations along the strands of DNA that make up chromosomes. These mutations occur at a more or less steady rate, so the greater the number of mutations that have accumulated, the more time must have passed. They act as a molecular clock, ticking away the time that has elapsed since any two groups of descendants diverged from their common ancestor. The bigger the differences in the genomes of species being compared, the further back in time they must have diverged. To get an absolute, rather than a relative, measure of time, the clock must be calibrated against fossils of known age, but then it can be used to estimate when groups of animals arose even if the fossil record is sparse or non-existent. In any case, the laws of probability – or at least the rarity of fossilization – make it highly unlikely that we'd ever find the earliest fossil of any group,* so these molecular analyses might be expected to give a more accurate picture of the timing of key evolutionary events. It's not an exact science since it depends on a series of assumptions about, among other things, the rates

* This is known as the Signor-Lipps effect, after Philip Signor and Jere Lipps, which states that, since the fossil record is never complete, neither the first nor the last organism of any group will be recorded as a fossil.

that mutations happen. So, trying to marry the fossil record with what molecular studies tell us still engenders plenty of lively debate.

Molecular analyses of living arthropods do point to their origin in the Ediacaran period, although they can tell us nothing about what that earliest arthropod might have looked like or whether any of the many described Ediacaran fossils are actually arthropods.[9] There are no unequivocal arthropods from the Ediacaran, either body fossils or trace fossils, such as *Rusophycus* from the early Cambrian. In any case, the Ediacaran seemed like a gentle world, with simple animals going quietly about their business – until the Cambrian exploded.

Within a short time, arthropods began to diversify and dominate their ocean world. They were both fuel and catalyst for the Cambrian Explosion. As sophisticated predators like *Anomalocaris* evolved, arthropods helped to escalate the arms race that drove a rapid expansion of life in the early Cambrian. By the time the Burgess Shale creatures were crawling and swimming through the ocean around 30 million years later, arthropods had consolidated their hold on the world. About a third of all species in the Burgess Shales are arthropods. By the end of the Cambrian, almost all the major arthropod groups had evolved.[10] Today they are still the most diverse group of animals on the planet – a distinction they've held, unchallenged, since the early Cambrian.

Adaptations that evolved to cope with the new, tougher world of Cambrian oceans would later prove just as effective in conquering dry land. Insects inherited these features from their most distant arthropod ancestors, but it's what they went on to do with them that catapulted them to far greater success than any other arthropod group. To understand how they've done that, we need to know a bit more about how insects fit into this most ancient of dynasties.

ARTHROPOD EVOLUTION

The detailed relationships of all the varied arthropod groups to each other have long been the subject of much wrangling among biologists. It has, however, at least been agreed for some time that the living arthropods can be broken into two large-scale groups: the *Chelicerata* (sea spiders, horseshoe crabs, spiders, scorpions and their relatives) and the *Mandibulata* (myriapods, insects and crustaceans).

This fundamental division of the arthropods is based on the mouthparts of the two groups. The 'fangs' of spiders are more correctly called *chelicerae* (singular: chelicera) and all the Chelicerata, including the horseshoe crabs I watched spawn on Delaware Bay, share this feature, although spider fangs are a specialized form of chelicerate mouthparts, hinged like a jackknife for stabbing prey and injecting venom. In most other chelicerates, these mouthparts form pincer-like feeding structures, easily visible in those horseshoe crabs that get flipped onto their backs by the waves.

This may not now come as much of a surprise, but members of the second group, the Mandibulata, are defined by their possession of mandibles, which evolved as slicing and crushing appendages, although many insects have taken this basic design and modified it beyond all recognition. Being able to remodel their mouthparts to cope with radically different diets has helped insects exploit a huge range of ecological opportunities. These differences in mouthparts might seem like a tiny point of distinction, of interest only to biologists poring over jars of specimens in their ivory towers. However, in reality, understanding this very early step in the evolution of arthropods answers a big part of the question as to why they have been so successful for so long.

The first arthropods, as illustrated by those 537-million-year-old resting traces, were long-bodied creatures with many repeated segments, each carrying a pair of legs. We don't really know what the animal that made those impressions looked like, but we can make a good guess. In fact, I've found animals living today that are probably not all that different from the very earliest arthropods.

One of the joys of spending time in a rainforest is searching through the thick leaf litter or rolling over rotting logs to see what might be lurking beneath. If you are really lucky, your efforts might reveal a pencil-length 'worm' with little stumpy legs that quickly hides itself away again. Its skin, which may be a startling pink, red or blue depending on the species, has the appearance of soft velvet, so these elusive animals are often called velvet worms. They belong to a group known as *Onychophora*, which closely resembles fossils called *lobopodians* from very early in the Cambrian period. Several lobopodians are preserved in the Burgess Shales and this group of segmented, worm-like creatures shared a common ancestor with all the arthropods, an ancestor that probably looked a lot like those fossil lobopodians, or even modern velvet worms, at least as far as having a body composed of lots of similar segments. As arthropods evolved, some or all of those segments became specialized to perform specific functions, but the same segments were able to specialize in different ways in different types of arthropod.

The giant whip scorpion or vinegarroon (*Mastigoproctus giganteus*) is a relative of spiders, scorpions and horseshoe crabs. Like these creatures it's part of the chelicerate branch of the arthropods.

The most fundamental illustration of this almost limitless flexibility, marking a deep division among arthropods, are the mouthparts of the two major groups. Chelicerae and mandibles may both serve as mouthparts, but they evolved from limbs on different segments. The segment whose appendages became chelicerae in the chelicerates evolved into antennae in the mandibulates, structures which could not be more different in form or function. Conversely, the mandible equivalent in chelicerates is their first pair of walking legs.[11] This set of specializations arose very early in arthropod evolution, which is why it now marks the broadest division within the group. But that basic multi-segmented arthropod body plan had the capacity for an almost endless set of such variations, exploited in distinct ways in each major group. Design flexibility is the hallmark of the arthropod body plan, a quality pushed to the limit in the most successful of the arthropods, the insects.

THE FIRST INSECTS

How and when insects began to tread their own path through evolutionary history isn't known exactly. The first insects are poorly represented in the fossil record, making their early story difficult to interpret. The oldest fossil of an insect that we currently have was preserved in a type of fine-grained rock called chert, near Rhynie in Aberdeenshire, in north-east Scotland. Discovered in 2004, *Rhyniognatha hirsti* dates to between 396 and 407 million years old, which places it in the Devonian period, although, since it seems to show several advanced insect features, the fossil suggests that insects arose earlier than this.[12]

However, controversy still surrounds this mere smudge of a fossil. Only the head of *Rhyniognatha* was preserved and more recent analyses, using sophisticated 3D scanning techniques, suggest it is not the head of an insect at all, but of a centipede.[13] Other, more obvious centipede fossils are known from the Rhynie chert, which is not surprising since this group, the myriapods, was probably the first of the arthropods to invade land. Myriapod trackways have been found in the Ordovician period, some 50 million years before the Rhynie chert was laid down.

Luckily, even in the absence of definitive fossils, it's possible to make some good estimates of the timing of major turning points in the arthropod story using the molecular clock. These techniques tell us that the three early terrestrial lineages (arachnids, myriapods and insects) arose sometime between the Cambrian and the Silurian periods, the myriapods possibly early in the Cambrian and the arachnids later in the

Cambrian, colonizing the land perhaps at the boundary of the Cambrian and Ordovician periods.[14]

The insect part of the story is now also becoming clearer following an intensive effort by more than 100 scientists working on the 1KITE (1K Insect Transcriptome Evolution) project. This project aims to study the transcriptome* of 1,000 insect species spread across the diversity of different living groups. So far, the results suggest that insects evolved back in the early Ordovician period (around 479 million years ago), as marine organisms, before invading the land sometime later.[15] Insects might have been the last of the ancient arthropods to set foot on land, but their late start doesn't seem to have handicapped them.

What could that most ancient of insect ancestors have looked like? A clue presented itself in 1979, when some very strange creatures were discovered in the Bahamas by Jill Yager from the Smithsonian Institution. Lucayan Cavern is a water-filled cave carved into the limestone foundations of Grand Bahama Island. It is connected to the Atlantic but runs far inland, so is fed by both salt and fresh water. Since the two types of water have different densities, the fresh water lies as a distinct layer on top of the denser salt water, separated by a sharp boundary called the halocline. In the pitch black, below the halocline in the saltwater layer, Yager collected animals that looked for all the world like swimming centipedes. In fact, they were even more bizarre and unexpected. They turned out to be a whole new class of crustaceans called Remipedia.[16] Remipedes have subsequently turned up in similar caves elsewhere around the Caribbean, in the Canary Islands and even in one cave in Western Australia. And they are very strange crustaceans indeed.

Like centipedes they have two fang-like jaws (called maxillules in remipedes) and also like centipedes, but unlike every other kind of crustacean alive today, those jaws deliver a venomous bite.[17] In the wild and in laboratory aquaria, remipedes rummage around in the silt on the cave floor and were first assumed to be filtering out small organic particles, but with those venomous jaws it looks as though they would be quite capable of overpowering other crustaceans that share these caves. Recently, I sent a film crew in pursuit of these creatures deep into a flooded cave on Mexico's Yucatán peninsula, where they managed to

* The transcriptome is different from the genome. The genome refers to all the DNA in a cell. However, not all of this is actively being used to code for proteins. Active sections of DNA are first transcribed into molecules of RNA that go on to guide the production of proteins. The transcriptome is the total of all these RNA molecules, which is a reflection of the active or expressed part of the genome.

film a remipede clutching a cave shrimp in its jaws. However, the most remarkable fact about remipedes had to wait until molecular studies of their genome were carried out. Remipedes turned out to be the closest living relatives of insects.[18]

Molecular techniques have transformed our understanding of the relationships among the various groups of arthropods. Before we were able look directly at the genetic code, scientists had to compare the anatomy of the different groups to construct the arthropod family tree, but different interpretations led to several alternative arrangements of how this mighty tree had grown and branched. One popular theory was that insects were closely related to myriapods, so the two were placed together in a group called Tracheata (for their shared possession of tracheal breathing tubes). Modern molecular phylogenetics has created more robust and reliable trees and instead indicates a close relationship between insects and crustaceans. Thanks to molecular phylogenetics, the Tracheata as a group has been consigned to history and in its place now stands the Pancrustacea – a group that includes all the crustacean and hexapod lineages. For those who lump spiders, scorpions or even millipedes together as 'insects', the message from molecular phylogenetics is that insects are far more closely related to crabs than they are to spiders or millipedes. Furthermore, the very close relationship between insects and remipedes allows a glimpse of what the first insects might have looked like.

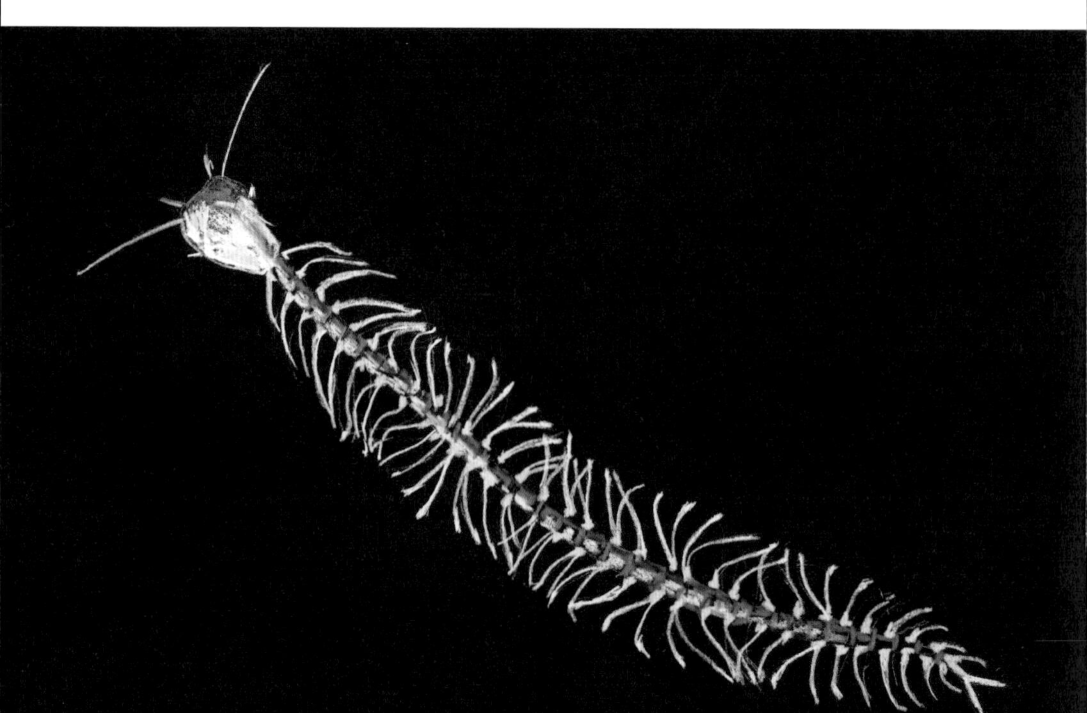

Although they share a common ancestor, remipedes did not give rise to insects, and because remipedes are aquatic and live in a very specialized habitat, they can't tell us much about what the first terrestrial insects looked like. However, it's not beyond the bounds of possibility that an ancestral marine insect would have looked something like a remipede. It probably had repeated similar segments, each carrying legs, much as remipedes, as well as centipedes and millipedes, have retained to the present day. But the first of the insect innovations that would set them on the path to success was to streamline this repetitive segmental pattern into the more ergonomic and adaptable system that we saw in the introductory chapter.

This basic ground plan that unites all insects is a body built around three distinct regions – the head, thorax and abdomen. These evolved by the fusion and subsequent specialization of segments of the ancestral multi-segmental body. Six of the original segments fused to form the head, and their appendages became mouthparts or antennae. The thorax is composed of three segments, each of which still carries walking legs. In most insects, the last two thoracic segments also carry wings. The last eleven segments formed the abdomen; most of them lost their appendages, apart from cerci, sense organs that protrude from the back ends of some insects.

These regional specializations reduced the number of legs to just six. At the other extreme, millipedes, in which each segment carries two pairs of limbs, may have several hundred. Literally, millipede means 'a thousand legs', though until very recently none were known with this many legs. The record stood at around 750,[19] which is still a pretty impressive feat of coordination when it comes to the simple act of walking. Then, in 2021, the record was smashed by a millipede with 1,306 legs – the first true millipede. *Eumillipes persephone*, from Western Australia, is like a strand of spaghetti with legs, less than a millimetre wide but nearly 10 centimetres long.[20] It lives underground, where its long, thin body and short but numerous legs facilitate the tricky business of travelling through twisting burrows.

Watching a giant millipede, perhaps 30 centimetres long, crawl along a branch is a hypnotic experience as each pair of legs in turn is lifted and moved forwards, creating a perfectly coordinated Mexican wave. But this arrangement limits what millipedes can do. Luckily, their ponderous perambulations aren't too much of a hindrance – millipedes feed on decaying vegetation that isn't too hard to outrun. On the other hand, their myriapod relatives, centipedes, are all carnivores and active

Remipedes are strange crustaceans that live deep in caves, and were only discovered in 1979. Genetic analysis shows them to be the closest living relatives of insects.

hunters. Centipedes have fewer legs than millipedes. They only have one pair of much longer legs on each segment, and they're a lot quicker on their feet. I have huge respect for giant centipedes. These monsters often reach 20 centimetres in length and can deliver an excruciating bite. They are lithe and fast enough to make even my longest forceps seem far too short when I have to handle them. Faster still are house centipedes, *Scutigera coleoptrata*, so called from their fondness for damp, dark cellars. They have only fifteen pairs of legs, but these are extremely long, and house centipedes move like lightning.

Insects have taken this reduction in legs further, to just six, the minimum number that allows a stable tripod to be formed by three standing legs while the other three are swung forwards. This hexapod arrangement seems to be the most efficient way of using legs to move on land. However, true to form, insects have been much more inventive with their legs than just using them for walking or running. They have shaped their legs into deadly traps or musical instruments to play songs of courtship. They serve as sensitive chemical detectors or semaphore flags, along with a host of other unexpected uses. We'll explore these in much more detail in the next chapter. For now, it's enough to know that having six highly adaptable legs is another reason why insects have become so diverse.

Despite the obvious efficiencies of this hexapod ground plan, insects didn't take over the world immediately. Even today, wingless insects, sporting a three-part body with six legs, are not very diverse, and such insects are remarkably scarce in the fossil record for many tens of millions of years after their likely origin.[21] Only six scraps of fossils are known from the Devonian period, including the disputed *Rhyniognatha*. No fossils at all are known from the early Carboniferous period.[22] This lack is so conspicuous that the period between 385 and 325 million years ago is known to palaeoentomologists as 'the Hexapod Gap'. Whether this represents a real pause in the conquest of dry land or simply a lack of fossil evidence is hard to say. Very few rocks of terrestrial origin exist for this period, so while the record of marine life is extensive, the sampling of life on land during this time is very sparse. We can only hope that further careful searches will reveal more ancient insects. In 2012, in a quarry in Belgium, just such a find was made. *Strudiella devonica* is remarkably complete, and unlike *Rhyniognatha* is undisputedly an insect. It has been dated to 370 million years ago, so has shortened the Hexapod Gap by 15 million years.[23]

However, a similar, although shorter, gap also exists in the vertebrate record. Known as Romer's Gap, after the American palaeontologist Arthur Romer, the paucity of fossil records during this period (360–345 million years ago) has long been a source of frustration for vertebrate palaeontologists. It spans the time that vertebrates were first emerging from the sea, and clouds our picture of how our own ancestors rose to the challenges of life on land. The coincidence of these two gaps raises suspicions that something other than just a lack of fossils is going on. There is some evidence that this was also a time of lowered oxygen levels, so perhaps these gaps in the fossil record reflect a real pause in the diversification of life on land,[24] as the stresses of breathing in the rarefied air took their toll.

Taking the timings derived from molecular data into consideration, one picture to emerge is of an early diversification of insects that was curtailed by adverse environmental conditions. Then, as oxygen levels rose again, insects really took off. In the late Carboniferous period, abundant and spectacular flying insects appeared as if from nowhere.

THE RISE OF INSECTS

Molecular studies suggest that insect flight evolved about 400 million years ago, not all that long after insects first learnt to walk. If so, and despite a lack of clear fossil evidence, the late Carboniferous flying insects must have had a long flying ancestry. There is only a little evidence from the fossil record to support this conjecture. Those who interpret *Rhyniognatha*, which dates to around this period, as an insect point out that its jaws exhibit features shared by all winged insects today and so regard it as an already advanced flying insect. Logically, then, flight must have evolved even earlier, perhaps in the Silurian period,[25] although we have yet to find convincing fossils to back this up.

More recent studies have cast some doubt on this tentative scenario. As more data have become available, it looks as though fluctuations in oxygen concentration through this period were far less dramatic than previously thought. So, perhaps the Hexapod Gap wasn't caused by low oxygen concentrations after all, but by the fact that insects remained relatively rare for the early part of their history, as they struggled to exist alongside the myriapods and arachnids that had been crawling around dry land for a lot longer. It looks as though the insect body plan wasn't a recipe for immediate success, although it was, nevertheless, a vital part of their eventual domination. This compartmentalized layout allowed insects to develop a thorax largely filled with muscles, while

the abdomen came to house a large and elaborate food storage organ – the fat body – which could provide all the energy needed to power those muscles.[26] In other words, the insects' basic body design was a prerequisite for flight and it was the later evolutionary innovation of wings that finally gave insects their decisive edge, as they became the only arthropods to conquer the air. This alternative picture suggests that flight evolved much later than suggested by molecular studies, and that the rapid appearance of abundant fossils in the late Carboniferous period reflects reality – it simply shows how successful this innovation proved to be.[27]

Whenever it happened, the evolution of flight was undoubtedly a major factor in the triumph of insects. They were the first creatures to take to the air and had the skies to themselves for perhaps 170 million years, before vertebrates finally joined them.[28] It's hard to overestimate the importance of wings in the insect success story, so we'll return to this topic in much more detail in a later chapter. Yet as auspicious as flight proved to be, it may not have been the smartest thing insects ever did.

Towards the end of the Carboniferous period, the first of the *endopterygotes* evolved.[29] These are the insects most familiar to people today – those with *complete metamorphosis* that entails a larval stage followed by a

The giant cave cockroach, *Blaberus giganteus*, is one of the largest roaches in the world. All roaches undergo incomplete metamorphosis, their wings growing as external wings pads until they reach full size.

transformative pupal stage, from which emerges an adult that often looks nothing at all like the preceding larval stage. The transformation from caterpillar to butterfly is so miraculous that in the past not all believed it was possible. An unfortunate German naturalist called Renous was arrested in San Fernando, Chile, in the early nineteenth century for uttering the heresy that caterpillars turned into butterflies. He later recounted this tale to Charles Darwin, who visited Renous while on his round-the-world voyage on board HMS *Beagle*. Darwin, who had been a keen entomologist since his university days, was so struck by this absurdity that he included the tale in the book of his journey, usually referred to with its shortened title *The Voyage of the Beagle*.[30]

Complete metamorphosis may have been heretical to the authorities in nineteenth-century Chile, but for insects it was a game-changing evolutionary innovation. More than 80 per cent of modern insects exhibit complete metamorphosis – that's 60 per cent of all animals alive on the planet today.[31] There's a strong argument to say it's the most successful strategy to have evolved in any animal – ever. How on Earth could complete metamorphosis, also known as *holometaboly*, confer such an extraordinary advantage? As yet there is no consensus, but an obvious starting point is to compare the life cycle of insects that undergo incomplete metamorphosis (hemimetaboly) with that of holometabolous insects and ask what advantages the latter might offer.

The distinction usually pointed out is that the young of hemimetabolous insects look like miniature versions of the adults, whereas the young and adults of holometabolous insects look entirely different. This, as we'll see shortly, is a huge oversimplification. A more important difference is the presence of a pupal stage in holometabolous insects, which dictates, among other things, how the adults of each type of insect acquire their wings.

A grasshopper is a typical example of an insect with incomplete metamorphosis. The young, which are usually called nymphs, look very much like the adults, except that they are smaller and wingless. The nymphs moult several times and at each successive stage their wings grow on the outsides of their bodies, which are easily seen as ever-larger wing-pads. By the last nymphal stage, the wing-pads are large enough to contain fully formed adult wings, neatly folded up inside until the adult crawls free of the last nymphal skin and can pump up its wings to full size with fluids from its body. I'll never tire of seeing adult dragonflies emerging at the edges of a pond at dawn and expanding their soft, translucent wings ready to harden in the first rays of the sun – an everyday miracle. Because

(*overleaf*) Hummingbird clearwings, *Hemaris thysbe*, demonstrate their mastery of the air as they hover in front of flowers with all the skill of their namesakes.

their wings grow externally through each nymphal stage, such insects are often called *exopterygotes*, meaning 'external winged forms'.

Among insects exhibiting complete metamorphosis, the young stages (in this case called larvae) must also moult to grow, but their wings are not visible in any of the larval stages. The last larval stage moults into a pupa and at this stage the wings grow rapidly from small discs of cells that were hidden deep inside the larva's body. In a butterfly or moth pupa, the wings-to-be are clearly visible as part of the surface sculpting of the pupa. Since in this case the wings grow inside the pupal body, these insects are called *endopterygotes*, meaning 'internal winged forms'.

What do these contrasting life cycles mean, at least for the examples I've chosen? Grasshopper nymphs look and behave much like the adults except that they can't fly. More importantly, they eat the same kind of thing that the adults eat. An army of locust nymphs marching across a field can do nearly as much damage to a crop as the adults. Caterpillars, on the other hand, munch their way through leaves, while adult butterflies sip nectar. Caterpillars occupy an entirely different ecological niche from adult butterflies, which means that, unlike grasshoppers and locusts, the young stages don't compete with the adults, allowing higher populations to exist in a limited area.

The fact they can cram many more individuals into a patch of land in this way certainly appears to give these insects the edge, but it's not quite that simple. I spent the early part of my career working on dragonflies, which – along with mayflies – are the most basal insects alive today, classified together in a group called the Palaeoptera* (ancient wings), which split off from other insects long before complete metamorphosis evolved. Yet they both do a pretty good job of occupying different niches as adults and nymphs. Adult dragonflies are fast-flying aerial predators of small insects, while the young stages live underwater, catching large prey by ambush. They do so with a unique structure, the labial mask, which is essentially a hinged gin-trap that shoots out nearly half the length of the nymph's body. Mayfly nymphs also live underwater, eating detritus or scraping algae from rocks, while adults fly above the surface, although they do so for such a brief time that they don't bother eating at all. It's as complete a divergence in lifestyles as between any caterpillar and butterfly.

---

* Molecular phylogenetics has done much to clarify the insect family tree, but the position of the Palaeoptera is still obscure. It may not be a natural – or monophyletic – group, in which all members share a common ancestor. But it is clear that both dragonflies and mayflies branched off early from the rest of the tree and therefore represent basal branches.

The nymph of an emperor dragonfly, *Anax imperator*, looks very different from the adult, with its hinged labial mask slung beneath its head.

Nor do the nymphs of dragonflies or mayflies look anything like the adults. The nymphs are well adapted to aquatic life, breathing through gills that, in mayflies, form plates or filaments along the sides of their abdomens and, in dragonflies, form six rows inside the hind gut. Dragonfly nymphs breathe by pumping water in and out of their rear ends. In the early 1980s, I worked extensively on the internal and external remodelling that takes place as the last nymphal stage of dragonflies prepares to take to the air. The elaborate gill chamber is reabsorbed and re-formed into the very different adult hind gut.[32] The labial mask is disassembled and replaced by a more conventional set of chewing mouthparts. This always seemed to me to be a rather complete metamorphosis for an insect that sits solidly among hemimetabolous kin. Clearly, simple niche separation isn't the whole answer.

However, neither dragonflies nor mayflies form a pupa, a stage that allows even more extensive remodelling between larva and adult. Caterpillars and butterflies could not be more different from each other, and perhaps this more extreme divergence in form is the key to endopterygote success. It allows larvae and adults to specialize in entirely different ways. The larvae have evolved to maximize efficiency of feeding and growth, while the adult is specialized for reproduction and, since it's the only

stage with wings, for dispersal. This idea has been around for some time. Darwin (who clearly did believe that caterpillars turned into butterflies) discusses the idea of separate specializations in larvae and adults in *The Origin of Species*,[33] though he credits the concept to the German-born biologist Fritz Müller, who worked for the National Museum in Rio de Janeiro as a 'Travelling Naturalist' (now, there's a job).

Evidence that caterpillars, maggots, grubs and other simple, cylindrical larvae are shaped to maximize their growth rate has accumulated since Darwin's day. Fruit-fly larvae can grow from hatching to pupa in just ninety-six hours. In the same time, honeybee grubs grow 800 times in size, fed 140 times by worker bees that definitely earn their name. The cylindrical shape of all these larvae makes it easy to supply oxygen to rapidly growing and respiring tissues; it also facilitates moulting.[34] Insects with incomplete metamorphosis have to extract complicated long legs and antennae from the old skin; many fail to free themselves completely and die. I've watched many of my stick insects moult and it's like watching some escape artist wriggle free of a straitjacket. It's a miracle any of them make it at all.

A simple cylindrical shape can also support itself with internal hydrostatic pressure. This means that such larvae don't need a thick exoskeleton. The cuticle of an insect's exoskeleton is made up of two separate layers – endocuticle on the inside and exocuticle on the outside. The exocuticle can be hardened by a process called tanning, which gives it strength and rigidity. Once tanned, the exoskeleton is hard to break down, which is great for protection but a problem if a larva, striving for maximum growth, wants to reabsorb and recycle this valuable material at each moult. The cast skin of insects consists mostly of this tough exocuticle and although a few insects do try to eat their old skins, they're hard to digest and are therefore largely wasted. Caterpillars pumped up with their internal hydrostatic pressure don't need a tough outer layer for structural support, so their cuticle can be made largely of endocuticle. This has two advantages. Firstly, endocuticle is more flexible than tanned exocuticle so it can stretch, at least to some extent, as the caterpillar grows rapidly between moults. Secondly, it can be broken down by enzymes as the moult approaches, so it can be recycled, allowing the caterpillar to devote more resources to growth. Partially digested and therefore softer endocuticle also makes moulting even easier.

Perhaps the advantage of complete metamorphosis is that an insect can wholly decouple its young stages from the adult one. The young stages are simple creatures, evolved to grow as fast as possible; then, at

the pupal stage, all that growth is converted into the much more complex form of the adult. Adults need to be more complex because they not only need to feed (although some don't), but also to find a mate, look after eggs or larvae, navigate and migrate, and perform a host of other sophisticated behaviours that ensure they can produce a new generation of fast-growing larvae. It's a compelling argument and may well be the reason for the extraordinary success of the higher holometabolous insects, but it's possibly not what originally drove their evolution back in the Carboniferous period.

Insects used regularly in laboratory studies represent a minuscule fraction of the diversity that exists in the wild. Those selected tend to be insects that are easy to rear in large numbers, but these chosen few can distort our view of insect biology. Snakeflies rarely make it onto the lab animal list, but they occupy a key position on the insect evolutionary tree as 'primitive' holometabolans. They have a distinct pupal stage but this has external wing-pads, like those of grasshoppers; apart from a lack of signs of wings, the larvae of snakeflies look very much like the adult and live in a similar kind of way. Snakeflies give us a glimpse of what the first endopterygotes might have looked like and they don't look like butterflies, flies, beetles, bees or any of the higher holometabolans. Despite being insects with a complete metamorphosis, snakeflies are nowhere near as diverse or successful as these other holometabolous insects. Obviously, just being holometabolous isn't a guarantee of success, and perhaps that was also true for the first insects to evolve complete metamorphosis.

There's some evidence from the fossil record to support this. Despite the total dominance of holometabolous insects today, the evolution

The small tortoiseshell, *Aglais urticae*, an insect with complete metamorphosis. The caterpillar could not be more different from the adult, so an intermediate pupal stage is needed to transform one into the other.

of complete metamorphosis didn't spark an immediate renaissance of insect-kind. It wasn't until much later that holometabolous insects began to diversify and take over the world. I think it quite likely that the ideas outlined above are a big part of what makes endopterygotes so successful today, but it took time for the first holometabolans to evolve the extreme specializations of larvae and adults that would really make the difference. However, evolution can't look ahead and plan for the future; if the above outline is true, then we still haven't discovered the true factors that drove the evolution of complete metamorphosis in the first place. Although we may now have a broad picture of the key events in the origins and evolution of insects, a great many mysteries remain.

Back on the shore of Delaware Bay, the tide is beginning to turn. The horseshoe crabs must soon return to the sea before they are left high and dry. The females have struggled a little way beyond the reach of the waves, often with one or more males in tow, each hanging on with powerful pincers until the female has dug a pit in the sand and laid her eggs. Now they have just enough time to fertilize the eggs before lumbering back to the safety of the sea. Not all make it. The beach is littered with dried horseshoe crabs from previous high tides, and there are plenty more now stranded from the spawning that I have just witnessed. Some seem disorientated and are heading further from the sea, others have been turned over onto their backs by the waves and are unable to right themselves.

As I watched, a young child dragged her reticent mother along the beach, frequently stopping to hurl stranded horseshoe crabs as far into the waves as her small arms allowed. It would have seemed churlish not to help and, much to the relief of her mother, who stopped to chat to the cameraman as he packed up the gear, my new friend and I worked our way along a few hundred metres of shoreline, rescuing as many crabs as we could.

The caterpillar of a silver-striped hawkmoth, *Hippotion celerio*. Such simple, soft-bodied, cylindrical larvae have evolved purely as eating machines.

Moving onto dry land is no easy step, at least to judge from a beach covered as far as I could see in dead and dying horseshoe crabs. But once the ancestors of insects made this transition, a world of new opportunities lay before them – opportunities that they were able to exploit in exuberant style with their ever-adaptable body plan. Their jointed legs, inherited from their aquatic arthropod ancestors, at first allowed them simply to walk or run on dry land. But eventually they would turn out to be a significant factor in their success. Insects, as we'll see in the next chapter, have found many additional uses for their limbs.

I paused in my rescue mission to take a close look at one overturned crab and to admire its waving legs and claws and its desperately flapping gills, like the pages of a book blown open in the wind. To spare this individual the indignity of being hurled by its long tail into the water, I carried it out to deeper water and let it go. Here, back in its element, I watched it trundle over the seabed as it headed back out into the bay, hopefully to return on future high tides and continue a tradition that has lasted for more than 400 million years.

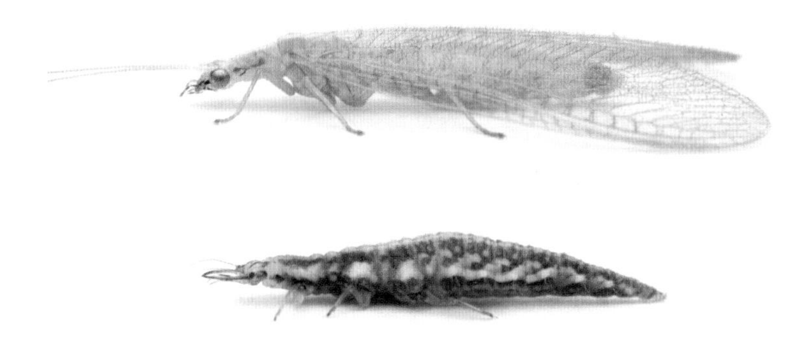

Common green lacewing, *Chrysoperla carnea*, a holometabolous insect, perhaps similar to the earliest holometabolans in which, apart from wings, adult (above) and larva (below) look very similar.

(*overleaf*) Emperor dragonfly, *Anax imperator*.

3

Hexapods
How evolution gave insects a leg-up

It's a matter of proportion, that's what it is; and when you come to gauge a thing's speed by its size, where's your bird and your man and your railroad, alongside of a flea? The fastest man can't run more than about ten miles in an hour – not much over ten thousand times his own length. But all the books say any common ordinary third-class flea can jump a hundred and fifty times his own length; yes, and he can make five jumps a second too, – seven hundred and fifty times his own length, in one little second – for he don't fool away any time stopping and starting – he does them both at the same time.

Mark Twain, *Tom Sawyer Abroad*, 1894

Fleas have been making their prodigious leaps for a lot longer than human beings have walked the Earth. Ancient fleas plagued dinosaurs from at least 150 million years ago and were, appropriately enough, huge.[1] Some reached 2 centimetres in length, twice the size of the very biggest fleas alive today, although these monster fleas didn't have the modern flea's spring-loaded jump. The ability of today's fleas to disappear from the floor and almost instantaneously reappear on your leg borders on the uncanny and, as Mark Twain points out, their predilection for hopping knows no bounds. From a standing start, a human flea, *Pulex irritans*, can jump more than 30 centimetres straight up and can keep this up every few seconds for seventy-two hours straight.[2] That energy has been harnessed, quite literally, by entertainers throughout history – in flea circuses. Tiny gold collars fitted over the fleas allowed them to use their extraordinary stamina to pull carriages or other contraptions many times their size. In fact, fleas are so small that often it looked as though the carriages were self-propelled.

FILMING THE HIGH JUMP OF FLEAS

The fleas' superpowers are all the more impressive when their jump is filmed at high speed to slow it down so that it can be both seen and appreciated. However, given the speed of the action and its unpredictability, filming a jump is a near-impossible task. Unfortunately, it was also one of the first jobs that came my way after I joined the BBC's Natural History Unit in the 1980s. Back then, we shot everything on film, which made this job even harder. Today's digital cameras record continuously to an internal memory, so all you have to do is wait for the action to happen, then press the button, and the camera conveniently saves the previous few seconds, which makes capturing unpredictable action a whole lot more predictable. On film, however, we had to predict the unpredictable and trigger the camera ahead of the action. Furthermore, it takes a mechanical film camera a certain amount of time to reach its running speed, so our predictions would have to be pretty much spot on.

Then there was the amount by which we needed to slow down the jump just to make it visible. Film cameras normally run at a speed of twenty-five frames a second, so a standard 400-foot load of film lasts about ten minutes. We worked out that the jump was so fast, we'd need to run at 10,000 frames a second, in which case our ten-minute roll of film would last about a second and a half – not much room for error. There was a camera, known as an 'E10', that could achieve this performance,

(*previous page*) A tiger beetle, *Anthia sexmaculata*, from North Africa.

but it runs the film through so fast that the last 20 feet or more is usually pulverized into tiny fragments as the tail of the film whips around inside the camera. This reduces the time available to record the action even further, as well as forcing us to spend a lot of time between shots simply cleaning out the camera.

The final technical problem was light – or rather the lack of it. To film something that small at that frame rate we were going to need a lot of light – so much that, when lit, the only way I could see our miniature set was to wear welder's goggles. Now we just needed the fleas to jump when I told them to, something even the best flea circus trainers couldn't do. However, I had noticed that in the flea green room where the stars were awaiting their call, they almost always climbed to the top of an object before they jumped. The scene we were filming was set inside a house, on a carpet, so in the tiny film set, a 1-centimetre glass cube made from the coverslips used on microscope slides, I arranged for a single tuft of the carpet pile to project above the others. That narrowed down the 'where' of the jump if not the 'when'. There was still no easy way to predict to a fraction of a second when a flea would jump.

My plan was to cover up the powerful filming lights with bits of card, so they didn't overheat the set between takes. I would then uncover the lights briefly when a flea was in position, and we'd see what happened. The first time, the flea simply dropped dead on the spot, so great was the light intensity, but on a few subsequent occasions, the shock served to stimulate the flea to jump. Once we had worked out the choreography of unveiling the lights and firing the camera, we ran off about a dozen rolls of film. Even then, we had no way of knowing whether the flea had actually jumped while the camera was running at the right speed. The only way to find out was to send the film off for overnight processing and project it in the viewing theatre the next day. On most rolls, we had to sit through ten minutes of a flea just perched there, thinking about jumping, or worse still, ten minutes of a tuft of carpet doing nothing at all after the flea had jumped ahead of the trigger. But on a few rolls, we caught the jump, slowed down 400 times, and watched a flea catapulting itself off the tuft with such force that the fibres rippled with the shockwave.

Fleas are masters of the high jump, leaping over 300 times their body length, the equivalent of a human easily clearing the Empire State Building in Manhattan.

I later found out that I wasn't the first person in the NHU to try this. When I joined the unit, I soon became friends with Barry Paine. His was the gentle voice of a great many Sunday evening natural history programmes on BBC2. He was also a producer, a thespian of considerable talent and a natural comedian whose anecdotes filled many an evening in the BBC Club bar. Shortly after hearing of my flea-filming exploits, he regaled me with tales of a film he had worked on several decades earlier, when he had teamed up with a renowned scientist, Miriam Rothschild, to make a film about one of her real passions – the rabbit flea, *Xenopsylla cunicularis.*

Miriam Rothschild was born into the wealthy Rothschild banking family and followed what seemed to be a family passion for natural history. Her uncle Lionel Walter Rothschild built a natural history museum at Tring in Hertfordshire (now part of the Natural History Museum, London), and her father, Charles, described more than 500 new flea species. Miriam inherited her family's wealth and her father's enthusiasm, as she stepped into his shoes to become a leading authority on fleas. The extensive Rothschild collection of fleas, started by her father but substantially added to by Miriam, is still housed at the Natural History Museum.

Making a film on rabbit fleas in 1966 took considerable technical ingenuity, not least from Eric Lucey of the Research Film Unit at the University of Edinburgh, who was tasked by Barry with filming the near-instantaneous flea jump. This university unit was set up to develop specialist film technology to aid research projects, but recording a flea jump tested the skills of the team to the limit.[3] In the end, they surmounted all the considerable technical difficulties to produce results that excited not only the film producers but scientists as well, because film looked as though it might offer a way of answering a question that had been nagging at biologists for a century – how can a flea jump so quickly and so far?

BUILT FOR SPEED

The surprising answer to that question will have to wait until a little later in the chapter, since to fully understand just how amazing fleas are, we first need to know a bit more about insects that don't skip around with such boundless energy, and explore how their versatile jointed limbs gave insects a critical leg-up in the survival game. We saw in the previous chapter that insects are *hexapods* – that is, they have six legs, one of their defining characteristics. But, as I pointed out in the introductory chapter, although all insects are hexapods, not all hexapods are insects. Three groups of tiny creatures, the diplurans, the proturans and the collembolans, all possess six legs and all used to be regarded as insects, but genetic studies have recently shown them to be related groups that branched off the family tree before true insects evolved.[4]

It's unlikely that any of these non-insect hexapods would be noticed by a casual observer, although the collembolans, also known as springtails, sometimes make their presence felt by sheer force of numbers. Tiny black *Hypogastrura nivicola*, sometimes called snow fleas, speckle newly fallen snow like ash from a nearby fire. Despite their name, they are not related to fleas, but like their namesakes, they can skip around in an energetic fashion. However, they don't use their legs to jump. They have a special organ, the furcula, attached to the tip of the abdomen, which catapults them into the air – springtails literally have a spring in their tails. Steely-grey *Podura aquatica* live in such numbers on the surfaces of ponds that they form conspicuous rafts that look like patches of grey scum, and they are small enough to skip and jump over the water without breaking through the surface film.

We saw in the previous chapter that arthropods appeared very early in the Cambrian period with the first blossoming of complex, multicellular life. Their ancestors were probably lobopodians (or something very like them), worm-like creatures that lived in the late Precambrian period but continued to live alongside the rapidly diversifying arthropods during the Cambrian period that followed. Distant relatives of these archaic monster worms survive today – the onychophorans or velvet worms – and share with lobopodians a long, multi-segmented body that carries a pair of tubular limbs on most of the segments.

As arthropods diversified, those simple, unjointed lobopodian limbs gradually evolved into the articulated legs that characterize all arthropods past and present, although this oversimplification hides the fact that the details of exactly how this happened are still being debated.[5] As we discovered in the introductory chapter, arthropod legs were originally double-pronged structures. It's likely that these structures originated by combining a lobopodian's tubular limbs with gill flaps that are found on some lobopodians. The earliest arthropods carried these jointed limbs on most of their numerous similar segments, as do modern remipedes or centipedes and millipedes. However, as we saw in the previous chapter, insects pursued a strategy of regional specialization in which different segments took on different roles. They lost all but six of their many legs

(*overleaf*) The elaborate, fan-shaped antennae of a cockchafer, *Melolontha melolontha*, are derived from primitive legs.

– or, to be precise, they lost them as functioning legs.

The appendages on some of their other segments, rather than just being lost, took on new roles. On the head, they became antennae for smelling or tasting the world, and three sets of mouthparts, an all-purpose cutlery set that gives insects such a wide choice of menu. At the tip of the abdomen many insects have a pair of *cerci* (singular: *cercus*), which are also derived from ancient legs; just like the antennae, in some insects the cerci serve a sensory function, but at the other end of the insect. In others, such as earwigs, they've become intimidating pincers. Some insects also have *styli* (singular: *stylus*) on their abdominal segments, also derived from highly modified abdominal legs.

There's no better example of how an insect benefits from all these segmental specializations than the much-maligned cockroach. The American cockroach, *Periplaneta americana*, is one of the most common pest roaches worldwide, aided in its all-conquering success by an array of useful appendages, all derived from primitive legs. Sensitive antennae guide a roach to anything edible, where its powerful biting *mandibles* work like knives to cut up whatever it finds into small pieces, which are then passed to the dextrous *maxillae* behind, which in turn fork the food into its mouth. But open a door into a room where roaches are feasting, and they'll scatter before you've even taken a single step. Their cerci

detect the slightest air movement, and by combining information from both cerci a roach can even tell the direction from which a threat is approaching, just by the patterns of disturbance in the air. Its six long legs soon propel it to safety.

American cockroaches are among the fastest insects around. At top speed they can cover a metre and a half in a second. In slow cruise mode, they use a standard insect gait, in which they rest on three legs, two on one side, one on the other, while moving the other three forwards. So, they always have a stable tripod to rest on – the advantage of keeping six of their ancestral legs. Stable it may be, but that doesn't mean slow. The frequency that a roach can move its legs is not far short of the frequency it flaps its wings in flight. As its speed increases, there's a period when all six legs leave the ground at once, just like in a galloping horse. However, to reach its top speed, a roach does something no horse can do. First it starts to run on just four legs; then, in turbo-boost mode, it lifts up onto its two back legs and runs bipedally.[6]

Yet as fast as they are, roaches are not the fastest things on six legs. A roach's metre and a half a second translates as fifty body lengths a second, a measurement that makes comparing different-sized insects a bit more meaningful. A tiny phorid fly, for example, just millimetres long, although nowhere near as fast in absolute terms, can nevertheless cover 112 body lengths a second. These flies are appropriately called scuttle flies due to the way they sprint and stop many times a second.[7]

Several kinds of desert ant have independently evolved long legs with speed to match. The most famous, having been pursued over burning sands by David Attenborough, are the Saharan silver ants, *Cataglyphis* spp., referred to by one scientist who has studied them for decades as the racehorses of the insect world.[8] These ants only emerge from their nests during the hottest part of the day, in an explosive exodus that often lasts for just a few minutes.[9] They can tolerate their body temperature reaching 50°C but must return to the nest before they get any hotter. Even so, this is more than hot enough to fry any other insect, and that's exactly why the ants wait until midday to leave their underground nest to forage. They're looking for insects or other invertebrates that have succumbed to the heat. There's no predicting where they might find such a meal, so they cross the desert in high-speed zigzags to search as large an area as they can, as fast as they can. They run at 70 centimetres a second, but they are much smaller than roaches, so in relative terms they are much faster. However, perhaps the most extraordinary thing about these ants is what happens when they find food. Despite criss-crossing

A green tiger beetle, *Cicindela campestris*, uses its long legs to run at high speed in pursuit of prey.

the desert in a seemingly random manner, an ant now makes its way directly back to the nest entrance, as straight as an arrow.

To do this, the ants use vector navigation, or what a pilot would call 'dead reckoning'. For this they need to know all the different directions in which they have travelled and the distance covered. They use the sun as a compass, to provide information on direction, and they have a built-in pedometer that counts their strides to estimate distance.[10] However, this neat device, called a stride integrator, could be a major source of navigational errors if it is not accurate, and even a hardy desert ant doesn't want to get lost out here. Unlike roaches, desert ants maintain their tripod-style gait across all their walking speeds,[11] and their stride length seems to be more constant than in other insects,[12] perhaps both ways of reducing errors in distance measuring. These ants can also remember their vector data. If they found a particularly good foraging site, they can make their way straight back there, having offloaded the booty from the first trip.

In other desert areas, evolution has crafted very similar-looking ants, long-legged, fast and extremely heat tolerant, which behave in much the same way as the Saharan silver ants. In the deserts of Arabia and the dunes of the Namib in southern Africa, hot rod ants, *Ocymyrmex* spp., fulfil this role, while in the deserts of Australia, species of the genus *Melophorus* have mastered the art of dead reckoning at high speed. However, in both

The huge mandibles of a male stag beetle, *Cyclomattus tarandus*, evolved from primitive legs, as did all three sets of an insect's mouthparts, although these jaws are now used for combat with rival males.

the Namib Desert and the baking flats of the Australian outback, these ants have some serious competition for speed – beetles.

Australia claims to be the home of the champion insect sprinter, a tiger beetle called *Cicindela hudsoni*. It scoots over dry lake beds at speeds of about two and a half metres a second – the fastest absolute speed of any insect. But there is a smaller tiger beetle, *C. eburneola*, living here that is even faster than the so-called record holder in terms of body lengths per second. This tiny tiger beetle covers an extraordinary 170 body lengths a second.[13] Comparing this to a human performance is at best spurious, since physics and the laws of scaling don't work like this. But – and I'm sorry – I can't resist a comparison that at least gives a sense of how fast these beetles are in their world – *C. eburneola*'s sprint is like Usain Bolt running 100 metres *in a third of a second!*

About thirty species of tiger beetle live in open saline habitats in Australia. Often these places are lake beds that remain dry for decades, scattered like desert islands across the outback. Nine species of these tiger beetles have become flightless, for the same reason that many oceanic island insects lose their wings. Their island homes are so remote from each other that flying anywhere is likely to dump them somewhere they can't live. Their flight apparatus is expensive to grow and maintain, so if it's not needed, natural selection soon does away with it. However, these beetles still need to get around as well as escape from enemies. So, they've traded flight for speed over the ground. The nine flightless species are the fastest Australian tiger beetles.[14]

Speed does bring its own problems, though. Imagine running 100 metres in a third of a second. For that brief moment, the world would be an undecipherable blur. The same is true for tiger beetles. Scientists at Cornell University in New York discovered that at their top speed, the local tiger beetles were effectively blind. Their eyes simply can't capture enough light to form an image. Yet tiger beetles are visual hunters, running down prey in a fast chase, more like cheetahs than tigers.

They get around this problem by stopping for just long enough to get a fix on their prey and to plot an intercept course. It only takes them a matter of milliseconds and they are off again, before once more stopping to re-acquire their target and readjust their course. Despite the stop-start chase, the beetles can run so fast that they soon overtake their prey, at which point they bring their wicked-looking, sickle-shaped mandibles into play to dispatch it quickly.[15] In Kruger National Park in South Africa, I once watched a giant tiger beetle, *Manticbora* sp., catching a large moth like this. It's not the fastest tiger beetle but, at 3 centimetres long and armed with massive slicing jaws, it's a formidable predator. A single bite from those jaws sliced the moth neatly in two.

On the dashes between orientation stops, tiger beetles still need to negotiate rough terrain and skip over obstacles in their path. How do they do this if they can't see? The Cornell team resorted to high-speed photography to find out. The beetles run with their antennae held out stiffly in front of them. The instant they touch an obstruction, the beetles react by tilting their bodies up and vaulting over the obstacle. Beetles with their eyes painted over were just as effective at negotiating the course, but those with clipped antennae ran headlong into every obstacle in their path.[16]

On the shifting dunes of the Namib Desert, a different family of beetles lives life in the fast lane. This part of southern Africa is home to hundreds of kinds of darkling beetle (Tenebrionidae), many of which live on open sand or sparsely vegetated dunes. Those living out in the open are the fastest of their kind, for all the same reasons that desert ants and tiger beetles have a need for speed.[17] In fact, one of their number vies for the title of the world's fastest insect. *Onymacris plana* is a black bullet of a beetle, with sleek lines and elegant flanges like an expensive sports car. It's active in the daylight hours, when its long, spindly legs hold its body well clear of the burning sand. It can cover 30 metres non-stop at a metre a second, but it's not just long legs that give it this remarkable turn of speed over loose sand.

Compared to related beetles, *O. plana* uses much less energy when running at speed. One suggestion is that its flanged body forms an

aerodynamic surface that generates lift when it's running flat out. This means that the beetle can expend less energy on supporting its body weight and put all its effort into generating forwards thrust.[18] The rapid passage of air over its body also helps keep it cool when running under a ferocious sun. Air-cooled and aerodynamic – they named the wrong kind of car after a beetle.

SKATING AND SWIMMING

It's not all about speed, though. From diggers and jumpers to climbers, their legs give insects all-terrain capabilities. They can tackle almost any surface, even the water's surface. Just as the hot desert sands incapacitate most insects, providing ample easy meals for the few hardy stalwarts that can cope, so the surface of a pond or lake acts as a lethal trap for most small insects, held in an unbreakable grip by surface tension. Again, there's a community of insects completely at home here, taking advantage of these endless free lunches. Many are true bugs (Hemiptera), relatives of the ocean striders, *Halobates* spp., that we met in Chapter 1.

Most familiar are the pond skaters, *Gerris* spp. and *Aquarius* spp. Late in the season they assemble in rafts thousands strong, skittering over the water's surface and creating hypnotic swirls of ripples. If a stiff wind gets up, they sometimes pile onto protruding sticks or stones, a dark, writhing mass of bodies several layers deep, as they avoid being blown across the lake. However, they're quite capable of powering themselves over the surface into a headwind if they need to. The secret to their pond-skating prowess is their ability not to get wet.

The body of a pond skater is covered with two layers of tiny hairs. The longer ones are responsible for an extremely effective waterproofing, while the smaller ones trap a layer of air around the insect should it suffer the misfortune of becoming submerged.[19] This waterproofing is critically important on the pond skater's legs, which must not break through the surface film. If they did, the pond skater would be as helpless as its prey. A pond skater usually stands on its four back legs, but the middle pair are the longest and serve as sculls to row it over the surface. Since these legs rest on top of the water's surface, and don't dip below like a rower's oar on the power stroke, it's not obvious how a pond skater generates thrust.

One idea, proposed nearly half a century ago, was that a pond skater creates tiny waves that its legs push against to drive it forwards.[20] More recent studies, again making use of high-speed cameras, have shown that pond skaters have two techniques for water walking, one used at slow

A pond skater, *Gerris lacustris*, uses its unwettable legs to stand on the surface film. Its long middle legs are used to scull over the water's surface.

speeds, the other when it needs to move faster. At slow speeds, some of the longer hairs make enough contact with the water for the rowing legs to leave a series of vortices just below the surface, so the legs do work a little like conventional oars. For higher speeds, it swings its rowing legs faster over the surface, which distorts and bends the meniscus into tiny ripples, technically called *capillary waves*.* Under such circumstances the surface behaves more like a trampoline springing the insect forwards.[21] In fact, pond skaters can even jump from the surface, which they do to escape their own predators or when taking to the air.

Water crickets *Velia* spp. are also water walkers and, living up to their name, also water jumpers. A water cricket is smaller and stouter than a pond skater but more brightly coloured, with two red 'go-faster' stripes along the top of its abdomen. I usually find water crickets on small, slow-flowing streams, which shows that they can even skate over running water. They do this by harnessing the surface tension in a different way from pond skaters. I've filmed these creatures on many occasions and

* Capillary waves are tiny waves, so small that their dynamics are dominated by surface tension.

often noticed that when we get one in frame it will suddenly scoot off with barely a movement of any of its legs. It almost looks supernatural, although it isn't. It's a phenomenon based on good old-fashioned physical science. It's something called *Marangoni propulsion*.

Camphor boats work in the same way. Attach a tiny piece of camphor to the back of a small model boat and it will dart forwards with no apparent means of propulsion. The camphor is a surfactant, a substance that reduces surface tension. So is the washing-up liquid you use on dirty dishes. It makes the dishes more wettable and therefore easier to clean, but dip one end of a cocktail stick in the washing-up liquid, then float it on the water's surface, and like the camphor boat it will scoot forwards. When the surface tension is reduced at the back of the boat or at one end of the stick, the stronger surface tension at the front will draw it forwards.

A water cricket operates on the same principle. It produces a tiny drop of surfactant at the tip of its proboscis, and when this touches the

Water measurers, *Hydrometra stagnorum*, are less nimble than pond skaters, stepping over the water's surface in a more stately manner.

surface, the cricket is rapidly propelled away.[22] An even smaller water walker, *Microvelia*, uses the same trick, but being smaller it's even more effective. Related to the water crickets, *Rhagovelia* spp., often called ripple bugs in North America, have an additional way of coping with flowing water. They have an expandable fan of hairs on the tips of their middle legs that dip into the water and make their rowing action even more effective.[23]

Much harder to spot on the surface are water measurers, *Hydrometra* spp. They're a little smaller than the water skaters but much thinner. With six long legs, they always remind me of tiny stick insects. They move much more slowly and deliberately than water skaters and water crickets and avoid the open surface, preferring to skulk around the shelter of marginal vegetation. Yet, in a sense, they are the truest water walkers – they don't row over the surface with one elongated pair of legs, but walk in a stately fashion using the standard insect tripod gait.[24] They sometimes look slow and ponderous compared to skaters, but they are also elegant, sliding over the water's surface with all the effortless grace of a professional ice-skater.

All of these bugs live on top of the surface film, but another group of insects lives *across* the surface film, half above the water, half below. Whirligig beetles (Gyrinidae) are supremely adapted to their strange double lives. Their compound eyes are divided horizontally into two, the upper half for viewing the surface world, the lower half aimed at the underwater world. Their antennae are flattened and lie along the surface, where they detect the tiniest ripples from other beetles or from prey trapped on the surface film. There is even some suggestion that a whirligig beetle can detect its own waves reflected off obstacles breaking the water's surface – a ripple-based sonar. These beetles are certainly drawn to each other and often aggregate in huge numbers. If disturbed, the whole swarm performs wild, high-speed gyrations, moving fast and turning quickly, as if showing off their mastery of the surface film.

True masters they are indeed. They're the fastest swimming animals in the world relative to size, and even at speed can turn in just one-quarter of their body length. Yet, from the surface, no propulsion system is visible. These shiny black beetles, just a few millimetres long, skitter over the surface as if animated by some magic spell. To really appreciate the evolutionary achievements of whirligig beetles, I brought a small swarm into our film studio in Bristol, where we had prepared a specially constructed filming tank. The tank was shallow and had an optically pure glass window in the base, so we could mount a high-speed camera beneath, looking vertically up. From below, and slowed down forty times, we could see that all the action took place beneath the water's surface. The beetles' middle and rear legs were working furiously. Like the archetypal duck, it's all calmness and tranquillity above the surface, but frantic activity below.

These legs each sport a large, foldable fan of overlapping plates that, when extended, increase the surface area of the legs by forty times. The fan extends on the backstroke, then folds on the forwards stroke to lower resistance, just as rowers feather their oars on the return stroke.[25] Curiously, the middle and hind legs beat at different rates. The back legs, the main source of propulsion, work at sixty strokes a second, while the middle legs move at only half this rate.[26] The action of these legs is the most efficient movement known in the animal kingdom – 84 per cent of the energy burnt by the beetle is transferred directly to thrust. While the hind legs provide the power, the beetle seems to use its middle legs to stabilize its course. In shape, whirligig beetles are very smooth ellipsoids, which makes them inherently unstable, particularly at high speeds, although that instability is the reason why they can twist and turn with great agility. However, if they want to swim on a less erratic course, they must make constant adjustments with their middle legs.[27]

In one respect, though, these beetles have the worst of both worlds. The submerged part of their body faces resistance from the water. When they move, they also create ripples on the surface, so the exposed part of their body experiences wave resistance. Calculations show that the combined resistance is most severe at certain intermediate speeds, so whirligig beetles avoid travelling at these middling speeds, instead moving either slowly or really quickly.[28] I see this pattern whenever I pause by a lake to be entertained by these beetles. At times, the whole swarm is performing gentle gyrations that then suddenly switch to frenetic pirouettes with barely any transition.

Whirligig beetles have one other trick. Approach them too closely and they dive beneath the water's surface. They do this by altering the stroke plane of their hind legs to produce a downwards force.[29] Once underwater, they find themselves in a world full of swimming insects. The ancestors of insects evolved in an aquatic environment, but all these modern aquatic insects had fully terrestrial ancestors before re-invading this watery domain. So, they had to adapt a terrestrial tool kit to cope with life underwater and, in particular, they had to transform walking legs into swimming legs. Luckily, the adaptable insect body plan is up

to the task, and any pond or lake is brimming with accomplished insect swimmers.

Diving beetles (Dytiscidae), like whirligig beetles, swim using their back legs, which are fringed with movable hairs that automatically fan out on the power stroke, then fold down on the recovery stroke to reduce resistance. The other four legs fit neatly into grooves on the underside of the body to maintain their streamlined shape. But unlike whirligig beetles, dytiscids reach large sizes. *Dytiscus latissimus*, found across northern Europe, can reach four and half centimetres in length. *Megadytes ducalis* is even bigger, the largest diving beetle in the world, or at least it was. It is now feared extinct. For a long time, this species was known from only one specimen in the Natural History Museum in London collected from an unrecorded location in Brazil. Recently, another ten specimens have turned up in dusty museum drawers with better location data attached. Hopefully, expeditions to these places might turn up one of these giants still alive and swimming.[30]

The true bugs (Hemiptera) are also at home underwater. Like the beetles, backswimmers use long, hair-fringed back legs to swim. However,

Backswimmers, *Notonecta* spp., use their long, hair-fringed rear legs to swim.

both beetles and bugs face a problem with life underwater. Part of their terrestrial legacy is that, as adults, they must breathe air (water beetle larvae are fully aquatic and breathe through gills on the abdomen). Adult water beetles store an air supply beneath their wing-cases, backswimmers in channels beneath their abdomens that are roofed with long, waterproof hairs. Unfortunately, the air makes them very buoyant. Backswimmers spend much of their time hanging off the underside of the surface film of the water, and they need to row extremely hard to dive. The instant they ease off even slightly they bob back to the surface like corks. The only way they can stay submerged without effort is to cling to underwater vegetation. However, they're not quite as hapless as they might at first appear.

Several kinds of backswimmer, such as *Buenoa* and *Anisops*, can achieve neutral buoyancy and hang in mid-water with barely a flick of their legs. They do this by carefully managing the size of their air bubbles. At the surface they not only replenish their air bubbles, but also store oxygen in their bodies in cells packed with haemoglobin, the same protein that carries oxygen in our blood. Haemoglobin is more unusual in insects (because they breathe through their tracheae), but makes sense for many air-breathing aquatic insects because oxygen concentrations in water are lower than those in air. Haemoglobin has such a strong affinity for oxygen that it can grab these molecules even from very low concentrations.

At some point during its dive, the bug will have used enough of its air supply to have caused the bubble to shrink to just the right size to offset the bug's weight. In other words, the bug is now neutrally buoyant. The trick is to prolong this state for as long as possible. The bubble is trapped so firmly that it can't shrink beyond a certain point, even as the bug uses up oxygen from within it. Instead, the pressure drops inside the bubble and, since gases always move from an area of higher pressure to one of lower pressure, oxygen from the water passes into the bubble, so maintaining the size and buoyancy of the bubble along with the bug's air supply. In addition, the bug can also release oxygen from its haemoglobin to sustain the bubble at its optimum size.[31] In this way, the bugs remain neutrally buoyant, not expending any energy while they wait for prey to drift within range.

Backswimmers and many water beetles have retained the power of flight in order to colonize new ponds and lakes. Backswimmers, as their name suggests, swim on their backs, but I've watched them float to the surface then roll over, at the same time breaking through the surface film. They rest briefly on the surface, then use their long back legs to launch themselves from the surface film as they open their wings and take flight. Their powerful back legs serve both for swimming and, in a much cruder way, jumping, but many insects, including the fleas with which we opened this chapter, have become expert jumpers.

HOW INSECTS ATTAIN THE WORLD RECORD FOR HIGH JUMPING

Common adjectives (at least, polite ones) used to describe cockroaches might be scurrying, scuttling, scampering or running. Few would suggest jumping, but at least one species does just that. Luckily, it's not one you'd find leaping onto high shelves around your kitchen. You'd have to search on Table Mountain in South Africa to find *Saltoblattella montistabularis* – as far as we know, the world's only jumping roach. It's a newly discovered wingless species but compensates for its lack of aerial abilities with impressive leaps. Its two hind legs are twice as long as the front four, and using these it can accelerate to take-off velocity (around 2 metres per second) in only one hundredth of a second.[32] This makes it even harder to catch than a roach in the kitchen.

Some insects have such long hind legs that they look more of an encumbrance than a help. However, when legs this long extend, they generate a lot of leverage, allowing an insect to jump further, in much the same way that pole vaulters can leap much higher with a long pole than they could unaided. In fact, many katydids (Tettigonidae), also known as bush crickets, have such long back legs that they look like walking pole vaulters. Even newly hatched nymphs, just a few millimetres long, can leap over half a metre, which they do at the slightest disturbance. I have a breeding colony of giant katydids, *Stilpnochlora couloniana*, from the Caribbean, and I dread cleaning out the cages of young nymphs. Open the cage door and it's like watching popcorn popping in an open pan. Even with the help of my enthusiastic cat, it takes an age to round them all up.

Cave crickets (Raphidophoridae) also have ridiculously long back legs, which make them almost impossible to catch on the cave walls where they live. They also have long cerci that are even more sensitive than those of roaches. They can detect any air movement caused by a curious entomologist and pinpoint its source exactly. Their reaction is near instantaneous. Without even turning, they can leap, forwards, backwards or sideways, directly away from the disturbance, wherever it's coming from.[33]

The long back legs of a great green bush-cricket, *Tettigonia viridissima*, provide a lot of leverage when they extend, allowing the insect to leap considerable distances.

Grasshoppers (Acrididae) belong to the same order (Orthoptera) as katydids and cave crickets, and are the quintessential leaping insects, although their legs are shorter in relation to their bodies than those of either katydids or cave crickets. Nevertheless, they are powerfully built, with muscles that, gram for gram, can deliver ten times the force of ours. Indeed, the only muscles in the animal kingdom more powerful are the adductors that close the shells of clams. Even so, grasshopper muscles are not capable of propelling them as far as they often leap. The same is true of many other insects, especially the jumpers with much shorter legs than those of katydids. To put some figures on this, the maximum power a muscle can produce is about 300 watts per kilogram. But a planthopper (a small true bug) can accelerate in less than a millisecond from standing to its take-off velocity of 5 metres per second. This requires a power output of many thousands of watts per kilogram.[34] How does it do this?

Take a closer look at those powerful-looking hind legs of a grasshopper. Insect legs are divided into several distinct segments with flexible joints between them. An insect version of the song 'Dry Bones' would run (from the body outwards): the coxa is connected to the trochanter, the trochanter to the femur (a kind of insect version of a thigh), the femur to the tibia, the tibia to the tarsus. The tarsus itself consists of several segments, often of pad-like structures, that serve the insect as a foot. Grasshoppers also have distinctly knobbly knees. The joint between the femur and the tibia is swollen and crescent shaped in outline. For that reason, it's called the semi-lunar process and it's the source of a grasshopper's superpowers.

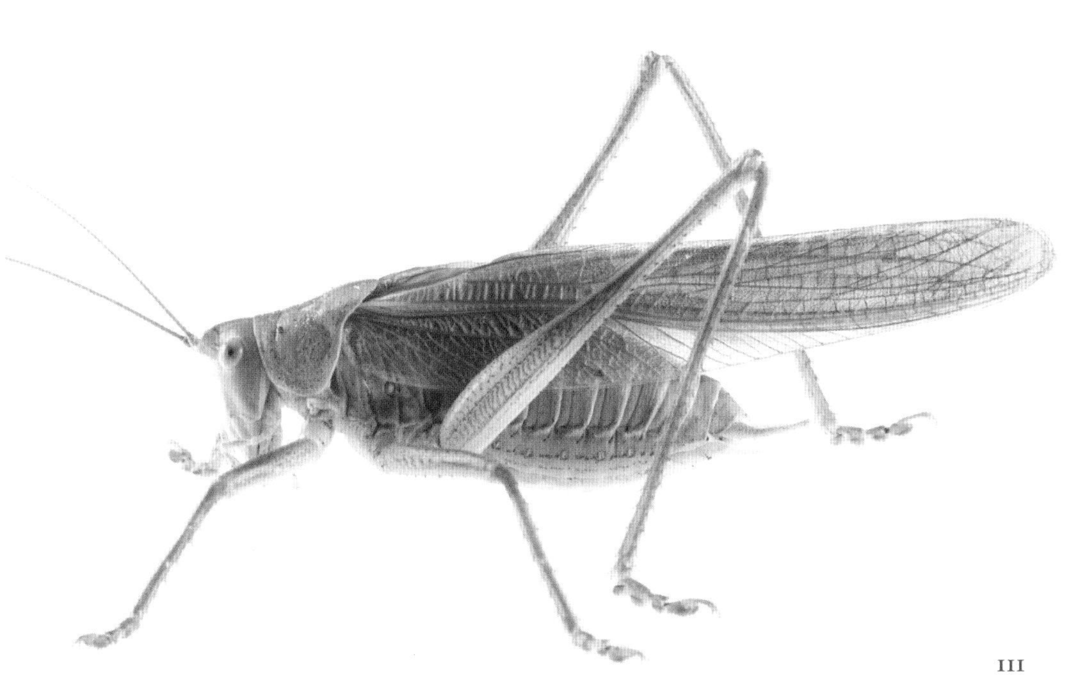

It consists of thick, stiff cuticle combined with layers of a super-rubber called resilin – which is remarkable stuff. It was discovered in 1960 at Cambridge University by the Danish zoologist Torkel Weis-Fogh in the wing hinges of locusts and dragonflies.[35] After this, it turned up in countless other places in insect bodies, where it was found to fulfil a variety of functions, from helping to pump venom through a bee sting, to allowing cicadas to sing at a deafening volume. Resilin is highly elastic and the most efficient rubber known. When stretched it springs back exactly to its original length, returning 97 per cent of the energy it stored when stretched. Moreover, it can be stretched like this through endless cycles without losing any pliability. It's a near-perfect energy-storage device, and energy storage is key to how insects generate the enormous power they need to perform their jumps. Insects therefore have built-in springs. But it's not quite that simple.

Because resilin is so easy to stretch, it can't store large amounts of energy. On its own it can't generate anything like enough power for a grasshopper jump. The cuticle of the semi-lunar process, on the other hand, is very stiff. It takes a lot of energy to bend it even a little, which means that with hardly any distortion it can store a great deal of energy. Unfortunately, being stiff also makes it more brittle, but the layer of resilin helps it to return to its original shape quickly and safely.[36]

When a grasshopper wants to jump, it has first to cock its springs. The tibia is raised to lie alongside the femur and the powerful muscles in the femur contract, bending the semi-lunar process in all three dimensions. Because loading the springs can be done relatively slowly, the muscles

The legs of grasshoppers, such as this stripe-winged grasshopper, *Stenobothrus lineatus*, are shorter than those of bush crickets, but have built-in springs that allow grasshoppers to jump further.

can pump a lot of energy into the spring. It's like an archer drawing back the bowstring. He can pull it slowly, storing more and more energy in the bow as it bends. Then he can loose the string and all that energy is returned in an instant. Likewise, when a grasshopper triggers its cocked legs, it shoots skywards. Malcolm Burrows, a Cambridge University zoologist who has studied a wide variety of jumping insects, has pointed out that such springs are the difference between how far you could throw an arrow compared to how far you could shoot it with a bow.

Apart from the long-legged katydids and cave crickets, together with some lacewings and a few bugs like flea-hoppers (Miridae), most jumping insects so far examined use springs of some description,[37] although their location and structure differ in different insects, showing that insects have evolved this trick on numerous separate occasions. And it was clear from the high-speed footage shot by Eric Lucey for the BBC's epic rabbit flea film that fleas must be doing something similar. That footage inspired Henry Bennet-Clark, then at the University of Edinburgh, to work with Eric Lucey to film a flea jumping at even higher magnifications, to see if they could work out the mechanisms involved.[38] Miriam Rothschild supplied the rabbit fleas and later went on to make her own studies of the jumping mechanism.[39]

Bennet-Clark and Lucey identified a pad of resilin on each side of the flea, derived from the wing hinge, which flightless fleas no longer needed, at least not for its original function. So, evolution repurposed it to become a catapult. Bennet-Clark and Lucey's calculations showed that if compressed enough, the resilin could store the energy needed to power the flea's high-speed jump. As it gets ready to jump, a flea first locks the joint between its topmost leg segment, the coxa, and its thorax. It then uses two massive muscles running right from its back to its hind legs to pull the leg up, rotating around the joint between the coxa and trochanter, the next segment down. This action also compresses the resilin pad. All the flea has to do is release the lock, and, like an arrow fired from a bow, it is catapulted upwards. It accelerates so fast that it pulls over 130g* on take-off. A well-trained combat pilot can handle 9g before blacking out.

* Our weight (as opposed to mass – the amount of stuff our bodies contain) derives from the acceleration due to gravity (about 9.8 metres per second on Earth). In shorthand, this is referred to as 1g. Rapid acceleration, such as in a fighter plane, produces higher g-forces and increases weight. Manoeuvring at 3g, for example, a pilot weighs three times as much as when sitting on the runway. The added acceleration also causes problems with blood circulation, which cause loss of consciousness. A pilot can survive sustained forces of 9g, but only with extensive training and a specially designed 'g-suit'.

Although the resilin pad is crucial to the spring mechanism it's possible that, as with grasshoppers, some of the energy is stored by bending the toughened cuticle around the resilin pad.[40] Wherever it is stored, though, once the leg is unlocked, the energy is released through the joint between the coxa and trochanter. This led to another enduring mystery surrounding the flea jump, one that wasn't resolved until 2011. Does a flea jump from its knees or its toes? Miriam Rothschild thought that, given how the spring mechanism works, the flea must jump by slamming its hind trochanters against the ground. Henry Bennet-Clark felt that it would use the whole leg for leverage and spring from its tarsus. A careful analysis by Gregory Sutton and Malcolm Burrows at the University of Cambridge showed that Bennet-Clark was right and fleas do jump from their toes.[41] The puzzle of the flea's extraordinary leap had finally been fully explained.

For a long time, fleas were considered the champion jumpers of the animal world, largely because no one had looked closely enough at all the other hopping, skipping and jumping insects. When Malcolm Burrows and his team at Cambridge University did just that, the flea was soon relegated from top spot. The new champion turned out to be an insect that will be familiar to many gardeners and country walkers. On a summer stroll through the countryside, it's not uncommon to find gobs of white froth on plant stems – *cuckoo spit* – made as a protective blanket by the nymphs of bugs called froghoppers, *Philaenus spumarius*. When fully grown, these creatures abandon their foam shelters, and turn into brown and boring bugs, undistinguished to look at, but still capable of outperforming a flea at the high jump. Hold one of these bugs on your hand and you can really feel the ping as it leaps to freedom.

A froghopper has a different kind of lock and spring that allows the hind leg to extend in under one thousandth of a second and launch it skywards at over 400g. Size for size, it can leap higher than a flea and is four times faster. Its hind legs are so specialized for jumping that it has to drag them behind as it walks on its front four legs.[42] But froghoppers didn't stay in the gold medal position for long. A few years later, the Cambridge scientists measured the performance of a little planthopper, *Issus coleoptratus*, and found that it pulled over 700g on take-off and could catapult itself over a metre.[43] To achieve this record-breaking performance, the power from the spring in each leg must be released simultaneously, so the two jumping legs need to be tightly synchronized. They hit the ground within three hundred thousandths of a second of each other, extraordinary precision for any biological system – and

Froghoppers like *Cercopis vulnerata* have such powerful, spring-loaded legs that they can outjump fleas.

indeed, they accomplish this with a solution that is more mechanical engineering than biology. Their back legs are linked together by a system of cogs, the first example of a true gear system found in nature.[44]

A few kinds of ant can also jump, and although their performances don't match those of planthoppers or fleas, they demand a lot of respect for other reasons. Working anywhere near a jack jumper ant nest is always nerve-racking. There are several kinds of jack jumper or jumping jack in south-east Australia and Tasmania, although *Myrmecia pilosula* is the one most often encountered. A related species, the hopper ant, *Myrmecia imaii*, is found only in south-west Australia. They all defend their nests with single-minded ferocity, and if disturbed, emerge leaping and bounding to face up to any threat. They move at great speed by alternating running with short jumps, described by pioneering ant biologist William Morton Wheeler in the early twentieth century as like a Lilliputian cavalry charging into battle. They can both bite and sting, and their venom causes far more allergic reactions in humans than that of any other ant – there have even been several fatalities in Tasmania. However, due to their animated leaps it's almost impossible to keep them off your legs.

Jack jumper ants also use their agility to pounce on unsuspecting prey, a trait they share with several other kinds of jumping ants. *Harpegnathos* species in south-east Asia and *Gigantiops* species in South America are both accomplished jumpers. *Harpegnathos* can jump more than 20 centimetres horizontally to escape a threat, although hunting jumps are much shorter. Their jumps are not particularly well controlled and their landings are

often undignified, so they can only guarantee hitting a target over short hops. These ants also have the curious habit of all abruptly jumping in concert. They may continue this frenetic behaviour for five minutes or more, although no one really knows why. One suggestion is that this exuberant communal bouncing serves to drive small invertebrates from cover so that they can be pounced on more easily.[45]

Gigantiops only jumps around 3 or 4 centimetres, but swings its heavy abdomen down as it jumps to help provide a little more thrust.[46] Other ants don't use their legs at all. Trapjaw ants, *Odontomachus* spp., can jump nearly half a metre – using only their jaws. These ants hold their long mandibles gaping wide open until they encounter something edible. Then the jaws snap shut on their victim in one of the fastest recorded movements in the animal kingdom. Again, as in the flea's leg, these speeds are only possible by using a spring-and-catch mechanism, but having hyper-fast jaws has another benefit for these ants. If threatened, the ants strike their closing mandibles against a hard object, and the force sends them spinning backwards and out of danger.[47] Their powerful jaw springs originally evolved to capture prey but, in yet another example of the remarkable adaptability of insects, these ants now use this capability as an escape mechanism as well. Only recently, trapjaw ants, at least those in Borneo, have also been seen jumping in a more conventional way, using their legs. This enables them to jump both forwards and backwards – using their legs to jump forwards and their jaws to jump backwards.[48]

In the world of insects, the more we look, the more we find. Trapjaw ants are not the only ants with ballistic jaws, and nor do they have the very fastest jaws of all. Dracula ants, *Mystrium camillae*, employ a different mechanism for jaw jumping. They build up tension in their closed jaws, pressing the tips of their mandibles together with ever more force. When enough force is applied, one mandible slides over the other, like a finger snap, but much, much faster. The movement is over in just twenty thousandths of a second, in which time the mandibles reach a velocity of 90 metres a second. Previously, the fastest recorded animal movements were the snapping mandibles of trapjaw ants, the claw strike of the Mantis Shrimp and the leaps of froghoppers and planthoppers. But the Dracula ant has them all beaten.[49] Incidentally, these ants get their name from the fact that the adults feed on the blood of their own living offspring. It's a gruesome but sustainable process, since the larvae survive to reach adulthood. Otherwise, there'd be no more Dracula ants to hold the world record for the fastest animal movement.

LETHAL LEGS

Insect legs can carry them at high speed over any terrain, but the trademark of insect evolution is their remarkable design flexibility – and legs can be used for a lot more than just getting around. Several groups of insects have turned their front pair of legs into traps as lethal as the high-speed jaws of trapjaw ants. The best known of these are the mantids.* The 2,500 species of mantids display an astounding variety of shapes and sizes, from creatures that look like the quartz pebbles that are scattered over the surfaces of the rocky deserts where they live, to those that are indistinguishable from dead leaves; from monsters, like *Idolomantis* and *Toxodera* spp. that grow to 17 centimetres, to the diminutive *Bolbe pygmea*, the world's smallest mantis at just a centimetre long.[50] Yet they all share one trait – the possession of raptorial front legs, armed with wicked-looking spines, with which they seize and trap prey.

Mantids are thought to have first evolved on tree trunks, and several families that show the most primitive characters still stalk such places today.[51] The six species of *Chaeteessa* scurry over tree trunks in South America and, unlike the spectacular diversity of forms displayed by their relatives, these little mantids could easily be mistaken for elongated cockroaches. They also differ from all other mantids in that they run on all six legs. Their raptorial legs are not as well developed and are still used for locomotion as well as for grabbing prey. In most other species, the front legs are so modified for catching prey that they aren't much use for anything else. Instead, the mantis holds them tucked under its body, or – when waiting in ambush – elevated out in front, in a position that gives these insects their epithet 'praying'.

As well as raptorial legs, mantids have huge eyes and highly mobile heads. There's something decidedly uncanny about the way a mantis follows you with its unblinking gaze as you move around it. Those sharp eyes can also lock on to prey, and the mobile head tracks its every movement, gauging the distance, waiting for its prey to come within reach of its front legs before striking at lightning speed. Their prey ranges from the tiny to the unfeasibly large. I recently watched a pygmy ground mantis, *Bolbe* sp., on fallen logs in the Queensland rainforest. It seemed to relish ants, particularly the introduced big-headed ants,

* There is some debate over the 'correct' way to refer to more than one of these insects – mantids or mantises. Originally, 'mantises' was more common in the US and mantids in the UK and Continental Europe. The word 'mantis' comes from the Greek word for 'prophet' and its Latinized plural should therefore be 'mantes'. 'Mantid' refers to members of the family Mantidae, to which all known species belonged until taxonomists inconveniently split the mantis order into several different families. So strictly speaking, neither is correct!

(*overleaf*) With its flattened abdomen and flanged middle and hind legs, a juvenile orchid mantis, *Hymenopus coronatus*, looks so much like a flower that it attracts other insects, which fly close enough to be caught by its lethal front legs.

Pheidole spp., which are becoming something of a nuisance in this part of the world. This little mantis just sat near a colony of ants, stuffing one after the other into its mouth like a kid with a sweet jar.

Others take prey almost their own size, and not always just insects. There are many records around the world of mantids catching birds, although most of these come from North America, where tiny, bite-size hummingbirds abound.[52] North America has its own native mantids, but some of the giants, like the Chinese mantis, *Tenodera sinensis*, have been introduced, either accidentally or deliberately for pest control, and these big species are easily capable of catching the local hummingbirds. There are even reports of them hanging around garden feeders where hummingbirds gather in large numbers to drink the sugary offerings. Mantids have also been seen catching and eating mice.

Even as an entomologist, with my soul firmly at home in the insect world, it's a little unsettling to watch insects feeding on warm and furry or feathery things. Perhaps more surprisingly, one mantis was observed catching and eating fish. A single individual of *Hierodula tenuidentata* was found stalking across water-lily leaves on a tiny artificial pond on a roof garden in Karnataka in south-east India. It perched on the edge of a leaf and waited until a guppy swam by beneath it, at which point it plunged its legs through the surface film and hauled its catch back onto the leaf where it could be eaten.[53] So far, this seems to be a unique observation, but it does show how adaptable mantids can be and how they can master new techniques to catch prey.

This rather shocking ability to tackle even quite large vertebrates has been known for some time. It was first reported by the German-Argentine entomologist Hermann Burmeister in the early years of the nineteenth century, when he recorded a mantis catching a lizard three times its own length.[54] However, these large vertebrate meals are the exception and mantids typically content themselves with a less grisly diet of invertebrates.

Mantids are masters of camouflage, which serves to conceal them both from their own enemies and from their prey. They're mostly sit-and-wait predators and adopt disguises that make them invisible in their chosen habitats. Dead-leaf mantids, *Deroplatys* spp., for example, look exactly like rolled-up dead leaves. Others sport bright green flanges that hide them among living leaves, while juvenile orchid mantids, *Hymenopus coronatus*, look like flowers. They are brightly coloured, off-white or pink, and sit with their abdomens curled over their backs, like petals. Their four rear legs grow large, petal-like flanges that complete their

flower disguise. The species was discovered in Indonesia in 1972 and at first was often portrayed sitting in an appropriately coloured orchid flower. Many wildlife filmmakers made the same mistake, posing their specimens on brightly coloured flowers. They reasoned that because it resembles a flower, a mantid could lurk unseen in a real flower and wait for unsuspecting pollinators to visit in much the way that crab spiders do.

In fact, the orchid mantis's disguise is so good that it doesn't need a flower to hide in. It simply pretends to *be* a flower. Insects are completely fooled and fly to the mantis as readily as they do to real flowers.[55] To add to its pulling power, this mantis also secretes two chemicals that oriental honeybees, *Apis cerana*, use for communication, and these bees often approach the mantis head on, responding to it exactly as they would to a genuine flower, usually with fatal results.[56]

A dead-leaf mantis, *Deroplatys dessicata*, well camouflaged as a dried leaf, waits in ambush.

Most mantids use their legs like jackknives, trapping their victims between the sharply spined femur and tibia. However, one species, *Carrikerella simpira*, only recently discovered in highland rainforests in the Andes, takes a different approach. The tibia has a clump of forwards facing spines and the mantis shoots its legs forwards to impale its prey on the spines. In this way the aptly named spear mantis can catch prey in awkward places that would be out of reach for the conventional swinging mantis strike. These 4-centimetre-long mantids catch a range of prey, from skipper butterflies over a centimetre long to 2-millimetre fungus gnats. One individual even spotted and impaled a tiny, half-millimetre mite, which demonstrates the precision of its targeting system and the accuracy of its strike.[57]

Since insects seem to manage perfectly well on four legs, a number of other groups have independently evolved raptorial legs. Mantispids or mantis flies are in fact a family of lacewings (Neuroptera), but they look so similar to mantids, with enlarged, well-armed front legs, that at first glance you'd be forgiven for thinking they really were mantids. Like true mantids some are also masters of disguise, although rather than favouring passive disguises like dead leaves, twigs or flowers that can hide them from predators and prey alike, mantispids go in for more aggressive mimicry to escape their predators. One North American species mimics paper wasps, which are boldly striped to warn all that they have excruciatingly painful stings. *Climaciella brunnea* has adopted the shape and colour of a paper wasp, and the habit of curling its abdomen up as if preparing to use its non-existent sting. All in all, it looks like a somewhat intimidating cross between a mantis and wasp, although it is actually neither.

An assortment of different true bugs (Hemiptera) also catch food with raptorial legs, though they're not always as well developed as those of mantids or mantispids. An interesting progression of ever more specialized front legs can be found among some of the assassin bugs (Reduviidae). Most of these bugs are predators that jump on passing prey and hold the victims with their front legs while injecting a lethal concoction of toxins and digestive enzymes through their sharp proboscis. In some, the forelegs are hardly modified at all, while others have legs that more closely resemble those of mantids. One group has even evolved legs that work in the same way as crab claws, with a movable tibia that closes like a pincer against a rigid extension of the femur, a unique structure in the insect world.[58] If that wasn't remarkable enough, one of these bugs has evolved tool use, an ability usually seen as the province of higher mammals and birds.

Although they look like mantids, mantis flies (Mantispidae) are not related. They are lacewings (Neuroptera).

A tool-using bug is too good a story to pass up on, and we were determined to be the first to film this behaviour for the series *Alien Empire* for the BBC, broadcast back in 1996. This astonishing discovery had been made a decade earlier by Elizabeth (Betty) McMahan, an American entomologist from the University of North Carolina at Chapel Hill, while she was working in Costa Rica. So, with Betty's help, cameraman Kevin Flay set off with a small team to try and capture the behaviour of *Salyavata variegata*.

This bug eats termites such as *Nasutitermes* that only emerge from their carton nests* under cover of darkness. It seemed that only the nymphs of the assassin bug were tool users and to make life harder, the nymphs conceal themselves, from both termites and cameramen, by covering themselves with bits of the carton nest. So, it looked like long nights in the rainforest trying to find and film a half-centimetre-long, invisible bug. Luckily, Betty had recorded *S. variegata* on nearly half of the termite nests she had surveyed.[59] Even so, it was many nights before Kevin finally saw a young assassin bug behaving suspiciously.

The bug had caught an unsuspecting termite on the outside of the nest and was feeding happily, sucking out a liquid meal after injecting enzymes into the termite's body. Once it was finished, it didn't just discard the drained carcass. Instead, it carried the termite over to one of the nest's entrance holes. Here it proceeded to dangle the carcass into the nest. Termites are programmed to pick up fallen comrades and dispose of them in a graveyard to avoid spreading disease through the crowded colony, so workers in the nest were simply unable to resist the bait being brandished before them. Like a master fisherman, the bug withdrew the dead termite, slowly enough that a few workers followed, still desperately trying to grab hold of this oddly active dead termite. As soon as the workers were far enough from the entrance hole, the bug dropped the dead termite and in one movement grabbed a living termite, which would in turn become first a meal, then the next bit of irresistible bait.

There is another group of assassin bugs that we'll meet again in Chapter 8, called resin bugs, which – it could be argued – are also tool users. Several unrelated kinds gather sticky resin from suitable plants

---

* *Nasutitermes* nests are built in tree branches or around the base of a tree trunk from a mixture of chewed wood and faeces, which forms a remarkably durable carton. Since termites dislike being exposed to the elements, they also construct covered runways of the same material from their nests down to the forest floor. They emerge to extend their nests with more carton at night.

and smear it on their front legs. In North America these bugs are often called bee killers due to their effectiveness at plucking bees from flowers with their sticky legs.[60] It seems that there's no end to the ingenuity of assassin bugs.

Yet other species seize spiders from their webs, which involves negotiating the spider's sticky snare, a lethal trap for most insects. One species, *Stenolemus giraffa*, which has been studied in detail, uses its front legs as precision tools to break each thread individually. The bug snaps a silken strand but holds each broken end taut in its front legs. It then releases the tension in each thread slowly to avoid sending any tell-tale vibrations through the web and alerting the spider. To make doubly sure that it remains undetected, this stealthy assassin often releases each half of the broken strand separately, waiting a couple of minutes between each.[61] In this way, it makes slow but steady progress across the web to the prize at its centre.

Although locomotion and feeding are two of the most obvious uses for insects' legs, we've barely scratched the surface when it comes to exploring some of the other bizarre uses that insects have found for their six legs. Some, like the kick-boxing chameleon grasshoppers, *Kosciuscola tristis*, of Australia, use their powerful back legs for fighting rivals, while others, like paddle-legged mosquitoes, *Sabethes* spp., use elaborate plumes on their middle legs to impress female mosquitoes. Sumo-wrestling 'nutcracker' cave crickets use their long, arched rear legs to squeeze rivals in a show of strength, or to hold females in a vice-like grip until they consent to mating. More courtly male orchid bees create an intoxicating perfume in channels in their enlarged legs to impress the ladies. We'll meet these insects in a later chapter as we discover how insects entice their mates. But sometimes we just have to stare in wonder, without knowing exactly what is going on. At least, that was all I could do when I first encountered a phantom cranefly, *Bittacomorpha clavipes*.

One late afternoon, in a sheltered forest in the Appalachian Mountains of North Carolina, I saw what I took to be a piece of thistledown blowing on the wind. It was only when I looked closer that I realized it was a phantom cranefly. It is aptly named since in the shadowy light of the forest, it flickered, ghost-like, in and out of sight. These insects have long legs, boldly striped in black and white, and each has a unique swollen section. The inflated part is filled with tracheae, so in essence is air filled, which makes the legs very light.[62] When it flies, its legs float out sideways in such a way that some have described it as looking like a drifting snowflake.

Quite what purpose these peculiar legs serve is not entirely obvious. They may act as sails to catch the breeze, or they may break up the insect's outline to make it less visible, or perhaps they just make it look nothing like an insect. The first one I saw certainly had me fooled, and if I had been an insect-eating bird I would have ignored it completely. If nothing else, the ethereal sight of one of these craneflies dancing on the wind emphasizes that we've barely begun to realize all the uses to which insects have put their six jointed limbs. They evolved in a distant aquatic ancestor but proved just as useful when insects emerged onto land. Yet despite their versatile limbs, it seems that insects remained scarce for the early part of their terrestrial invasion. Real success depended not just on occupying the land but on occupying the air above as well. Insects are the only arthropods ever to have evolved wings, and with that unique evolutionary innovation, they began their rise to unprecedented domination.

A stick insect, *Parapachymorpha zomproi*, is more leg than insect; its long legs aid its disguise as a dead twig.

4

First in Flight
How insects conquered the air

The desire to fly is an idea handed down to us by our ancestors who, in their gruelling travels across trackless lands in prehistoric times, looked enviously on the birds soaring freely through space, at full speed, above all obstacles, on the infinite highway of the air.

Wilbur Wright, aviation pioneer, 1867–1912

The Outer Banks of North Carolina are a chain of thin barrier islands looping out from the coast, protecting the Pamlico and Albemarle Sounds and the fractal shores of the mainland. It's a wild and remote place, often battered by Atlantic storms, yet it has played a pivotal role in world history on several occasions. In the lee of the barrier islands lies Roanoke Island, the site of the first attempt by the British to colonize North America. It was so remote and isolated from England that the colony failed and both it and its inhabitants mysteriously disappeared, but the 'lost colony' was only a start. It gave the British a taste for what the New World offered and a few decades later they were back, a little further north on Chesapeake Bay, where Jamestown, Virginia, became the first settlement in what would eventually become the United States of America.

A few kilometres to the north and east of Roanoke lies another island where, a little over 300 years after the settlers at Roanoke vanished, the first powered aeroplane took to the air, launching an era in which very few parts of the globe can be considered truly remote. Back in 2002, I found myself wandering around this island, over dunes and grassy flats humming with insects. I was looking for Carolina locusts, *Dissosteira carolina*, insects that, despite their name, are not usually prone to erupting in biblical plagues. In fact, they're almost impossible to spot. Sitting motionless on an open patch of gravelly sand or stony ground, their mottled colour blends in so perfectly that they're invisible. I can stare straight at one and not see it. Get too close, however, and it uses its powerful back legs to spring into the air, then spreads startlingly black wings that carry it safely beyond the reach of a curious entomologist. An insect miraculously appearing from nowhere, only to vanish again when it lands and folds its conspicuous hindwings beneath camouflaged forewings, confuses any pursuer, even a large-brained ape that knows exactly how its trick works.

The evolution of wings gave insects a huge advantage in the race to colonize the globe, not just by granting them the power of flight but, as amply demonstrated by those Carolina Locusts, in the countless other ways that they can be used to give insects an edge. Apart from escaping predators and rapidly colonizing new habitats, wings can be employed for camouflage, solar heating, warning displays and even singing love songs – to name but a few such additional uses that the ever-adaptable insects have come up with.[1]

Flight evolved just once in insects. All those beetles, dragonflies, butterflies and bees that are buzzing around me in the summer dunes of

(previous page) The wings of an oak-leaf butterfly (*Kallima inachus*) serve as a very convincing form of camouflage as well as for flight.

Carolina saddlebags dragonfly, *Tramea carolina*. Insects are the only invertebrates to have conquered the air.

North Carolina's Kill Devil Hills are descended from a common ancestor that flew perhaps as long as 400 million years ago. Insects then had the sky to themselves for around 170 million years before vertebrates began their own conquest of the air – plenty of time to explore every advantage that wings could offer. On their vehicle licence plates, North Carolinians proudly boast their claim to be 'first in flight', but they were beaten to it by a considerable margin.

The first vertebrates to achieve powered flight were pterosaurs, followed by birds and then bats, although several groups of theropod dinosaur other than birds either evolved powered flight or came very close to it,[2] and the bats may have taken to the skies on three separate occasions from a common ancestor with webbed hands.[3] For all these groups, flight proved extremely successful. There are more kinds of bird than any other terrestrial vertebrate, and one in every five species of mammal is a bat. Among mammals, bats are second only in number to the swarming hordes of rodents. We've already discovered that flying insects outnumber all of these other aeronauts by many orders of magnitude.

The last of the vertebrates to lift off was *Homo sapiens*, a mere century or so ago – on the very spot where I was chasing Carolina locusts. A white obelisk at Kill Devil Hills, near the town of Kitty Hawk, marks the

site where, on 17 December 1903, Orville Wright became the first human to ascend in a heavier-than-air, powered flying machine. Later that day, his brother Wilbur also flew the 'Wright Flyer', a precarious-looking contraption of struts and wires. Orville's flight covered 37 metres, only the wingspan of a modern jumbo jet, but after this momentous event our skill in the air developed at a breathtaking rate. A little over a decade after the first Wright Flyer hung uncertainly over the dunes of Kill Devil Hills, we were using planes to kill each other in the aerial battles of the First World War. A mere six decades after that first 37-metre hop through the air, humans flew a quarter of a million miles to the moon and back. Flight proved just as successful for humans as it had for all those other flying animals.

In the summer of 2002, I was roaming the dunes of the Outer Banks for a film celebrating the 100th anniversary of the Wright Brothers' first flight, by illustrating the parallels and differences between natural flight and human flight. Aviation pioneers before and after the Wright Brothers drew both their aspirations and their inspiration from nature, although largely from birds. Studying the wings of birds taught the

builders of the first flying machines a lot about aerodynamic shapes and about how to manoeuvre in flight. The Wright Brothers themselves made close observations of the turkey vultures that soared over their home in Ohio, and which can turn in impressively tight circles while scanning the ground below for carrion. They do this by twisting and tilting their outstretched wings, and later versions of Wright Flyers incorporated similar wing-twisting mechanics so that they could bank and turn, although with considerably less grace than a turkey vulture.

However, the main reason why large gliding birds proved a valuable source of inspiration for human flight is that birds are big, at least compared to insects. Gliding vultures live in a world dominated by the same kinds of aerodynamics that keep planes in the sky. For tiny insects, air behaves very differently. It feels much denser, like flying through molasses for the very tiniest flyers. In their world, viscous forces dominate, and they depend on a different kind of aerodynamics to stay in the air. Understanding how insects fly wouldn't have been much help in designing the first planes, but it's invaluable today in designing micro-aerial vehicles (MAVs) – tiny drones, like RoboBee and DelFly – that we'll meet shortly.

INSECTS GET THEIR WINGS

Before we explore how insects fly, there's an even more fundamental question to ask. Where did the wings of insects come from? Insects stand apart from all of nature's other flyers in that their wings are novel structures. Birds, other theropod dinosaurs, pterosaurs and bats all had to sacrifice their front legs to evolve wings, but at least that means we understand the anatomical origin of vertebrate wings. Insects, on the other hand, have kept all six of their legs and simply added wings to their inventory from scratch, which gives them much greater evolutionary flexibility than other flyers, but leaves us pondering how evolution created the wings of insects in the first place. This problem has occupied the minds of several generations of biologists; over that time, their deliberations have boiled down to two theories, each staunchly defended by its supporters.

The physical origin of wings is intimately bound up with what their precursors were used for. Evolution has no forethought. Every intermediate stage of a structure must give its bearer an advantage or natural selection will eliminate it. So, what were proto-wings used for before they were any good for powered flight? What were the steps on

Blanchard's tree nymph, *Idea blanchardi*, has huge wings on which it flies slowly through Indonesian forests.

the road to flight? This is a question, incidentally, that's just as crucial to answer for all the vertebrate flyers as it is for insects.

For all these animals, there are really only two possible answers, usually formulated as top down or ground up. In other words, did the ancestors of flying animals live in trees, leaping from branch to branch, where any factor that improved their performance as gliders would give them a distinct advantage? Or did these ancestral creatures live on the ground and use the precursors of flapping wings for something else – perhaps for giving them extra thrust or for turning at speed, for propelling them up inclined surfaces or for catching prey, all ideas that have been suggested for vertebrate flyers by proponents of the ground-up hypothesis?

The exact route map leading to powered flight is likely to be a little different for insects in their small, viscous world. Whether they followed the 'ground-up' or 'top-down' path also depends in large part on where those enigmatic wings came from, and that too falls into two well-defended camps – usually referred to as the paranotal theory and the gill theory.

Intriguingly, the origin of insect wings was being discussed among natural historians long before Darwin published *The Origin of Species* in 1859, albeit in the context of special creation.[4] The question they were trying to answer was how God had decided to fashion insect wings. In the post-Origin world, the debate took on a more modern perspective. In an 1870 book on animal anatomy, the German scientist Karl Gegenbaur suggested that insect wings evolved from gills, such as those still found along the sides of the abdomen of mayfly larvae.[5] These gills take many forms. In some species they are tufted filaments lying along the back of the abdomen. In others they lie protected beneath a toughened gill cover formed from the first pair of gills. However, one common species, *Cloeon dipterum*, which I often find when out pond-dipping, best illustrates why Gegenbaur saw parallels between gills and wings.

This species, known to fly-fishermen as the pond olive, often lives in small, stagnant ponds where oxygen levels may drop drastically, especially in hot weather. To combat this, the pond olive larva sports a row of gills down each side of its abdomen – large, flat plates containing a silver tracery of air-filled tracheae. Oxygen dissolved in the water passes through the thin cuticle of the gill plate and into the tracheae, from where it can diffuse into the tracheal network to supply oxygen to the rest of the mayfly's tissues. When oxygen levels in the water drop, the pond olive beats its gills to create a circulation of water and increase

The long tail streamers of a male Madagascan moon moth, *Argema mittrei*, twist and turn when the moth is in flight and are thought to confuse the sonar of hunting bats.

its absorption of oxygen. Thin, flat plates with hinges and muscles that allow them to flap – it's easy to see why Gegenbaur felt that gills were the precursors of wings, and they clearly had their own critical function before they became big enough to be useful as wings.

A couple of years later, another German biologist, Fritz Müller, suggested an alternative hypothesis. While studying termites he noticed that the nymphs of some species developed flattened outgrowths from the top of the thorax. At the time, many scientists believed that as an animal develops through its embryonic stages, it conveniently replays its evolutionary history.* After all, even embryonic humans start off looking almost indistinguishable from embryonic fish. By extension, Müller suggested that, although only queen and king termites have wings, these nymphal lobes, or paranota, replayed the ancient evolutionary origin of wings. Perhaps paranotal lobes originally helped insects glide a little better if they jumped out of trees.

The paranotal theory thus goes hand in hand with the top-down theory, while the gill theory neatly fits with the ground-up (or at least water surface-up) theory. Later scientists picked up the batons of both Gegenbaur and Müller, and the debate about the origins of wings and the route to powered flight rattled backwards and forwards for the next 150 years as convincing evidence stacked up on both sides. This long-standing puzzle was not easily laid to rest. No insect equivalent of *Archaeopteryx* has ever been found. Unlike the now well-documented fossil history of early birds, flying insects popped up in the mid-Carboniferous period already fully winged, with no earlier clues as to where wings came from. Throughout the debate, evidence remained stubbornly ambiguous, so much so that one of the most eminent entomologists of the twentieth century, Sir Vincent Wigglesworth, changed sides, having first considered the paranotal theory proven almost beyond doubt.

Subsequently, and despite Wigglesworth's influential change of heart, the paranotal theory gradually gained in prominence – until James Marden of Pennsylvania State University discovered some curious stoneflies. Like mayflies, stoneflies have aquatic larvae that bear gills, although in stoneflies the gills are like clumps of sausages at the bases of the legs rather than flat plates. Again, like mayflies, adult stoneflies have wings and most of them can fly. But many species emerge from the water in the

* This was known as the Theory of Recapitulation or Embryonic Parallelism, summed up by Ernst Haeckel as 'ontogeny recapitulates phylogeny'. But since we now know that embryos evolve as much as other life stages, the idea has been relegated to an interesting footnote in the history of biology.

depths of winter when the air temperature is too low for flight. They can, however, still make use of their wings in a way that got Marden thinking.

One species of stonefly, *Taeniopteryx burksi*, is commonly out and about as an adult in Pennsylvania's chill February air. The stoneflies that emerge onto stones breaking the surface in the middle of a wide river have somehow to make their way ashore so that they can clamber through vegetation in search of the opposite sex. Marden watched these stoneflies use their wings as oars to row over the surface with remarkable alacrity. Following up with experiments in which he carefully trimmed the area of the stoneflies' wings, he showed a direct correlation between the size of the wings and the speed of rowing.[6] Here was a plausible scenario for the evolution of wings, in which all the stages of proto-wings had their own advantages, and in which natural selection would favour enlargement of the oars for greater speed until they were big enough for the insect to take to the air.

However, stoneflies are using fully evolved wings to row over the surface. They are not a living insect *Archaeopteryx* demonstrating an intermediate stage. They simply use their wings as oars because it's too cold to fly – but it turns out that quite a few kinds of insect use their wings for skimming over the surface of water. *Cheirogenesia* mayflies are found only in Madagascar. Unlike most mayflies, they're flightless. Their wings are still long but much reduced in width, so their surface area is not large enough to support the insects in the air, but they can provide enough thrust to power the mayflies quickly over the water's surface. The insects stand on the tips of their feet, minimizing contact with the surface to reduce drag, and use their wings to skim over the water, just like tiny hydrofoil boats.[7] Again, these mayflies are not true intermediate stages on the road to flight. They have secondarily lost their ability to fly, probably because there are no fish in the Madagascan rivers where they live, so they have no need to fly – and nature abhors a waste of energy.

I recently came across a similar example, not in exotic Madagascar but in a gravel pit just 20 miles from my home. I was there to gather pond skaters, *Gerris* spp., to film on a pond set we had built in our studio, but among my collection were some small leaf beetles, *Galerucella nymphaea*, which live on the leaves of water lilies. The plan was to film a sequence on how pond skaters capture their prey with the help of long, water-repellent legs that allow them to stand on the surface film. As we discovered in the previous chapter, their middle legs are especially long and used as sculls as they skate at high speed over the surface. Any

(*overleaf*) When threatened, an Io moth, *Automeris io*, reveals conspicuous eye spots on its hind wings. It looks enough like a staring owl to startle most predators.

insects not so well equipped are trapped by the surface film, where their struggles are soon noticed by the predatory pond skaters. Hairs on the pond skaters' legs are finely tuned to the ripples produced by struggling insects, giving them an exact fix on lunch, and they quickly home in on their helpless victims.

While introducing the pond skaters to their temporary home in the studio, I accidentally dropped a water-lily leaf beetle, *Galerucella nymphaeae*, onto the surface and noticed that it managed to avoid becoming stuck. It could stand, if somewhat precariously, on the surface film but, without the pond skaters long middle legs, it could do little more than totter across the surface – surely an easy victim for the speedy pond skaters. However, as the pond skaters began to circle, the beetle revealed its secret weapon. Like a James Bond car packed with gadgets and tricks, the beetle slowly opened its wing-cases and unfolded a pair of wings. The wings began to whirr but instead of taking off, the beetle raised its middle pair of legs, leaving it standing on four, just like the Madagascan mayflies, then scooted off across the surface. A beetle that waterskis – even faster than a pond skater can skate.

Scientists from Stanford University in California have examined this remarkable beetle in more detail, and discovered that it could skim at half a metre a second. Translate that into body lengths per second, and it's the equivalent of a human waterskiing at 500 kilometres an hour. Curiously, waterskiing takes more energy than aerial flight, but lift-off is more difficult from the water's surface than from land since surface tension tethers the beetle firmly to the surface film.[8] So perhaps, on balance, it's their best option when they get blown off a lily leaf and onto pond skater-infested waters.

None of these skimming insects are pre-aerial ancestors of flying insects. They have all adopted this mode of travel from ancestors that were fully capable of flight.[9] Indeed, water-lily beetles can still fly very well, as I discovered while chasing my specimens around the studio. Even so, the widespread existence of water-skimming insects is seen as good evidence that the earliest ancestors of flying insects first rowed then skimmed over the surface – but could functioning gills, evolved for aquatic respiration, survive the transition to air?

Shortly after his *Taeniopteryx* studies, Marden and his colleagues described an even stranger stonefly from cold, fast-flowing streams in Chile. *Diamphipnopsis samali* is unusual in that the adults retain their larval gills along their abdomens. *D. samali* is yet another insect that skims over the surface film and as it does, its gills often dip into the water, so they

may still work as gills even for the aerial adult. Importantly for the gill theory of wing evolution, this obscure little stonefly shows that it's perfectly possible for an adult to have both wings and abdominal gills, as perhaps the ancient ancestor of flying insects did when some of its gills began to be used as oars and later as true wings.[10] In fact, a few fossils have been found from the upper Carboniferous and lower Permian periods that look as though they have both wings and gills, which is seen as further support for this idea.[11]

There is also some evidence from genetic manipulations that bolster the gill theory. There's a mutant strain of fruit flies that develops the early stages of wings along the abdomen in the same place that the gills of mayflies grow,[12] suggesting a common genetic origin for both wings and gills. However, this theory makes a big assumption – that the ancestor of winged insects had aquatic, gill-bearing larvae. At first glance, this may seem obvious. The larvae of the most 'primitive' insects alive today, mayflies and dragonflies, live underwater (with a few highly

The lantern bug, *Pyrops candelaria*, has brightly coloured hindwings that surprise predators when it takes flight.

specialized exceptions), as do those of stoneflies. But the most recent genetic studies aimed at clarifying the basal branches of the family tree of winged insects show that the distant common ancestor of all flying insects almost certainly had terrestrial larvae.[13] Despite all the evidence in favour of gills becoming wings, this looks like a body blow for that idea. So, what about the rival paranotal theory?

In recent years this idea, along with the associated top-down hypothesis, received a boost from a fortuitous discovery. While studying the different ecologies of tree-living and forest floor ants in Peruvian rainforests, Steve Yanoviak, of the University of Louisville in Kentucky, noticed that one particular arboreal ant readily threw itself from branches when approached, even though the trek back up from the floor perhaps 30 metres below would be both long and dangerous. A closer look revealed that these ants, *Cephalotes atratus*, never actually reached the forest floor. Instead, they initially fell headlong as if to certain doom, then suddenly pointed their bodies towards the trunk and began to glide, albeit more like a controlled plummet, gaining a foothold on the tree just a few metres further down.[14]

High-speed video footage shot in a vertical wind tunnel showed that falling ants spread their legs and steered with their abdomen as they glided backwards towards the tree. Intrigued by this unexpected behaviour, Yanoviak looked at other arboreal ants and found a few related species that could also glide, although not as well. A few years later, in the rainforests of West Africa, he found a different range of species that could also glide, even though not one of these ants had any physical adaptations to help them.[15] More striking still, Yanoviak later found a bristletail in Peru that was equally adept in the air.[16] Ants are flying insects and although the workers don't have wings, the reproductive queens and drones have large wings and fly well – if only briefly. So, like the surface-skimming of stoneflies and mayflies, gliding is a trick they evolved long after they learnt to fly. For bristletails, on the other hand, that's not the case.

Bristletails branched off very early from the insect tree, well before flying insects evolved, and have retained many features of those first insects. In many respects, they are the closest things alive today to the ancestral pre-winged insect, and that includes a lack of any semblance of wings at any stage in their lives. Bristletails have never flown, but that hasn't stopped them from becoming effective gliders. All bristletails have three long tail filaments, which – for the arboreal, gliding species – serve as rudders that allow them to steer back to the trunk in nine out of ten

A dead-leaf mantis, *Deroplatys dessicata*, is almost invisible in its camouflage but, if threatened, has a dramatic startle display, revealing eyespots that are normally hidden beneath its wing covers along with boldly patterned hindwings.

cases.[17] The existence of these remarkable, if undistinguished, creatures is a real shot in the arm for paranotalists. It shows that, due to their small size, insects can glide using just behavioural modifications before there is any sign of structural modifications – and a top-down hypothesis fits very neatly with the latest understanding that the ancestors of winged insects were terrestrial rather than aquatic.

It's not hard to imagine that following such a behavioural change, any physical adaptations, such as flaps or flanges, that improved gliding performance would give an insect a big advantage, so would be strongly favoured by natural selection. To test this, scientists built insect-sized cylindrical models, decked out with short winglets in different places, and demonstrated that several different configurations stabilized the glide, especially in the larger models – although winglets situated at the front end, in roughly the position where thoracic wings eventually evolved, were particularly effective when combined with long tail filaments, like those of bristletails.[18] There's some evidence among fossil insects that nature experimented in a similar way.

Palaeodictyopterans were abundant and diverse during the late Carboniferous and Permian periods, when high levels of oxygen in the

atmosphere allowed them to grow much bigger than insects today. They flew on two pairs of large wings that, as in all modern flying insects, grew from the last two thoracic segments. However, they also had a set of winglets on the first thoracic segment, the prothorax. Some have described these insects as six winged, although to me that's a bit of a stretch. However, the prothoracic winglets did probably stabilize the flight path of the faster flying paleodictyopterans,[19] and show that early in the evolution of flying insects, nature had played around with just the sort of outgrowths of the thorax postulated by the paranotal theory.

In the 150 years since these two ideas were suggested, evidence has accumulated to support both ideas, although both still have serious weaknesses. So, this long-standing puzzle remained unresolved – until recently. In 2010, detailed studies of the genetic control of wing development showed that two distinct pathways are involved; one is also involved with the development of gills and the other also controls outgrowths from the thorax.[20] Convincing evidence had built up for both sides of the argument because in some ways both were right. A closer examination of the larvae of those ancient palaeodictyopterans also suggests that their wings are a fusion of tissues from the sides of the thorax, where gills would form on other segments, and the top of the thorax. To grossly oversimplify, the lower tissues contribute the hinge mechanism, while the top of the thorax provides the wing aerofoil.[21]

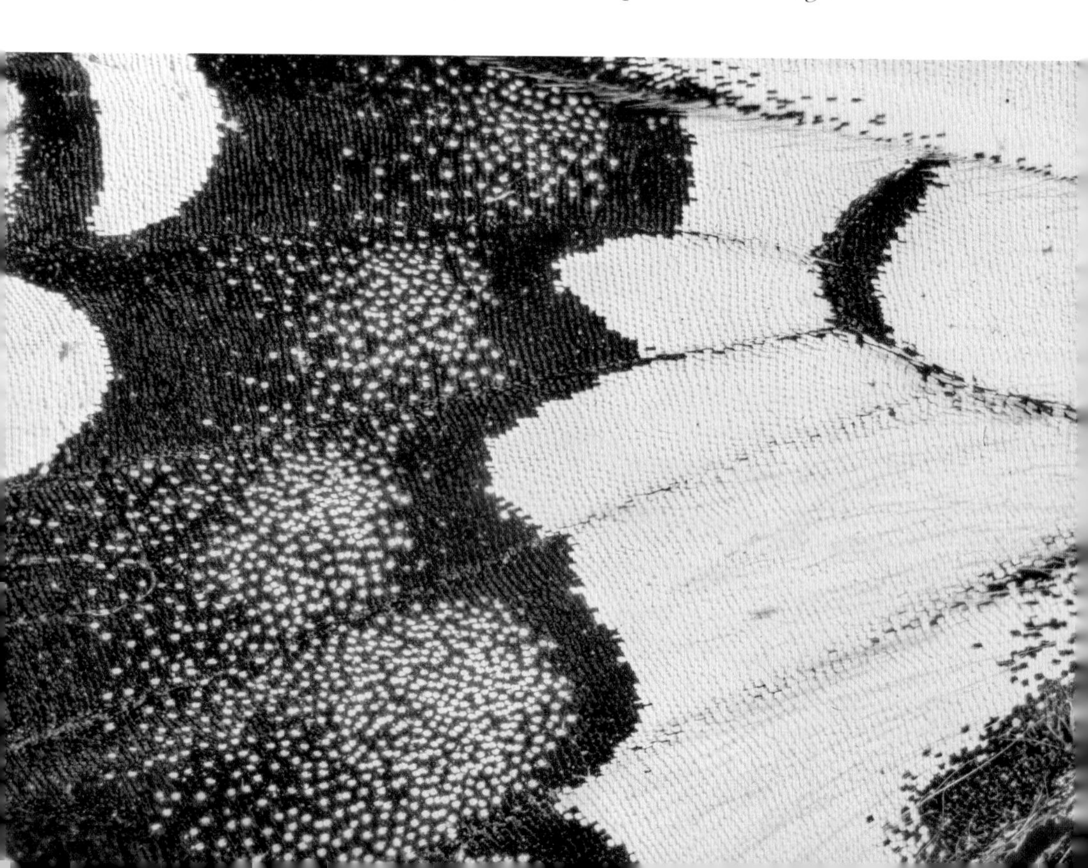

If the structural origin of insect wings is now more clearly understood, that still leaves the question of whether insects gained their aerial skill from the ground up or from the top down. Genetics is less helpful here, but I find the terrestrial ancestor of winged insects and the remarkable gliding performance of primitive bristletails very persuasive of a top-down route. It also seems likely that vertebrates followed just such a route, since in each group this hypothesis offers the simplest explanation. If I needed more convincing, a visit to the rainforests of south-east Asia did the job, because they're home to an unexpectedly eclectic mix of gliders.

The south-east Asian forests contain some mighty trees, taller than in rainforests elsewhere. Dipterocarps soar 70 metres above the ground, ramrod straight with no branches at all lower down, which might explain why such an unlikely assemblage of creatures has taken to the air here. It's by far the easiest way to get around. There are gliding frogs that really do no more than slow their descent and steer clumsily using large, webbed feet that work more like parachutes than wings. There are gliding lizards that spread a wing surface, supported by enlarged ribs, from the side of the body, on which they glide as far as 8 metres. There are flying snakes, the most improbable of creatures to take up gliding. Yet by flattening themselves and sucking in their bellies they can turn their long bodies into aerofoils. They stabilize their flight and steer by serpentine undulations, and they can stay aloft for a long time, covering up to 100 metres on each flight. It's the most absurd sight I've ever witnessed – a snake slithering through the air 20 metres above my head. And lastly there's the colugo, a mammal related to the primates. Colugos have a membrane that stretches between fore legs and hind legs, and, when they leap, they transform into large, square, furry flying magic carpets. The huge surface area of their gliding membranes carries them for 150 metres through the canopy with hardly any loss of altitude.

Surrounded by so many diverse gliders it's easy to believe that insects, initially without the need for similar structural modifications to keep them airborne, would have followed a similar route. I recently asked Steve Yanoviak whether he had searched these Asian rainforests for gliding insects. It turns out that neither he nor anyone else has yet looked – but I'm prepared to put money on the fact that some extraordinary discoveries await an entomologist intrepid enough to scale a 70-metre dipterocarp.

As we've already seen on many occasions, the insect body plan exhibits endless flexibility and that's no less true of the wings. They

The wings of butterflies and moths are usually covered in tiny, flattened scales that either contain pigments or reflect coloured light, to create intricate patterns like those on the wing of this European swallowtail, *Papilio machaon*.

occur in a bewildering variety of shapes, sizes and textures, although all are similar in their underlying structure and differ from the wings of all aerial vertebrates in many fundamental ways. One critical difference is that insect wings contain no muscles. The shape of bird and bat wings is controlled by muscles within their wings, so these animals can make continuous fine adjustments to the aerodynamics of their wings to adapt to flying conditions and to make precision manoeuvres. Yet watch a pair of dragonflies engaged in a territorial dogfight or try to swat a fly helping itself to a meal off your plate and it's pretty clear that insects don't seem at all hampered by having no muscles in their wings. That's all down to the remarkable structural complexity of insect wings.

In all insects, the wings grow as intricately folded structures, either as external wing-pads or internally within the pupa in the case of the vast majority of insects that undergo complete metamorphosis. On emergence as an adult, the wings must be pumped up to their final size and shape by blood pressure. Unlike vertebrates, with their complex network of veins and arteries, insects have an open system in which all their organs are simply bathed in blood, more correctly termed haemolymph in insects. They do, however, have a muscular tube, the dorsal aorta, an insect version of a heart, which creates a circulation of haemolymph around the body, a bit like an aquarium pump. They also have various accessory pumps to assist the heart, some of which are placed so they can pump fluid into the wings to expand them.

At this stage the wings consist of an upper and lower membrane of chitin, secreted by cells lying like the filling in a sandwich between the chitin layers. Once the wings are at full size, these cells are broken down and pumped out of the wings by the accessory organs. The upper and lower membranes now collapse and fuse to each other, and the wing is held in shape by a scaffolding of veins that form an intricate branching network. Although an insect wing looks like a dead, dried structure, that's far from the case. The veins contain nerve fibres and sensory structures, and since these living tissues need oxygen, tiny tracheae also pass along the veins.

To work at peak efficiency, the wings must also be kept clean. Chemical sensors on the wing surface detect foreign substances and trigger a bout of wing cleaning.[22] Butterfly wings also have heat sensors that trigger the wings to close if they become too warm, and even glands that produce pheromones. These are stored in special scales on the wings of certain male butterflies and dusted onto a female's antennae as a male courts her. So, despite appearances, insect wings are dynamic, living structures.[23]

While shooting a film on dragonflies recently, we had tanks full of larvae in the studio to film their predatory behaviour and then, in the spring, to time-lapse their emergence from the water and their wonderful transformation into masters of the air. Once their wings were pumped up and dry, we took the opportunity to try out a new super-macro lens that greatly magnifies the image. Insect wings might look like simple flat sheets – but look closer. Using a precision rail system to move the camera slowly along the length of the wing, I was amazed at the structures we could see. I've worked with dragonflies for nearly half a century, yet in that time I had never thought to look so closely at the wings from that particular angle. In cross-section, the wings have complex corrugations, and all the veins are bristling with stout, pointed hairs, the whole structure resembling a twisted barbed wire fence at this magnification. The corrugations running along the wing make it resistant to bending along its length just as a flat sheet of paper will buckle if held by one edge, whereas the same piece concertinaed into corrugations will remain stiff.

At the bases of the wings, however, things get really complicated. Here, the wings are joined to the thorax by a complex hinge that incorporates a collection of tiny, hardened plates. Some of these plates are associated with the base of the main wing veins and have individual muscles that control them. In some of the bigger dragonflies, there are seventy-one separate muscles in the thorax involved in wing mechanics.[24] As these muscles contract in different combinations, they pull on the bases of the wing veins, which are so arranged that they can warp and twist the wing, a little like the design that the Wright Brothers came up with for their later Wright Flyers.

Far from being passive flaps of cuticle, insect wings are as dynamic as those of birds or bats, in no way hindered by their lack of internal muscles. However, they are completely different from the wings of vertebrate flyers, and insects are also very much smaller than birds and bats. So, do they use their wings in a different way? That's another question that took a surprisingly long time to answer.

BEES CAN'T FLY – A WHOLE NEW BRANCH OF AERODYNAMICS

It's commonly said that, according to the laws of aerodynamics, bumblebees can't fly. Of course, whatever we might think, bumblebees do fly – and they've been happily going about their airborne business for as long as there have been bumblebees. This curious idea began,

as far as anyone can tell, back in 1934 in a book, *Le vol des insectes*, in which the French entomologist Antoine Magnan and his mathematician colleague Andrė Sainte-Laguë tried applying the known laws of fixed-wing aerodynamics to flapping wings. Their calculations showed that bees, and presumably other heavy-bodied insects, were unable to generate enough lift to support their bodies – hence the long-standing 'pub fact' that bumblebees can't fly.[25] There is, however, a kernel of truth at the heart of this myth. It's not the bee's problem – it's ours. The known laws of aviation definitely do not work for bees or any other flying insects, and it took a lot of research and ingenious technology to find out why.

The obvious difference between the wings of planes and those of flying animals is that those of planes are fixed, while those of animals flap up and down. In addition, the wings of animals need to generate both lift and forwards thrust, whereas planes can rely on propellors or jet engines to provide the power to move them forwards through the air. All their wings have to do is keep them up. This makes the aerodynamics of animals a lot more complicated. Even before the pioneering flight of the Wright Brothers, engineers were beginning to understand some of the principles behind fixed-wing aerodynamics, and our knowledge grew exponentially in the early decades of the twentieth century. Yet no one really understood how flapping wings worked right up until the early 1970s, by which time fixed wings were carrying us around the planet at supersonic speeds.

Using standard models of aerodynamics, calculations showed that insect wings could not produce enough lift to support their bodies. Yet the air is full of insects – so there had to be entirely novel mechanisms at work that enhanced the lift produced by the wings of insects. However, uncovering the secrets of insect flight would mean thinking about aerodynamics in very different ways. The University of Cambridge in the UK was a centre of such innovative thought, and it was here that Torkel Weis-Fogh, a Danish zoologist and pioneer of insect flight studies, suggested one neat trick. He used high-speed photography to examine the wing movements of a wide variety of insects and saw that some, like butterflies with their relatively huge wings, clapped their wings together over their backs at the top of the upstroke, then peeled them apart at the start of the downstroke, starting with the leading edge. He called this technique clap-and-fling and suggested that, as the wings peel apart, they induce spinning masses of air, vortices, over the wing's leading edge, which increase lift.[26]

For a bumblebee to fly, its wings must create more lift than is available through conventional aerodynamics. We're only just beginning to understand how insects do this.

Bigger flappers, like pigeons, use a similar mechanism to generate increased lift for a quick take-off, and the clap-and-fling is much easier to see (and hear) in these larger animals. Since its initial discovery, the aerodynamics of this mechanism have been refined and even now new subtleties continue to emerge. One recent study has shown that silver-washed fritillaries, *Argynnis paphia*, large and magnificent butterflies that flit through woodland glades across Europe, operate an enhanced version of clap-and-fling. As the wings close above the butterfly, the leading edges come together first, but the wing behind doesn't remain flat. It flexes into a slightly cupped shape, enclosing an air pocket. As the wings clap, some of this air may be forced backwards like a little jet to give the butterfly some forwards thrust.

Just as important, the cupped gap between the wings holds a bigger volume of air than if the wings were just flat surfaces. As the wings fling apart, this causes bigger vortices to form and therefore create more lift,[27] so these butterflies can make a quick getaway from predators. Their turbo-charged mode of clap-and-fling probably also explains why silver-washed fritillaries can be almost impossible to photograph on a hot, sunny day. These butterflies illustrate how those beautifully complex wings, with all their struts and braces, allow insects to harness a wide variety of unconventional aerodynamic tricks, many of which we are still discovering.

Had he not died far too young, at the age of fifty-two, Torkel Weis-Fogh would have doubtless gone on to further revolutionize our understanding of insect flight. However, one of his students at Cambridge, Charles Ellington, took up his mantle. In 1984, Ellington published a seminal series of papers on insect wing structure and function so extensive in scope that together they took up a whole volume of the prestigious *Philosophical Transactions of the Royal Society of London*. He started by taking a step back. Weis-Fogh thought that the bumblebee paradox applied to a few particular insects, but Ellington's wide-ranging analysis showed that, in theory, no insects can fly, at least according to conventional aerodynamics.

These papers led to the development of a new field of aerodynamics, called vortex theory, and to the realization that these mini-tornados are critical for insect flight. It turned out that the leading edge vortices that Weis-Fogh saw during his studies of the clap-and-fling mechanism form over the wings of a wide variety of insects. The same is true for the wings of birds and bats, but insects operate in a tiny world where air is more viscous and where these turbulent air flows play a much greater role. As

Many large-winged butterflies clap their wings together at the top of the upstroke, to generate vortices that enhance lift. Recently, we've discovered that the silver-washed fritillary, *Argynnis paphia*, operates a 'go-faster' version of this.

an insect flaps its wings, a vortex forms on top, just behind the leading edge. Vortices like these often form on wings of any size, then spin off the wing to leave a turbulent wake. Vortices shed from the wings of a big jet, for example, are visible as trails behind the wings if the plane is landing through misty rain. However, unusually, an insect's leading edge vortex remains attached to the wing during the whole downstroke. This was seen as a key mechanism for enhancing lift and one that is not accounted for by conventional aerodynamics. However, these kinds of turbulent airflow remain fiendishly complicated to understand. Recently, it has been suggested that the role of the leading edge vortex is not so much to generate lift but to prevent the wing from stalling at very high angles of attack.[28]

Watching from the window of an aircraft, it's obvious that the wing surface stays pretty much parallel to the airflow, which consequently remains smooth over the wing. The wing is working at a low angle of attack. The smooth or laminar airflow over the wing generates lots of lift to keep the plane in the air – and all the passengers can relax. During take-off and landing, the wings are a little more inclined to the airflow – operating at a higher angle of attack. If that angle gets too high, the airflow can separate from the wing and become turbulent, destroying the lift. In other words, the wing has stalled. Fortunately, aircraft wings are equipped with devices such as flaps and leading edge slots, which prevent stalls at higher angles of attack, so there is still no need to panic.

One thing that is clear from high-speed cinematography of flying insects, from Weis-Fogh's studies to the present, is that the terms 'upstroke' and 'downstroke' are a bit misleading. Many insects fly with strokes that are orientated more fore and aft than up and down, and often each wing is twisted almost perpendicular to the airflow at what looks like an unfeasibly high angle of attack. But the leading edge vortex keeps the airflow attached to the rear part of the wing, so it continues to generate lift.

This is only one way in which insects stay aloft. As well as flapping, insect wings rotate through large angles at the end of each stroke, a movement that creates yet more spinning eddies of air that add to the lift produced by the wings.[29] Such rotational movements of the wing seem to be critically important in keeping mosquitoes, with their very shallow wingbeats, airborne, although these insects also generate trailing edge vortices over their wings, as well as the more widely used leading edge vortices.[30] In other words, mosquitoes have three separate methods of generating lift from the complex way that air moves over their rapidly beating wings.

The very smallest insects of all face additional problems in moving through air that for them is as thick as syrup. We met feather-winged beetles in Chapter 1, where we realized how many significant sacrifices they've had to make to exist at such minuscule sizes. They can fit many fewer neurons into their tiny brains, and some have even had to move part of their brain into their thorax. Yet they can still fly – and fly very well. Larger insects generally fly faster than smaller ones, but feather-winged beetles can match the speed of beetles three times their size. The secret of the feather-winged beetles' unexpected prowess is yet another new mechanism for generating lift. As their name implies, feather-winged beetles don't have the membranous wings typical of most insects. Instead, their wings consist of thin bristles arranged along a central strut. In flight, the beetles move their very unaerodynamic-looking wings in a figure-of-eight, in a manner more reminiscent of the way that tiny swimming crustaceans move.[31] This novel approach to flying is perhaps not so surprising since air at this scale behaves more like water.

The structural complexity of insect wings and their equally complex movements, combined with their small size, means that there is no easy way to understand nature's first flyers. The literature is now replete with mathematical models and long equations – all well beyond my understanding – yet there is still a lot we don't know about how insects fly. We do, however, know enough to emulate at least a little of their aerial prowess. Micro-aerial vehicles, or MAVs, are attracting a lot of interest from many different sectors of society, although inevitably funds flow more freely for military applications, driven by concepts of tiny spy flies that can buzz their way unseen through the smallest of gaps.

Some time ago, for a series of films on bio-inspired engineering, I visited the Technical University of Delft in the Netherlands, home to DelFly and to a team of scientists that have extraordinary imagination and a sense of humour to match. I first met a passive walking robot that, given a slight incline to walk down, will do so with no power and with an incredibly realistic human gait. This robot is called Denise, because, as her inventors explained, the secret is all in 'de knees'. DelFly is a flying robot, kept aloft on very insect-like wings modelled on those of a dragonfly. The arrangement of the wings, however, is very un-dragonfly-like in that the two wings on each side are stacked one above the other like those of a biplane, so they flap towards and away from each other on

every cycle. However, the principles that keep DelFly in the air are much the same as for insects, including the creation of a leading edge vortex over the wings. DelFly can fly with remarkable precision, even hovering like a real dragonfly.

More recently, I filmed a robot that was even more closely modelled on a dragonfly. Esslingen am Neckar, in southern Germany, lies in a district famous for German engineering, with factories for Audi, Mercedes-Benz and Porsche all close by. I was here to visit Festo, a global engineering firm whose headquarters are in Esslingen. It is involved in a wide variety of industrial automation and control projects, but over the years has also built a series of robotic animals, as showcases for its skills and as testbeds for innovative technology. Its lighter-than-air manta ray and jellyfish are uncannily realistic, and among the other amazing creatures it has produced are hopping kangaroos and robotic ants. However, I had come to put a robo-dragonfly through its paces.

The day got off to a good start. As we assembled our kit in a huge glass building, one wall of which was lined with an enormous, fossilized seabed carpeted with sea lilies, a large trolley was wheeled up to us, laden with coffee, cakes and pastries. I enjoy those tough days soaking in rainforests or baking in deserts trying to find animals that are doing their best to avoid you, but I could get used to this – high-quality-location catering and animals that perform when you hit the 'on' button. Robo-dragonfly is a close match in design to the real thing, though a lot bigger, with nearly a half-metre wingspan. As it took to the air over that ancient seabed, it struck me that this was the same size as those giant *Meganeura* that flew in the late Carboniferous and Permian periods. It was the closest I've ever come to appreciating how impressive those ancient insects were.

The robo-dragonfly is controlled remotely by a pilot, who later confessed that it is nearly impossible to fly (and so, no, I couldn't have a go). Its wings twist and warp and each is powered and controlled independently, giving the pilot a total of nine different parameters that he has to worry about, with only a fraction of a second reaction time. It's a lot harder than flying a helicopter. Yet real dragonflies control even more parameters and fly and manoeuvre with much greater precision, all governed by a brain the size of a pinhead. As much as robo-dragonfly gave me a visceral appreciation of ancient giants, it also heightened my admiration for modern dragonflies. Often regarded as 'primitive' insects, dragonflies have become true masters of the air on gossamer wings that are actually aerofoils of incredible sophistication, controlled by miniature on-board computers that would be the envy of any fighter pilot.

The wings of dragonflies and damselflies, such as this banded demoiselle, *Calopteryx splendens*, are corrugated to give them strength. Corrugations also aid in the formation of lift-generating vortices.

MASTERY OF THE AIR – INSECT AVIONICS

The Royal Veterinary College doesn't sound like the sort of place to find out more about dragonfly flight – after all, it doesn't look after sick insects. However, at its Hawkshead Campus near Hatfield in Hertfordshire, among decidedly terrestrial horses and cows, Richard Bomphrey and his team at the Structure and Motion Laboratory are probing ever further into the mysteries of insect flight. In one lab, he generates fine plumes of smoke in a wind tunnel to visualize the patterns of airflow over a dragonfly's wings. In the lab next door is a large white arena with nine high-speed cameras arrayed around it. Dragonflies take off from a central perch and their flight manoeuvres are captured from every angle, slowed down eighty times. The images are broken down into data that a computer can handle and sent to a colleague at Leeds University who rebuilds an accurate virtual dragonfly in 3D, complete with all the same precision moves that the real one made as it flew. This can then be used to compute the airflow over its wings. Together, these techniques have allowed Richard and his team to understand a lot more about how those intricate wings work.

The complex corrugations along the wings that I'd seen earlier when shooting close-ups played a role in aerodynamics as well as wing strengthening. Additional vortices form in each valley of the corrugations, which help the wing develop slightly higher forces and make it even less prone to stalling.[32] Many other insects have similar corrugations on their

wings, presumably for similar purposes, but dragonfly flight is unique. It's quite unlike that of other insects in several important respects.

Dragonflies, along with mayflies, belong to a group called Palaeoptera – 'ancient wings' – which branched off early from the other flying insects (the Neoptera or 'new wings'). One characteristic of palaeopterans is that the muscles that power the wings are attached directly to the wings themselves. This allows each wing to operate independently, just like in Festo's robo-dragonfly. The main flight muscles can adjust the amplitude of each wing, while other muscles attached to the long veins alter camber, twist and angle of attack, with the result that a dragonfly has a lot of options – more than any other insect. Incidentally, mayflies, the other palaeopteran order, live for only a day or so as adults and are much weaker flyers, though they have similar direct flight muscles.

In cruise mode, dragonflies fly with their wings out of phase. That is, as the forewings are beating downwards, the hindwings are performing their upstroke. This maximizes horizontal thrust and as each wing moves through the wake left by the counter-stroking wing, there is also a small energy saving. For a dragonfly needing to get somewhere efficiently, this is the optimal pattern. However, if the dragonfly needs to move or manoeuvre quickly to escape a predator or catch a rival, it can switch to flapping its wings in phase. Now the two wings on each side act as a single aerofoil producing a much larger leading edge vortex, which increases lift, though at the expense of using more energy.[33] Add to that the ability to control all these wing parameters entirely independently for each wing and it's clear why dragonflies can turn on a dime, hover and fly backwards, all with equal ease. Robo-dragonfly eat your heart out.

However, it's no good having such a versatile and powerful motor without the sensory apparatus to keep up – the equivalent of a state-of-the-art, on-board flight computer. Dragonfly eyes are huge, covering most of the head and giving them virtually 360-degree vision at much higher resolution than the eyes of most other insects – they are easily capable of picking out the tiny flying insects that make up much of their diet.

A dragonfly's head is attached to its body by a very thin neck, which means that it can swivel its head through large angles to bring its sharp eyes to bear on its prey. When it takes off in pursuit, its mobile head serves as a kind of autopilot. The back of the dragonfly's head and the front of its thorax are bristling with hairs – sensors that monitor the position of the head with respect to the thorax. Feedback from these sensory hairs allows a dragonfly to monitor the position and movements

A green-eyed hawker, *Aeshna isosceles*, in economical cruise mode – beating its fore and rear wings out of phase. For greater speed, it can switch to beating its wings in phase.

of its body and to keep its head level, so its gaze remains fixed on its prey.

This system operates when the dragonfly is in steady or gliding flight, but if it needs to manoeuvre more violently, its thin neck gives it a major problem. The neck joint is so thin that it is mechanically weak. On take-off, a dragonfly can pull 4g and in a quick turn, 9g, so it needs to do something about avoiding a severe whiplash – and there are times when that thin neck puts a dragonfly in danger of decapitating itself or its mate.

The list of features unique to dragonflies seems endless and another is the way they mate. A male dragonfly spots a female in flight and uses his aerobatic skill to intercept her and grab her with a pair of hooks at the tip of his abdomen. He latches on to the top of her eyes and the pair fly like this, in tandem, for some time. As they manoeuvre, the strain on the female's head reaches dangerous levels. Now another unique dragonfly feature kicks in – the head-arrester system. This consists of tiny hairs called *microtrichia* on the back of the head and a series of plates, or *sclerites*, behind the head on the thorax. When muscles attached to these plates contract, the plates change their configuration, which brings them into firm contact with the microtrichia. Strong friction between the plates and the densely packed tiny hairs safely immobilizes the head.[34] The head-arrester system is also engaged during extreme flight manoeuvres. Even when perched and feeding on larger prey a dragonfly needs to stabilize its head. As powerful mandibles tear at the victim, the head arrester is triggered, to prevent the insect from beheading itself while eating.

Combining all these features has given dragonflies aerial skills that border on superpowers, including the ability to vanish from enemy radar. Insect eyes are exceptionally sensitive to movement, so a good way to sneak up on a rival is to appear not to be moving. Dragonflies do this using a trick called motion camouflage. They fly a complex trajectory towards a rival in such a way that they imitate how a stationary object would appear to the eyes of the other insect as it flies along, a feat which takes almost unimaginably precise flight control and positional sensing.[35]

Dragonflies make full use of their four independent wings, but throughout evolution insects have experimented with ways to work with a less complicated arrangement. Butterflies and moths retain four wings, but the forewings and hindwings on each side are linked together either by a simple lobe from one wing that overlaps the other, or a more complicated system involving a tiny hook on the hindwing that loops into a retaining structure on the forewing. This makes butterflies and moths effectively two winged.

Ruddy darters, *Sympetrum sanguineum*. When flying in tandem like this, there is enormous strain on the thin neck of the female. To prevent damage, dragonflies have a 'head arrester system' to safely immobilize the head.

Flies have gone a stage further and done away with their hindwings altogether, at least for generating lift. The hindwings have been reduced to tiny stalks with a knob on the end. These structures are called *halteres* and they are in large part responsible for a fly's uncanny ability to avoid being swatted. The halteres beat up and down just like the forewings, and the heavy knob at the end acts like a spinning gyroscope, which resists any attempt to tilt it off its axis. Likewise, the halteres resist alterations to their direction of movement as a fly manoeuvres. The base of the stalk is bristling with sensor arrays that when seen in a scanning electron microscope are things of extraordinary beauty.[36] They look like precisely aligned mosaics of tiny scales, each only a few microns (one thousandth of a millimetre) across with equally tiny hairs emerging between them.

As it resists a fly's movements, each haltere twists at the base of its stalk and the sensors measure the resulting strain that builds up. This provides the fly with continuous feedback on its flight attitude and allows it to compensate instantly for any perturbations to its flight path. The system also helps a fly to keep its head level, so it can better see its world and dodge any danger.[37] Indeed, as long ago as the early eighteenth century, the English clergyman and natural theologian William Derham showed that flies could not fly properly if their halteres were removed. I'm ashamed to admit that, as a young child avidly learning anything

I could about nature, I performed a similar cruel experiment on some craneflies I found in the garden… and Derham was right.

The halteres are easy to see on more primitive flies, like craneflies, although in these flies, they are less sophisticated than those of the more familiar house flies and bluebottles. More advanced still, the halteres of predatory flies, like robber flies, are packed with a dense array of sensors, allowing precise high-speed manoeuvres. Robber flies (Asilidae) lurk in the vegetation, keeping a lookout for unsuspecting prey flying overhead. As soon as a target is detected, the robber fly launches itself upwards at high speed. It flies an accurately plotted intercept course, deftly avoiding any obstacles in its path, then flips upside down at the last minute to grab its prey with long, bristly legs, before tumbling to the ground and injecting its victim with deadly toxins.[38] It's all over in a second.

Robber flies are so fast that they're almost impossible to film, although using a high-speed camera to slow down their attack would be

(*above*) A robber fly, *Efferia* sp., waits on a perch, ready to launch itself in pursuit of any prey that flies overhead.

Flies fly on just their forewings. The hindwings of flies are reduced to tiny stalked knobs (halteres) that work a bit like gyroscopic stabilizers.

a dramatic way to reveal just how adept they are in flight. So, a few years ago, I found myself on the high plains of Utah, where large robber flies are common in the short brush. To give us at least a chance of capturing the behaviour, I tethered a smaller fly onto a very finely teased out strand of dental floss, so thin it was invisible to the camera. I then attached this to a fishing rod and was able to move the prey steadily over the ground while the cameraman tracked it with a high-speed camera. As soon as a robber fly noticed my offering, it leapt up and grabbed a meal. Success! Slowed down forty times we could watch the aerial agility of the robber fly in all its glory – as it anticipated every turn and twist of its prey and even flipped upside down at the last minute so it could grasp its prey with its long legs. And we gave a whole new meaning to the term fly-fishing.

Wings revolutionized life on Earth and catapulted insects to unprecedented success, but they are also a real nuisance. Although they are much less delicate than they look, designed a bit like rip-stop material in which any tears are halted by the network of fine veins, they still get damaged. Late in the season dragonflies and butterflies with tattered and broken wings are a frequent sight. Thin aerofoils sticking out from the body are not just awkward, they also limit an insect's options. Dragonflies can't push through dense vegetation, let alone dig in the soil or swim through water. Yet many insects do exactly that. Almost more important than the evolution of wings themselves was the evolution of a mechanism for folding them out of harm's way. This innovation gave rise to the Neoptera (new wings), distinct from the palaeopteran dragonflies and mayflies, and with wing folding came a dramatic surge in the success of insects.

NEOPTERA – THE NEW FLYERS

Initially, wing-folding mechanisms allowed insects to simply fold their wings tidily over their backs. Stoneflies, for example, fold their wings into a curved shape that fits neatly around their long bodies, so they can scurry around in the vegetation without any hindrance. Beetles have taken this a lot further, by combining the advantages of wing folding with the efficiencies of using just two wings.

While flies fly with their forewings only, beetles fly with just their hindwings. The forewings have been converted into wing covers or *elytra* (singular: *elytron*) and their hindwings have become masterpieces of insect origami. Not only do they fold over the beetle's back at the hinge

mechanism, but the wing itself is criss-crossed by suture lines that allow it to fold over itself automatically, several times over, to be stowed like a well-packed parachute beneath the elytra.[39] This means that beetles can fly, if sometimes clumsily, and are still able to adopt the lifestyle of wingless creatures. Rowing on feathered back legs, water beetles are as at home underwater as they are in the air – the best of both worlds. Their streamlined shape also allows beetles to exploit dung produced in prodigious quantities by both wild and domestic herbivores. In fact, were it not for the sheer numbers of industrious dung beetles, we would be up to our necks in… well, you get the picture.

Australia provided a salutary example of just such a situation. Native Australian dung beetles, adapted to cope with the dry pellets of marsupials, found the slop produced by domestic cows utterly unpalatable. Dung accumulated on fields in vast quantities, un-recycled, where it provided breeding grounds for billions of flies, which began to swarm through the bush in such numbers that they forced Australians into hanging those iconic corks from their hats to lessen the nuisance. The only solution was to import Old World dung beetles, happy to swim through cow dung, then eat or bury it. Now, many millions of these

Beetles, like this cockchafer, *Melolontha melolontha*, have converted their forewings into tough wing cases. They fly using only their hindwings, although outstretched elytra may contribute a small amount of lift.

beetles are bred and released to keep Australia dung free. Ultimately, this vital service is only possible due to the evolution of wing folding and of protective elytra.

A few beetles have turned wing folding into an art form. Rove beetles (Staphylinidae) range in size from the impressive, 2-centimetre devil's coach horse, *Ocypus olens*, to tiny beetles, just millimetres long, that scurry through leaf litter and soil. Their elytra are tiny, barely covering the thorax, yet, somehow, they manage to fold full-sized wings into this tiny space. They achieve this by first laying the left and right wings together, then folding them simultaneously. Each wing must have a complementary folding pattern, so they stack neatly together. In fact, each wing has both the 'left' and 'right' folding pattern built into it, so either wing can adopt either pattern, which means that it doesn't matter which wing ends up on top when the beetle lands.[40] Once the wings are tucked away, a rove beetle is not hampered by the stiff wing-cases that cover the whole body of most beetles. It can twist and squirm through tiny spaces as if it had no wings at all.

Despite the wings being tightly packed into a small space, they are quick to deploy, thanks to that insect super-rubber, resilin, which we found in the leg springs of fleas and other jumping insects in the previous chapter. The hinged joints of a beetle's hindwings are also reinforced with resilin. The elasticity of resilin allows the wings to fold easily at the hinges, but the energy stored during folding can be released to spring the wings open again like an automatic umbrella.[41] In addition, a beetle can pump haemolymph into the wing veins to unfold and stiffen the wing.[42]

The evolution of elytra is one of the fundamental reasons why beetles are the most diverse order of insects – they have all the advantages of flight but can still occupy virtually any niche open to an insect. It's probably no coincidence that rove beetles, with even more sophisticated wing folding, are among the most successful of all beetles. However, a few other insects have also come up with ways to stow their wings into impossibly tiny spaces.

Earwigs are familiar creatures in the garden, yet few people realize that they can fly. They appear wingless, with flexible bodies that allow them to congregate in tight spaces, for example between the petals of your prize dahlias (though not in your ears, despite their name). Their wings are tucked away under tiny wing covers, a little like those of rove beetles – but earwigs fold their wings in a different way. First, they fold like a fan, then lines of flexion allow the folded fan to be folded again both transversely and longitudinally.[43] Earwigs are not as slick as rove beetles and their wings are slower to deploy, a process that needs help from their pincers. Even so, it's an impressive achievement to have wings that can fold so many times yet don't embarrassingly fold up when the earwig is flying. Once open and under tension, the wings bow in such a way that the lines of flexion are blocked by other structures on the wing, so they stay open automatically, until the earwig releases the tension. In this way, earwigs have managed to optimize their wings for both folding and flight.[44]

THE POWER OF FLIGHT

The evolution of the Neoptera also involved a change in how the wings were powered. In dragonflies and mayflies, the wings are attached directly to the flight muscles, but in other insects the main power muscles attach to the walls of the thorax, one set running front to back, the other from side to side. In contrast to the direct flight muscles of dragonflies and mayflies, these are called indirect flight muscles, because when they contract, rather than directly moving the wings, they distort the thorax, which flips into a different shape. It's the same as pressing the slightly domed lid of a tin box. At some point it clicks loudly from a slightly convex shape to a slightly concave one. It can be reset to its original shape by pressing from the inside, a phenomenon I discovered as a young child, much to the annoyance of my parents. The insect thorax, with its chitinous exoskeleton, behaves in much the same way. The longitudinal muscles work to flip the thorax shape one way and the transverse muscles then flip it back the other way. The wings are attached to the top of the thorax in such a way that when the shape distorts in one way, the wings flip down; then, when it clicks into the alternate shape, the wings flip up again.

This innovation goes hand in hand with a change in how the flight muscles work. When the direct flight muscles of a dragonfly receive a nerve impulse, they contract, pulling the wing one way. When the antagonistic muscles receive their electrical signal, they contract, pulling the wing the other way. In other words, one nerve impulse causes one muscle contraction, much as is happening in my hands as I type these words. Because of this one-to-one relationship, these are referred to as *synchronous* muscles. The indirect flight muscles work in a different way. A single nerve impulse turns them on, but they are subsequently stimulated to contract every time they are stretched. As the thorax flips to one shape driven by, for example, the longitudinal muscles, the transverse muscles

are stretched. This causes them to contract, which flips the thorax back to the alternative shape, so stretching the longitudinal muscles, which in turn immediately contract… and so on. There isn't any one-to-one relationship between nerve impulses and contractions, so these muscles are referred to as *asynchronous*.

Asynchronous muscles don't differ greatly in microscopic structure from synchronous ones, and it seems that different groups of insects independently evolved asynchronous flight muscles many times in their history.[45] They are yet another innovation that has played a major part in the insect success story because they enabled the miniaturization of flying insects.[46] Since they are driven mechanically at a rate that depends solely on the size and shape of the thorax, asynchronous muscles can achieve astonishingly high frequencies, far faster than could be generated by sequential nerve impulses firing synchronous muscles. They can therefore produce a much higher power output than equivalent-sized synchronous muscles.

To generate enough lift, tiny insects need to beat their wings very fast, at much greater frequencies than is possible with *synchronous* muscles. Asynchronous muscles, on the other hand, are quite capable of driving the wings of mosquitoes, for example, at 100–1,000 beats a second, a frequency responsible for that whine of misery that plagues many a summer picnic. Because these muscles operate at high frequencies and therefore generate more power, insects can produce enough lift to fly from a smaller wing area, which in turn has freed flies to modify their hindwings into gyroscopic stabilizers and beetles to transform their forewings into protective shields.

Asynchronous muscles allowed the miniaturization of flying insects and the ecological success that stemmed from that, but they are less critical for larger insects with slower wingbeat frequencies. The one highly diverse order not to use asynchronous muscles is the butterflies and moths (Lepidoptera) with their relatively huge wings, coupled into one large aerofoil on each side. Yet the Lepidoptera rank alongside the beetles (Coleoptera), flies (Diptera), and ants, bees and wasps (Hymenoptera) in their numbers of species, so we'll have to look elsewhere for the secrets of their success. However, for flies with fast and powerful wings combined with a sophisticated gyroscopic navigation, or beetles, with their ability to transform from winged creatures to burrowers and swimmers, the evolution of asynchronous muscles opened up a multitude of opportunities that catapulted these orders to mega-diverse success. As the participants at a 2009 symposium in Boston on 'Insect Evolution' somewhat prosaically concluded: 'Much of insect diversity today derives from the biomechanical consequences of flight at a small body size.'[47] Put another way, the whine of mosquitoes or the annoying buzz of flies might be all too familiar, but we should listen with new ears. It's the sweet sound of success.

Common green bottle flies, *Lucilia sericata*, are masters of the air. Using asynchronous flight muscles, flies can harness lots of power and, with their gyroscopic halteres, they can precisely control their flight path.

5

Wings over the World
Insect migration

Bearing up close to Cuba, they saw turtles of vast bigness and in such numbers that they covered the sea—and the next day such immense swarms of butterflies as even to darken the sky.

Christopher Columbus, 1494[1]

As I gazed out of the window of Lufthansa flight 498, eight long hours after leaving Frankfurt, Mexico City finally lay below us. It stretched further than I could see, which through the ever-present blanket of haze, wasn't all that far. Yet, we'd been passing over grey urban sprawl for what seemed like an eternity as the plane slowly made its decent into Benito Juárez International Airport – a clear demonstration that Mexico City is vast. Today it's a crowded home to nearly 22 million people, though this whole stretch of the Valley of Mexico has long been a centre of overpopulation, from well before Hernán Cortés and his army of conquistadors came here in 1519. Even back then, Tenochtitlan, a sophisticated Aztec city in the middle of Lake Texcoco complete with botanic gardens, an aquarium and the palace of Emperor Moctezuma, was a bustling metropolis. It would soon fall to the Spanish newcomers, sowing the seeds of the vastly more populous city of today. North along the valley lay the city of Teotihuacan, more ancient still and once one of the largest cities in the Americas. However, my intention was to spend as little time as possible in this claustrophobic sea of humanity.

Once we finally cleared our thirty cases of camera gear through customs and loaded it onto a couple of minibuses, we headed west through the modern city that has all but obliterated the long pre-Columbian history of the region. The journey seemed to drag on forever through crawling traffic, until finally we began to climb into the mountains, towards a destination just 70 miles from the city and, thankfully, a hotel in which to succumb to jet lag.

Next morning, I awoke refreshed and with keen anticipation. Hummingbirds buzzed round the flower beds of the hotel and unfamiliar bird calls from the bushes had me flicking my binoculars from branch to branch. However, it wasn't the birds that had drawn me here with a specialist 3D film crew. We had come to capture one of the greatest spectacles the natural world has to offer, but one that wasn't even known to science when I began my undergraduate degree in the early 1970s. It was something I'd wanted to witness for the last forty years and now fulfilment was just a few hours away. Our journey took us through the beautiful village of Angangueo to the tiny hamlet at El Rosario, centred on a vast dirt square surrounded by a selection of rustic bars and restaurants that would keep us well fed and watered for the next two weeks. In one corner of the square, a path led past a cluster of buildings where we met up with a group of local guides and assorted kind souls who had offered to help us carry the ridiculously large pile of kit that's needed to shoot a film in 3D. Just as well. El Rosario nestles 3,000 metres

(previous page) Monarch butterflies, *Danaus plexippus*, overwinter in a few groves of fir trees high in the Sierra Madre Mountains of Mexico. On warm days, tens of thousands descend to drink along streams.

up in the Sierra Madre Mountains, and once you start climbing a hill laden with heavy cameras you feel every one of those metres.

The path up the mountain passes through a large archway that welcomes visitors to this special place, with bold paintings depicting the unmistakable black-veined orange wings of monarch butterflies. This patch of mountain forest is where most of the monarch butterflies from North America end up for the winter. There are twenty-two tiny groves of oyamel firs in this area, the only known overwintering sites for northern monarchs, *Danaus plexippus plexippus*, from across almost the whole North American continent. Their summer range is 12 million square kilometres,[2] but in winter they're all crammed into a single square kilometre.

THE MONARCH GROVES OF MEXICO

Over the years, I've noticed that as I get older the hills, for some reason, get steeper. On the ascent, I felt the frequent need to pause, to seek out some bird I'd spotted, a thinly veiled excuse to gasp a few lungfuls of thin air. As on all my film trips, I carry a field guide to the local birds, but every time I stopped to flick through the pages, several of our local helpers gathered round me to admire the pictures. I soon lost the book to my new birdwatching friends as they enthusiastically worked out which of the birds lived in these forests. This turned out to be very helpful since, by narrowing down Mexico's bewildering variety of birds, they made it a lot easier for me to identify birds seen in fleeting glimpses. Eventually, despite these distractions, I made it to a shady grove of oyamel firs.

There are tens of millions of monarchs in this one grove alone, though first thing in the morning they can be surprisingly hard to spot from a distance. I had already noticed what I took to be heavy, drooping branches of fir trees but as I drew closer, I realized they were great masses of butterflies clustered in tight roosts, the less vibrant undersides of their wings making them inconspicuous in the dappled light. Then, from behind I heard what struck me as the whole forest gasping in awe. I turned to see an orange and black cascade of butterflies pouring out of the crown of a fir tree that had caught the warm morning sun. This triggered a nearby cluster to erupt into flight, and then another and another. Soon the air was full of butterflies, hundreds of thousands of fluttering wings. I'd never have believed that you can hear a butterfly's wingbeats, but amplified by so many individuals they sound for all the world like a distant waterfall. If that's not wonderful enough, then

(*overleaf*) During the winter, monarchs, *Danaus plexippus*, hang from the boughs of Oyamel Firs in clusters that contain tens of thousands of butterflies.

consider that some of these butterflies had flown nearly 4,000 kilometres to reach this tiny patch of trees.

I've read a lot about insect migration over the years, but standing in that grove on a Mexican mountainside, the sheer scale of the journeys made by such tiny, fragile creatures was brought home to me in a way no amount of academic understanding ever could. I also finally grasped the emotions that a sixty-five-year-old Fred Urquhart, the first scientist to witness this spectacle, must have felt when he staggered up this mountain in 1976. He described his exaltation – after nearly four decades of trying to uncover the mysterious life cycle of this butterfly – in a lavishly photographed article in *National Geographic* magazine later that year.[3] That was just as I was beginning my PhD in entomology.

Seeing those astonishing pictures of trees draped in butterflies further fuelled my enthusiasm for the extraordinary world of insects, although, like Fred Urquhart, it would take me forty years to witness this phenomenon first-hand. Even as a child, Fred had watched groups of monarchs in the autumn, heading off with a clear purpose to destinations unknown. In the late 1930s, working as a zoologist at the University of Toronto, he decided to unravel this mystery once and for all, never dreaming it would take him most of his academic life. Together with his wife, Norah, they began by trying to attach uniquely numbered sticky labels to the butterflies' wings, so individuals could be identified if recaptured and their journeys mapped. Unfortunately, all they managed to do was create a gluey, tangled union of paper, butterfly and fingers. It was not until the 1950s that they finally solved the problem.

Have you ever been frustrated trying to peel a price label off a glass or plastic item you've just bought? It clings with a tenacity that no amount of scraping or scrubbing will erase. For the Urquharts, though, this was a revelation. These pressure-activated adhesive labels clung to Monarchs' wings with that same tenacity, through rain or shine, and the mission to track monarchs to their mysterious destination could begin in earnest.

As the Urquharts recruited volunteers across the country to tag monarchs and to keep an eye open for those that had previously been tagged, the Insect Migration Association was born. By the start of the 1970s around 600 of those volunteers, called 'citizen scientists' by the Urquharts, had helped to build up a picture of monarch movements across North America, plotted on a huge map on Fred's study wall. Across the continent, lines converged on Texas, although there was no evidence at all that hundreds of millions of butterflies were holed up for the winter anywhere in Texas. However, the Lone Star State is, as Texans invariably point out, a big place, with canyons and valleys so remote that, in the early twentieth century, they managed to hide the last remnants of the great southern herd of buffalo. Maybe a large swarm of butterflies could also hide here, but despite searches no 'lost valley of the butterflies' was ever discovered. The most likely explanation was that the butterflies had flown on, across the Rio Grande and into Mexico.

With this possibility in mind, Norah wrote articles about the project for Mexican newspapers, one of which caught the attention of Ken Brugger, an American living in Mexico. He volunteered his help and after he later married Cathy, a Mexican, the pair scoured the countryside for any sign of the missing monarchs. In late 1974, when he came across heaps of dead monarchs by the roadside in the Sierra Madre Mountains, Ken knew he was getting close. However, these mountains are vast and dotted with remote peaks and valleys. How could they pinpoint where the monarchs were headed?

On 2 November, Mexicans celebrate the Day of the Dead, honouring departed friends and relatives. Cathy recalled that, in this area, the day is also marked by clouds of butterflies drifting through the cemeteries, the souls of departed children according to local people. By questioning these locals about the direction in which these departed souls were flying, the Bruggers were able to narrow down their search and, early in 1975, as they climbed the slopes of Cerro Pelon above Angangueo, they bumped into local woodcutters who were very familiar with the biological treasure hidden in these mountains. They pointed the way to a small grove of firs and later that day, Ken relayed their discovery to the Urquharts in a breathless phone call. The mystery of the monarchs had been solved, although this wasn't the most remarkable part of the story.

Early in 1976, Fred and Norah made the journey to the Sierra Madre themselves, and while wandering around the grove Fred noticed a monarch with a paper tag bearing the number PS 397. It turned out that this butterfly had been tagged in Minnesota the previous autumn, 2,000 miles (a little over 3,000 kilometres) distant. Among the tens of millions of butterflies swarming through the grove, Fred had stumbled on the one individual that proved beyond all doubt that monarchs undertook migrations every bit as spectacular as those of birds.

Before the Bruggers' and Urquharts' remarkable discovery, the true scale of these journeys wasn't even guessed at, though Fred knew full well from his childhood that monarchs were clearly migrating somewhere. Observations elsewhere of butterflies and other insects travelling in large numbers in the same directions also hinted at the scale of migratory movement in the insect world. In the late 1930s, as Fred Urquhart was beginning his lifelong study of northern monarchs, William Beebe of the New York Zoological Society was observing mass movements of the Southern Monarch, *D. p. megalippe*, over Rancho Grande in Venezuela. Previously, this subspecies, along with other tropical insects, was assumed to be sedentary. Yet Beebe witnessed this butterfly, along with a wide variety of others, heading purposefully across the landscape.[4]

In 1958, English biologist Carrington Bonsor Williams, known to fellow entomologists as 'C. B.', wrote a pioneering book, *Insect Migration*, published in Collins' classic 'New Naturalist' series,[5] summarizing the current understanding of insect movements. In the Colonial Service in the early years of the twentieth century, C. B. witnessed flights of hordes of butterflies in the West Indies – butterflies that were clearly on their way somewhere else. In North Africa he saw similar flights of painted lady butterflies and, further south, plagues of African locusts flew in swarms of truly epic proportions in a never-ending search for fresh food. Closer to home, C. B. knew that large numbers of silver Y moths, *Autographa gamma*, appeared each spring on the south coast of England. It was obvious to him that insects were capable of more than just a few short dispersal hops, but how far did these insects travel and were these movements true migrations in the sense that migration was understood in birds?

Even recently, defining just what migration really means for an insect has been fraught with difficulties.[6] When a bird migrates, it travels perhaps thousands of kilometres (or tens of thousands of kilometres in a few cases) to its destination, then later reverses its journey and travels back to its original point of departure, often with an accuracy of a few metres or even to the same nest. Each individual may repeat this journey many times in its life. With a few exceptions, insect migration is different.

BUTTERFLIES AND MOTHS ON THE MOVE

Migration for an insect means a journey undertaken by successive generations in which no single individual ever completes the whole round trip. This is certainly true of monarchs. A single individual, like PS 397, does make the impressive journey south, but it will never again see the place where it was born. That journey will fall to future generations. I'd timed my visit to the Mexican monarch groves to witness the preparations for spring departure. In the warming weather, uncountable numbers of butterflies drank at seeps of water in the forest or nectared on colourful flowers along the forest edge, taking on fuel for the journey ahead. Clouds of them fluttered low over an open meadow on the hilltop where the ground was littered with mating pairs. These butterflies will shortly head north and some will make it as far as 2,000 kilometres from their wintering grounds.[7] Others pause soon after crossing into the States, where they seek out milkweed plants* among the gaudy displays of Texas and Oklahoma spring wildflowers.

Here they lay their eggs, and with that the baton is passed to their sons and daughters in the relay race north. Once adult, this new generation moves further north still, breeding again as the butterflies follow summer across America and into Canada. Eventually, the great-grandcaterpillars of the butterflies that left Mexico hatch in the late summer and turn into a very different kind of butterfly. They don't mature sexually, but instead store lipids in their bodies, fatty chemicals to tide them through a long winter, and they fill their tanks with flight fuel in the form of sugary nectar. These are the super-butterflies that will make the long journey to a tiny grove of oyamel firs on a Mexican mountainside that sheltered their great-grandparents.

Because monarchs are common and familiar in North America – and because they happened to migrate across a continent where people had the time and resources to be curious about where they were going – their long-distance journey was one of the first to be mapped. It remains one of the most spectacular examples of insect migration. However, as our eyes were opened to the true measure of insect migration, many other impressive examples have emerged. In North America, monarchs face the problem of a cold winter. In India, butterflies face the seasonal problem of the torrential monsoon. Many species here have solved that problem by migrating.

Three species of butterfly, the dark blue tiger, *Tirumala septentrionis*, the common crow, *Euploea core*, and the double-banded crow, *E. sylvester*, make a regular journey between the Western and Eastern Ghats, two lines of mountains that run parallel to the west and east coasts of India. On their journey between the two regions, migrating dark blue tigers have been recorded flying in their millions around Bangalore.[8] This species belongs to the same group as the monarchs (Danainae) and, like monarchs, individuals preparing to migrate switch off their reproductive system, which allows them to both divert energy to flight muscles

* Monarchs belong to a group of nearly 160 species of butterfly (Danainae), often called milkweed butterflies because many feed exclusively on milkweeds of the genus *Asclepias*. The milky sap that gives these plants their name is highly toxic to many animals but clearly not to milkweed butterflies.

and to live a lot longer. This switch happens between April and June in the Western Ghats, and is triggered by the approaching monsoon. These butterflies must be responding to some sign of the impending downpours, because the exact timing of their migration varies from year to year, tracking whether the monsoon is early or late.[9] The Western Ghats is the first region to intercept the South-west Monsoon, at its most torrential as it sweeps in off the warm Indian Ocean. Once clear of the worst of the rains in the Eastern Ghats, the butterflies breed and this new generation then heads back to the Western Ghats between October and December after the rains recede. Other observations in India suggest that around 250 species of butterfly and moth do the same and migrate just before the monsoon hits.[10]

In Taiwan, four crow butterflies, *E. tulliolus koxinga*, *E. mulciber barsine*, *E. eunice hobsoni* and *E. sylvester swinhoei*, travel south to the area around Maolin to escape the cold winter in the north. Here, in a few sheltered valleys, they swarm in great numbers, not quite on the scale of monarchs in Mexico but nonetheless impressive. In Central America, a spectacular day-flying moth, the urania swallowtail moth, *Urania fulgens*, is a little less predictable. Every few years great swarms of these insects are seen on the move, passing observers in groups of fifty or more every few minutes. The scale of the migration in 2005 was massive; from July to September it was witnessed across the whole of Costa Rica, which gives some idea of the sheer numbers of migrating individuals in just this one species.[11] In some years they even reach as far as the Gulf states of America. Urania moths migrate not to escape adverse weather but to find new feeding areas. Their caterpillars feed on a number of related vines (*Omphalea* spp.), which react to millions of chomping mandibles by slowly pumping up the level of toxic chemicals in their leaves. Over a few generations of moths, the plants in one area become so toxic that they're inedible and the moths have to set off to pastures new, where the naive vines have no idea what's about to hit them.[12]

A moth from the same family (Uraniidae) in Australia is also well known for its mass movement. On the *East Coast*, zodiac moths, *Alcides metaurus*, head north in the Australian winter into subtropical Queensland, where they festoon trees in brightly coloured masses. Queensland is also the favoured winter resort of another danaid butterfly, the blue tiger, *Tirumala hamata*. Swarms in their millions are so conspicuous that they attract press attention each year and were spectacular enough for Captain Cook to have made a note of them in his journal 250 years ago, one of the very first natural history observations made on the great southern

Bogong moths, *Agrotis infusa*, line the walls of crevices high in the Australian Alps, where they spend the summer, before returning to the lowlands in the autumn.

continent. Many of these butterflies end up on Magnetic Island, off Townsville, a wonderful sight as bright blue and black butterflies flutter in their thousands around flowering shrubs.

In all of these examples from around the world, it takes several generations to complete a migratory cycle, but one Australian moth bucks this trend and behaves more like a bird. Bogong moths, *Agrotis infusa*, are nothing to write home about, at least in terms of looks – just brown and boring. But the numbers that spend the summer in the Australian Alps put even the Mexican monarchs in the shade. All over south-east Australia, as far as southern Queensland, bogong moths emerge from their pupae in early spring. Instead of breeding, all these moths make a beeline for the alpine regions of New South Wales, the Australian Capital Territory and Victoria, where they can escape the blistering heat of summer in the Australian lowlands.

The moths fly by night, in some years perhaps four and a half billion of them. Many, like the proverbial moth to a flame, are distracted by the bright lights of Canberra or Sydney, where they create a moth blizzard in streets and gardens. People with less empathy for this extraordinary event call them a nuisance; I call it one of the most mind-boggling phenomena I've ever seen. Those that escape the lure of the lights converge on a few favoured summits, where they line the walls of cool crevices between massive granite boulders. These narrow clefts are a tight squeeze for a human, but well worth making the effort to stick your head inside. Up to 17,000 moths are packed into every square metre, giving the impression that the cave walls have been tiled with an intricate mosaic.[13]

Here they enter a state of aestivation and pass the summer months in torpor. However, they are not entirely inactive. As evening approaches a few moths begin to scurry over their sluggish neighbours, then more and more, until the walls and floors of the cleft are a seething mass of moths. Now they pour out of the crevices and into a sky still bright with the afterglow of sunset. They fill the air in spectacular swarms, flying until well after dark. No one really understands why they do this. It seems like a waste of energy, burning up the fuel they'll need for their return journey to the lowlands in the autumn. However, bogong moths are known to change resting sites during the summer, sometimes moving to higher and cooler elevations as the season progresses, so it may be related to this behaviour. In any case, it's by far the best time to appreciate the scale of their extraordinary migration.

As summer turns to autumn, the moths rouse fully and fly back to where they hatched, now ready to breed themselves. This behaviour wasn't really investigated by scientists until the 1950s, although it had been well known for perhaps 10,000 years before that. Four and half billion moths, packed with enough nutritious fat to see them through the summer and turning up each year in a few convenient caves, was too good a chance for local Aboriginal tribes to miss. 'Bogong' is an Aboriginal word for 'mountain' and several tribes converged on these mountainsides each summer, to conduct local politics, initiation rites and corroborees and, of course, to feast on moths.[14]

Many local creatures also home in on the caves to make the most of the summer windfall. It's possible that Aboriginals used feasting little ravens, *Corvus mellori*, or other birds to help them find the aggregations. Bogong moths also dominate the summer diet of a very rare marsupial, the mountain pygmy possum (*Burramys parvus*), a creature only known from the fossil record until a living one turned up in 1966 at a ski resort in Victoria. There are probably only about 2,300 of these tiny possums, living in an area of just 6 or 7 square kilometres, dependent for part of the year on this impressive insect migration.[15]

A similar round-trip migration, though over a much smaller distance, occurs in North America. In late summer, convergent lady beetles, *Hippodamia convergens*, head to cool woodlands where they assemble in red and black masses, carpeting logs, stumps and boulders. They come to these places to escape the heat that dries up plants and kills their aphid prey. This beetle occurs widely across North America but the biggest aggregations I've seen are in California, in woods on the hills around San Francisco Bay. Deep in the shade of giant coastal redwoods they blanket the ground in a seething mass, another great insect spectacle. Yet their name doesn't come from the fact that millions of these beetles converge on these spots, but from the fact that two white lines on their thorax converge towards their head. You just have to love the pragmatism of entomologists.

Like monarchs, the lady beetles use the same spots from year to year. When their food dries up, for example in California's Central Valley, each lady beetle rises vertically into the air until it intercepts high-altitude winds that whisk it towards the mountains. In the aggregation, each lady beetle leaves a small spot of scent where it rested. Hundreds of thousands of similar spots last long enough to create a powerful homing beacon that new generations can track when it's their turn to head for the hills. Sheltered in these forests, the lady beetles will also spend the winter here ready to head back to the Central Valley when spring returns. They probably choose these traditional sites because they provide the best

shelter for the winter months. The same is true for monarchs in Mexico. The few favoured groves of oyamel firs provide just the right range of temperatures so that the butterflies don't freeze to death, but are not so warm that they remain too active and burn up their winter fat stores too quickly.

There are also northern monarchs down under. They turned up in Australia in 1870 and established populations there, some of which now undertake their own migrations, though shorter and less spectacular than those in North America. It's likely that monarchs rode into Australia on the back of a cyclone from islands in the Pacific,[16] which they had conquered by the second half of the nineteenth century. They now occur on islands from Hawaii westwards through Micronesia and Melanesia. At the same time, they crossed the Atlantic to establish outposts on the Canary Islands and in mainland Spain and North Africa.[17] The more remote island populations have, very sensibly, given up migrating.

In 1993, the entomologist Richard Vane-Wright suggested an intriguing hypothesis to account for the monarch's sudden conquest of large parts of the globe.[18] He reasoned that the butterfly's only food, various kinds of milkweed, are plants of open ground and must have spread

dramatically as European settlers and later Americans cleared the continent of much of its forests. Clearances peaked in the second half of the nineteenth century and Vane-Wright argued, in what he called the 'Columbus Hypothesis', that monarchs must have likewise dramatically increased in number, compelling them to migrate in the spectacular numbers we see today and even forcing some to abandon their native continent.

A more recent analysis of the monarch's spread across the Pacific suggests a more mundane hypothesis.[19] If emigrants were flooding out of a North America that was becoming overcrowded, their colonization of new areas should follow a broad east–west pattern. Instead, historical documents suggest a few human-assisted introductions, perhaps to three different spots across the Pacific, from where the butterflies then island hopped, either on their own very capable wings or wind assisted, to reach as far as Australia. Rather than being a recent phenomenon, other evidence suggests that the great migration of monarchs to Mexico is far more ancient than the European conquest of the Americas.[20] As we've seen, many danaids are migratory, including other subspecies of *D. plexippus*. Yet, as impressive as the globe-conquering achievements

Convergent ladybugs, *Hippodamia convergens*, gather in vast numbers in sheltered spots where they can sit out the summer heat or the cold of winter.

of monarchs are, they're no longer considered to be the long-distance record holders that they once were. That honour has now passed to a common butterfly familiar on both sides of the Atlantic.

THE EPIC JOURNEY OF THE PAINTED LADY

Back in the spring of 2009, I was crawling around the nature reserve at Park Gate Down in Kent, in the far south-eastern corner of England. I'd travelled there to photograph monkey orchids for a previous book, and had spent most of the morning on hands and knees, completely focused on finding the best angles of the best specimens. When I finally paused to pour a cup of coffee, I was startled to find myself surrounded by clouds of hundreds of butterflies. I soon realized they were painted ladies, *Vanessa cardui*. Kent is the first landfall for many migratory species crossing the English Channel, both vertebrate and invertebrate, and I knew straightaway that these butterflies must be new arrivals to our shores. What I didn't know at the time was that this was the vanguard of a particularly impressive influx of these butterflies. Painted ladies arrive here from the south every year, but in 2009 reports were soon coming in from across the country of astonishing numbers of these butterflies.

Painted ladies are truly cosmopolitan. The same species lives on all continents except South America and Antarctica.[21] It has even established a permanent outpost on the sub-Antarctic Amsterdam Island in the southern Indian Ocean, a home it shares with albatrosses and penguins.[22] In North America it winters in the deserts of Mexico and, like the monarchs, follows summer as it rolls north across the continent. However, unlike monarchs, it has no winter rest. It breeds on thistles and other plants throughout the winter, building up its population ready to track fresh plant growth as more northerly regions warm up again.

The painted lady follows a similar pattern in the Old World, where in years like 2009, with wetter than normal winters, phenomenal numbers build up on the flush of plant growth in north-west Africa, in an area known as the Maghreb. These butterflies then make their way across Europe, breeding as they go, to reach as far as Iceland and even beyond the Arctic Circle to Svalbard. This is a butterfly that shares its world with polar bears at one extreme and penguins at the other. But the true scale of what this butterfly can achieve had to wait for a more sophisticated way of tracking it than just gluing sticky labels to the wings – one that relies on understanding a bit of basic nuclear physics.

As spring approaches in the monarch's overwintering sites, they descend in droves on to flowers to refuel on nectar, ready for the journey north.

Many atoms come in a number of different forms, called isotopes, that differ in the number of neutrons in their nucleus. Some isotopes, such as those of uranium, are unstable and soon break down, releasing a lot of energy in the process, but others are stable and last an eternity. Hydrogen, the simplest of elements, normally has just one proton in its nucleus, but another form, deuterium, also has a neutron in its nucleus. Scientists usually denote these two isotopes as H^1 and H^2. Oxygen, with a bigger and more complicated nucleus, normally has eight protons and eight neutrons (O^{16}) but several other stable isotopes exist, including O^{18}. So, what has all this to do with migrating insects?

The ratio of these different isotopes of hydrogen and oxygen conveniently varies from place to place and, as they are absorbed by growing vegetation, the plants preserve that ratio – more or less. So too do the insects feeding on those plants. The ratio can be measured in chitin from pieces of insect wings, even from museum specimens, and provided you have a map of how isotope ratios vary geographically you can tell where an insect grew up. This approach to tracking migrations has led to some startling revelations about the painted lady.

It's been known for some time that painted ladies are absent from north-west Africa in late summer, but arrive from Europe in great clouds in October and November, when enormous numbers crowd into oases and along verges.[23] Stable isotope analysis reveals that some of these butterflies cross the Sahara to reach the tropics of East Africa and the

Ethiopian Highlands.[24] Given that their descendants eventually reach back into northern Europe, these butterflies have an annual cycle that covers 15,000 kilometres, the longest of any butterfly so far discovered. While some of the European butterflies head to the Maghreb in north-west Africa to lay eggs on the abundance of thistles, common mallow and desert nettles, others cross the Sahara to take advantage of a flush of winter growth in the Sahel, then travel on to the Kenyan and Ethiopian highlands in midwinter.[25] It's hard enough to imagine the journey endured by small birds that must cross this barren desert, let alone a tiny butterfly.

Some of these tropical butterflies then move back to the Mahgreb in late winter, ready to recolonize Europe in the spring. Like many migrating birds that move through the tropics, the butterflies are following an atmospheric phenomenon known as the Inter-Tropical Convergence Zone (ITCZ). At sea this is more widely known as the doldrums, the

Every year, painted ladies, *Vanessa cardui*, arrive along Britain's southern coasts. In some years, the influx is so great that dozens can be seen together. When you next watch one feeding in your garden, consider that this insect makes the longest migration of any butterfly – over several generations they travel as far as the plains of East Africa.

windless zone near the equator, but it is part of a much bigger pattern of global atmospheric circulation. It is the zone where the north-east and south-east trade winds converge. Here, around the equator, where the sun is hottest, humid air carried by the trades is warmed and rises high into the atmosphere. As it cools at high altitudes, the moisture precipitates to form rain. The zone moves north and then south of the equator over the course of each year, bringing seasonal rains that trigger fresh new plant growth. Painted ladies track this growth northwards on their way back to North Africa. This is an insect that depends on both temperate and tropical ecosystems for its survival, across half the planet.

MONARCH FLIGHT PATHS

Stable isotope analysis has also helped flesh out the story of monarchs in North America, as has a similar technique, discovered by another key figure in the monarch story. The American entomologist Lincoln Brower spent six decades delving in depth into the life cycle of this butterfly, not just its migrations but its chemical ecology. He was a founder of the International Society of Chemical Ecology, which seeks to understand the role of chemistry in the ecology of plants and animals, and the monarch proved to be a perfect study animal. Its caterpillars feed on milkweed plants, which do their best to protect themselves from grazers by filling

their leaves with toxic, foul-tasting chemicals called cardenolides. Monarchs, along with a few other insects, have, in evolutionary terms, worked out how to get round this and even store these chemicals in their bodies, using them for their own defence. To advertise the fact that they are distasteful and best avoided, monarch caterpillars are conspicuously coloured in white, black and yellow. These chemicals survive the pupal transition to the adult butterfly, which is therefore also protected by its foul taste. So, the adult butterflies, too, flaunt themselves in dazzling orange and black livery.

It has always been assumed that bright colours serve as a conspicuous and easy-to-remember signal to predators. To show that this is indeed the case, Brower and his wife came up with an experiment now known as the 'barfing blue jay assay'.[26] The unfortunate blue jays were fed with Monarchs and if they hadn't seen one before, they greedily ate it, only to be violently sick a few moments later. However, the easily recognizable wing pattern only took one lesson to learn, and after their first experience these blue jays wouldn't go near another monarch. Brower also showed that if the caterpillars were reared on non-toxic food plants, the jays would happily feast on the adult butterflies for as long as they were offered. Brower's interest in monarchs soon expanded to the rest of their extraordinary lives, though he had first to find the overwintering sites in Mexico for himself.

After their discovery in 1976, Urquhart and Brugger kept the exact location a tight secret. After all, a whole continent's population of butterflies crammed into a few hectares makes the monarch very vulnerable, even to tourists keen to admire the spectacle for themselves. Lincoln Brower, together with a colleague, Bill Calvert, had to piece together the location from various clues in the published pictures and other sources. Luckily, oyamel firs are not too widespread and Calvert soon found the groves. Brower then used his knowledge of chemical ecology to unpick the northwards part of their journey.

He had discovered that different milkweeds in different areas had unique combinations of cardenolides, patterns that, like the stable isotopes of hydrogen and oxygen, were preserved in the butterflies. The chemicals are only ingested by the caterpillars so, by measuring this 'cardenolide footprint', Brower could tell where each butterfly was born and raised. In this way he showed that the northwards migration took place over several generations, each breeding successively further north.

However, the complexity of the life of this extraordinary insect only becomes apparent when the late-summer super-butterflies begin

The northward migration of monarchs takes place over several generations. The butterflies stop and breed en route and, once adult, the next generation will continue the journey.

their journey south. Alongside chemical and isotopic footprints, citizen science has continued to be critical in unpicking this part of the story. Following Urquhart's pioneering use of a curious public, several long-running projects, such as Journey North and Monarch Watch, still provide valuable data that helps plot the butterflies' travels in detail. We now know that there are two broad flyways for Monarchs that bred east of the Rockies. One runs down from central Canada through the American Midwest, into Kansas, Missouri, Oklahoma and Arkansas, from where the butterflies funnel into Texas. It used to be thought that this super-butterfly generation went all the way to Mexico, but stable isotope analysis has recently revealed that some of these butterflies stop off in Texas to breed and their offspring then continue the journey south.[27]

A second flyway passes south along the eastern side of the Appalachians, through eastern and Atlantic coastal states. Butterflies using this flyway lag behind those on the main route and they seem to have a reduced chance of making it to Mexico.[28] However, this hides another layer of complexity. Although there is a much lower recovery rate in Mexico of monarchs tagged in the east, it's the coastal route that is particularly risky. Those flying east of the Appalachians but away from the coast do seem to have as much chance of reaching their wintering sites as those on the central flyway. But those tracking the Atlantic coast south are eight times less likely to make it.[29] Perhaps these butterflies expend too much energy battling autumn storms and winds along the exposed coast and run out of fuel before making it to Mexico – but they may not all be doomed.

Some butterflies probably make it to Florida, where they join a population that lives permanently in the Sunshine State. Others might even get as far as Cuba. Recent studies using stable isotope analysis show that about half the monarchs in south Florida belong to a sedentary resident population, while the other half are migrants from the north.[30] Residents can breed year-round in southern Florida's balmy climate, while migrants have paused their breeding to divert resources to migration. Monarchs here have adopted one of two very different strategies. Furthermore, the stable isotope ratios among the migrants suggest a much more complex pattern of migration than I have outlined above. There's still a lot we don't know about this very familiar butterfly.

On the other side of the country, there is yet another population of this butterfly that doesn't migrate to Mexican mountains. To the west of the Rockies, monarchs make the much shorter journey to one of around 400 wintering sites scattered along a 1,000-kilometre stretch of Pacific coast from Mendocino County, California, to Baja California in Mexico. Here they assemble in native pines and cypresses as well as introduced *Eucalyptus* trees. As with so much of the life of this butterfly, though, it's never quite that simple. The Continental Divide running down the Rockies that separates the central and eastern populations from those of the west doesn't seem to be a complete divide. Some butterflies from the west turn up in the Sierra Madre mountains, and the pattern of movements within the population to the west of the Rockies is still a long way from being fully understood.[31]

INSECT NAVIGATORS

Just the fact that all these insects make journeys of such magnitude is impressive enough and that's before we even ask how they do it. Many insects make clever use of winds to assist their journey. The silver Y moths that migrate north through Europe in the spring somehow know when high-altitude winds are blowing in the right direction to carry them along.[32] Painted ladies moving south in autumn probably do something similar, ascending to around a kilometre in height to ride high-speed winds. This makes their autumn migration much less obvious than their spring arrival and could be one reason why the true scale of their record-breaking migration wasn't recognized for some time. It used to be assumed that the spring migrants bred in northern Europe, only to die as the weather turned cold in autumn.

If the wind direction is not exactly right, some insects can compensate, at least to some extent, by flying at an angle to the wind, and if the wind veers well away from their intended flight path, they drop back to the ground and wait for better conditions. In this way, insects can cover the ground as fast as many birds. One monarch tracked in a recent study was able to cover 143 kilometres in a single day.[33] Long-distance migrants also have other tricks to help their journeys. Many butterflies, monarchs included, glide and soar as much as they can to save energy, and some migratory insects have a larger wing-surface area than their more sedentary relatives, to help them glide more effectively. This is true even within the same species. In Florida, monarchs from the migratory population have longer wings than the residents.[34] Even Monarchs from migratory populations have different wing lengths. Measurements of monarchs in the Mexican groves, combined with stable isotope analyses, show that the Monarchs that travel the furthest have the longest and largest wings.[35]

To find its way to its destination, an insect needs to know where it's going. Again, many of the fine details have been studied in the ever-popular monarch, although the findings are applicable to other species as well. Monarchs follow good aviation practice and have two separate navigational aids, so one can act as a back-up. Like birds, they primarily use the sun to guide their journeys, though since the sun appears to move across the sky each day from dawn to dusk, to use it as a compass, a Monarch also needs to know the time of day. Many animals have a circadian clock that ticks away, measuring the time from dawn, usually in part of the brain. The monarch's clock, however, is at the base of its antennae and since it has two antennae, it always has a back-up chronometer.[36]

Monarchs can also use the Earth's magnetic field to orientate, and it has been suggested that this butterfly even has a map sense, like a built-in satnav. Knowing where you are on an east–west axis is a particularly hard navigation problem to crack. For human explorers, knowing which line of longitude they were on had to wait until the 1720s, when John Harrison invented a reliable and accurate marine clock that allowed sailors to plot the position of the sun where they were, then to compare it to where it would be at Greenwich time and thus to work out how far they were to the east or west of Greenwich. Yet many animals that have been experimentally moved to a new location instantly adjust their course to get back to where they want to be, probably by referring to some kind of internal map. Recent experiments with monarchs, however, have failed to show that they have such a map.[37]

Painted ladies also use a sun compass to navigate, though in their case their target destinations are broad geographical zones. Monarchs, on the other hand, are aiming for a few very specific groves of trees hidden in the vastness of the Sierra Madre Mountains. The geography of the

Silver Y moths, *Autographa gamma*, make use of tailwinds to aid their migration. If the wind direction becomes unfavourable, they drop back to the ground to wait until conditions improve.

mountains helps funnel butterflies towards their destination after they cross the border, but it's likely that they switch to navigation by smell as they get closer, following the distinctive scent of oyamel firs as it wafts on the wind.

In Australia, migrating bogongs are faced with a harder problem. They fly at night. Like monarchs, they can sense the Earth's magnetic field, but they also have acute night vision. They can see landmarks along the horizon as well as stars and the moon. On clear nights, they can navigate by the stars and the pattern of the Milky Way. If they can't see enough stars, they can even pinpoint the moon's position by the brightness of the sky when the moon itself is still below the horizon. Their built-in magnetic compass allows them to judge which landmarks to use as guides to travel in the direction they want to go. They then switch off the magnetic compass (which is energetically more expensive to use) and just use appropriate landmarks to navigate. At intervals, they switch the magnetic compass back on to cross-check that the landmarks and the magnetic field still correlate. If an experimenter has artificially altered this relationship since the last cross-check, the moths become confused and disorientated.[38] Like monarchs, bogongs must also aim for a very precise target, the boulder fields on the mountain peaks, and they probably use the same trick, switching to smell for the final leg of their journey. Since granite boulders don't smell much, it's been suggested that it's the odour of dead moths – those that failed to make it through the previous summer – that guides them to their final destination.[39]

I've always been impressed by migrating birds that can return to the exact tree where they nested in the previous year after a journey of perhaps thousands of kilometres. However, it seems that insects are just as good or perhaps even better at this, because most have no first-hand knowledge of their hibernation or aestivation sites. They are the descendants of those that left the site last year. A few insects are even able to pass on their flight plans to their fellow travellers. Honeybees have legendary navigation skills and spatial memories to match. They are also famous for swarming as they move a part of their colony to a new site. This is the way that honeybees found new colonies, usually close to the old one, but the swarms of one species, the giant honeybee, *Apis dorsata*, undertake a much longer annual migration.

The giant honeybee nests at high altitudes in the Himalayas, where it builds massive combs, suspended from overhangs on inaccessible cliffs. Each colony builds a single comb that may be a metre in length; often there are dozens, occasionally hundreds, of colonies close together on a suitable cliff face. To film these impressive bees for the *Alien Empire* series involved dangling a cameraman on a rope from the clifftop. Suspended several hundred feet from the ground, it was no easy task to capture the behaviour of these bees on camera. For one thing, they've put a lot of effort into building their combs, and they defend them with vigour. Yet every year they abandon the combs and migrate in a swarm to the lowlands.

The bees spend the dry season on the high cliffs but head to the lowlands during the monsoon, where the rains encourage a profusion of pollen and nectar-bearing flowers. The journey may be 200 kilometres long, at the end of which the bees build a new comb below the branch of a tree or even on a building. In transit, though, the bees have no time to build combs, but shelter in bivouacs made of the living bodies of bees. They seem to choose stopover sites that provide plenty of food and water, and stay in these bivouacs for several weeks at a time. They also seem to know exactly where they're going when they leave on the next leg of their journey.

Bees communicate directional information to other colony members by dancing. The famous waggle dance involves a bee performing a figure-of-eight pattern in which it waggles its abdomen each time it crosses the centre of the eight. The dance cleverly encodes both directional and distance information. The bees dance on a vertical surface, and the direction is given by the angle of the waggle walk to the vertical – that's the angle to the sun you'll need to fly. The distance is proportional to the length of the waggle walk. Bees use this to direct other workers to good sources of food, but giant honeybees also use it to tell the migrating swarm where they're headed next. For an hour or so before lift-off, bees dance on the outside of the bivouac. In this way, 40,000 bees all know the plan and stick together on their journey ready to form a new bivouac when needed.[40]

On their epic journey south, monarchs also gather in large groups at the end of a hard day's flying. Tracking the position of these roosts is one way that projects like Journey North have mapped the Monarchs' flyways. Sometimes these aggregations build up into impressive swarms. Point Pelee is the southernmost point of Canada, a long, thin triangle of Ontario projecting south into Lake Erie, a jumping-off point for Monarchs crossing the lake. I trekked out along this 14-kilometre spit of land one autumn, only to watch heavy clouds building up overhead and feel a cold wind whipping off the lake. I didn't much like the look of the weather for the next few days – and neither did the monarchs.

On their journey south, monarchs often gather in roosts as night falls. Sometimes, if the weather turns against them, they'll wait in sheltered spots, where impressive numbers build up.

Monarchs funnel along this peninsula as they fly south from Ontario since it offers them the shortest and safest crossing of the lake – but it's still risky, so they wait for the perfect weather conditions. Over the next few days, their numbers built up in the shelter of scrubby woodland until the glades were full of fluttering wings. Then, one morning, the bad weather finally broke, and as I stood on the tip of Point Pelee, I watched hundreds of frail-looking butterflies fly out low over the water and disappear over the horizon – with a mere 3,000 kilometres left to fly.

Butterflies using the more arduous Atlantic coastal route face similar problems. Cape May is a long peninsula that juts out to the south-west across the mouth of Delaware Bay, towards Cape Henlopen on the south side of the bay. It's a migration hotspot for insects and birds, as is the whole of this vast bay. I've watched the hypnotic swirls of great flocks of red knots as they descend on the beaches in spring to feast on the eggs of horseshoe crabs, a vital refuelling stop on their journey from Patagonia to Arctic breeding grounds. In late autumn, spectacular flocks of snow geese, heading south from the high Arctic, call in at Chincoteague, on the west side of the bay. These birds are pure white with jet-black wingtips and their long skeins flying overhead look like strings of pearls hung across a crisp blue autumn sky. Monarchs, along with many other insects, also make their way down Cape May as they head south in the autumn and, in 2006, they could count among their number a dragonfly called Dave.

DRAGONFLIES – THE RECORD BREAKERS

Dave was a green darner, *Anax junius*, and he was one of several of these migratory dragonflies that had been fitted with a radio transmitter. These 300-milligram devices were as small as they could be, but even so amounted to a quarter of the dragonflies' body weight. Nevertheless, Dave seemed unhindered, and over the twelve days that he was tracked, covered a very respectable 145 kilometres a day. Dave and his fellow travellers were followed by ground vehicles and even by light aircraft. I can only imagine what Dave thought when his huge, all-seeing dragonfly eyes spotted a Cessna tracking his every move. One intriguing discovery from this remarkable achievement (of both dedicated researchers and heavily laden dragonfly) was that the migratory behaviour of dragonflies bears a close similarity to that of birds using the same routes.[41]

Small birds flying down Cape May often baulk at the prospect of almost 20 kilometres of open water to reach Cape Henlopen on the other side of the bay's wide entrance. Instead, they backtrack north along the eastern shore of the bay until it narrows enough for them to hop across. Dave did exactly the same thing and, like many birds, he had distinct stopover periods between bouts of migration so he could top up his fuel reserves.

During the autumn migration, green darners fly south, hundreds to thousands of kilometres from where they hatched, and some make it all the way from southern Canada to Mexico.[42] However, like monarchs, their full annual migratory cycle involves several generations. More recently, stable isotope tracking has allowed scientists to follow the details of this annual cycle without having to burden dragonflies with heavy radio backpacks. The generation that migrates south breeds when it arrives on its warm winter grounds. The adults that emerge here also breed and, in the balmy southern weather, this generation quickly goes on to emerge and breed themselves. By the time this third generation emerges, spring is burgeoning across the continent and these dragonflies can head north again.[43] So, the annual migratory cycle in this species involves three generations.

Several other dragonflies in North America are long-distance migrants, although in most cases their journeys are not known in anything like as much detail as those of the green darner. These include the black saddlebags, *Tramea lacerata*, the spot-winged glider, *Pantala hymenaea*, the variegated meadowhawk, *Sympetrum corruptum*, and the wandering glider, *Pantala flavescens*. Black saddlebags, along with green darners, often build up in dramatic numbers at migratory hotspots like Cape May and Point

In some parts of its range, the migrant hawker, *Aeshna mixta*, undertakes a southward migration in the autumn, making use of tailwinds to speed its journey.

Pelee, and variegated meadowhawks have been seen migrating in large swarms down the Pacific coast from Washington to California. Many of these must get blown out to sea because this species sometimes fills the stomachs of salmon caught off the coast of Oregon.[44] However, it's the last of these migrants that now holds the record as the champion long-distance migrant. The wandering glider is also – far more aptly – known as the globeskimmer.

Globeskimmers have conquered even more of the planet than monarchs or painted ladies and they too have made it as far as Amsterdam Island, although it's probably too cold for this dragonfly to establish a permanent base camp here. But there is a resident population on equally remote Easter Island.[45] It doesn't breed in New Zealand but has even turned up there from time to time, either by riding winds from breeding sites in Australia, an ocean crossing of 2,000 kilometres, or by being swept up in tropical air masses moving south from established colonies on Pacific islands, an even more impressive open-ocean journey of at least 4,000 kilometres.[46]

It's been known for some time that on the Maldives, in the Indian Ocean, globeskimmers descend in swarms so vast that they defy imagination. The odd thing about this is that there is no standing water in the Maldives, so nowhere for them to breed. These insects were obviously on their way from somewhere to somewhere else and just dropped in here for rest and refuelling. We now know that the Maldives are a tiny part of a much greater picture – the longest insect migration yet recorded.

Stable isotope analysis of the wings of these dragonflies shows that they originated in northern India, rode the winds of the ITCZ to

southern India, then struck out over the ocean. The dragonflies call in on the Maldives and the Seychelles, before heading on to East Africa. They use much the same route as many birds that migrate to East Africa from Asia. Amur falcons even follow the migrating swarms of dragonflies, so they can refuel in mid-air to power their own journeys. Globeskimmer larvae develop very quickly, in as little as thirty-four days, so they can breed wherever they find small temporary pools of water. They return to India on the winds of the South-west Monsoon. If this wasn't astonishing enough, it's possible that some dragonflies originally came from north of the Himalayas. If so, their long journey also crosses the tallest mountain range in the world. Their annual cycle covers up to 18,000 kilometres, beating the record of 15,000 kilometres set by the painted lady and pushing the monarch further down the league table.

Many other dragonflies are also powerful flyers with enough stamina and endurance to travel long distances. The vagrant emperor dragonfly, *Hemianax ephippiger*, lives in the tropics of the Old World, across much of Africa and south-east Asia where, like the globeskimmer, it uses the ITCZ to follow the seasonal rains that provide it with breeding opportunities. Its normal migrations are not quite in the same league

as the globeskimmer's, but it does make occasional excursions as far as northern Europe. It is becoming an ever-more frequent visitor to Britain, where it might turn up at any time of the year. An adult dragonfly flying around in the depths of winter will very probably be a vagrant emperor dragonfly.

Sometime around 2000, this species even managed to cross the Atlantic to discover the New World. Like Columbus, this insect explorer landed in the Caribbean, and its descendants are now island hopping their way throughout the Caribbean, from where it might soon reach the North American continent. Helped by climate change, it could easily become a new addition to North America's dragonfly fauna.[47] The vagrant emperor's transatlantic crossing was a one-way trip, probably as a result of one or more of these dragonflies being blown out to sea by storms. It is not a migration in the strictest sense, but the vagrant emperor's colonization of the New World illustrates that, despite all the associated dangers, long-distance travel opens up a world of new opportunities.

These long-distance record holders are certainly impressive, but it's the sheer scale of insect migration around the world that is mind-boggling. Some 600 different kinds of butterfly are known to migrate.[48] No one has any idea of how many migratory beetles, flies, bugs or any of the other kinds of insect there are. Many of these don't cover such vast distances, but the numbers involved defy imagination – trillions of tiny bodies all travelling with determination to reach their own destinations. In 2014, the US National Weather Service tracked what it thought was a weather front moving over Albuquerque in New Mexico, until it realized that it was moving in a non-random way. It turned out to be a swarm of grasshoppers on the move, a swarm the size of the city itself.[49]

Using radar specifically designed to detect insect movements, scientists in the UK estimated that three and a half trillion individual insects migrate annually above southern Britain.[50] Many of these are flies. Another study calculated that 1–4 billion hoverflies migrate in and out of Britain each year. These insects carry out vital pollination services, while their predatory larvae devour perhaps 10 trillion aphids each year.[51] And these are figures for just one small part of the globe.

MIGRATION – A RISKY BUSINESS

It seems as though most of the insect world is on the move. Migration has allowed insects to occupy places that may not be suitable for the whole year. It lets insects exploit temporary abundances or escape

Eastern amberwing dragonflies, *Perithemis tenera*, are not, as far as is known, long-distance migrants, but with such large wings in relation to their small body, they wander great distances searching for new habitats.

intolerable heat or cold and, in broader terms, to make the best use of their environment. However, it comes with its own risks. These journeys test insects to their limits, and changes in the modern world are pushing many beyond the brink of their endurance. Migrants are often more vulnerable than sedentary insects (the same is true of birds), because they depend on critical resources throughout their entire journey. They need to find secure places en route where they can rest, and they rely on refuelling stops, like motorway service stations, at convenient distances along their route. Not surprisingly, migrant insects have experienced some severe declines over the last few decades, but one of the most recent is also one of the most alarming.

In 2019, scientists in Australia reported a staggering and thoroughly depressing 99.5 per cent decline in the population of bogong moths at their summering sites in the Australian Alps. Their numbers have been in slow decline for many years, driven by habitat changes in their winter breeding grounds, the overuse of pesticides (Australia still allows the use of neonicotinoids that are now banned in the European Union) and light pollution that disorientates migrating moths. A severe drought in 2017 caused a further catastrophic loss, but the sheer scale of the decline after this has caught conservation bodies completely by surprise.

These anonymous, boring-looking moths are actually quite famous down under. In times past, they swarmed into Canberra in such huge numbers that they covered the white walls of Parliament House and fluttered through the chamber of the House of Representatives. They don't, perhaps, reach the status of a national icon, but the severity of this unexpected decline might resonate strongly enough for some of the factors linked to falling numbers to be addressed. A female bogong moth can lay 2,000 eggs when she returns from her mountain sojourn, so if conditions improve, it wouldn't take long to rebuild their impressive historical population. However, like all insects, they face some problems that can only be tackled on a global scale. Due to climate change, nowhere will escape wholesale habitat changes, which will affect insects in countless ways.

For example, changing patterns of rainfall will have massive effects on monarch butterflies in North America. Feeding on well-watered plants, monarchs can store up to 80 to milligrams of lipids in their bodies, but if forced to feed on drought-stressed plants, they can only lay down half of these energy reserves, not enough to fuel their long migratory journeys.[52] Climate already has a major impact on monarch populations, which vary dramatically from year to year. monarch numbers in Mexico

during the winter of 2021–2022 showed a welcome increase of around 35 per cent on the previous year, at least in terms of the area of oyamel firs that they occupied. This was also higher than the average for the whole of the preceding decade. However, this small piece of good news is set against a backdrop of a steady decline since the wintering sites were first discovered.[53]

Western monarchs overwintering in California had an even more impressive revival of fortunes in the same period. In the winter of 2020–2021, an alarmingly low number of butterflies turned up – a total of just 2,000 monarchs were counted across the whole state and there were none at all in many of the traditional overwintering sites. The following year saw almost a quarter of a million butterflies flock to more than 200 wintering groves, a hundred-fold increase in numbers.[54] Scientists are still unsure what drove such a dramatic comeback. While the vast increase was more than welcome, such huge swings in population are also a cause for great concern, because they leave western monarchs very susceptible to extinction during their low ebbs – and we should remember that since the 1980s, western monarchs have declined by 95 per cent.

Migrants have always faced many hazards on their journeys, but crossing the modern world brings new levels of jeopardy. For Monarchs traversing the crowded and bustling United States, this means crossing

Hoverflies, *Syrphus ribesii*. Billions of hoverflies migrate to Britain each year, where they carry out vital pollination services, while their larvae devour uncountable numbers of aphids.

endless highways, where worryingly large numbers of these butterflies meet an untimely end.

Half a million monarchs die each week on roads in Illinois alone, and that's just a fraction of the distressingly large number of 20 million other butterflies and moths that also perish each week in the same way in that state.[55] It gets even tougher for monarchs as they head south. Crossing Texas, the terrain channels the butterflies into a narrow corridor called the Central Funnel. Here, in some years, more than three and half million monarchs are killed on the roads that cross their migration route. Averaged out, around 2 million are killed every year – or about 3 per cent of the population that overwinters in Mexico.[56] Simple speed restrictions during the migration season would go a long way to reducing these appalling figures.

As spectacular as the Mexican winter sites are, they are where the monarchs are most vulnerable to a broad range of threats. The oyamel fir groves are increasingly hemmed in by expanding agricultural land. Smoke from fires set on this land drifts up the mountainsides and into the groves, which causes the butterflies to erupt in panicked, frenzied flight. This uses up their stores of fat, which they'll need both to survive the winter and to begin the journey north again in spring.[57] The butterflies chose these sites in the first place because the conditions are just right for them to achieve the tricky balancing act of eking out their resources over winter, with enough energy left to mate and fly north in the spring. Anything that disrupts this balance could have disastrous consequences, but that is probably what will happen as a result of climate change.

Temperature and rainfall patterns are changing rapidly in the mountains. Climate modelling suggests that it's likely that three-quarters of the monarch groves will no longer be fit for purpose by 2030, and by the end of the century there will be virtually no suitable wintering sites for monarchs.[58] Even today, although they are much diminished, these sites remain one of the greatest wonders of natural history. The weeks I spent in the Sierra Madre are still etched with crystal clarity on my mind, in part because of the spectacle, in part because I knew how fragile this phenomenon was, but in largest part because of the people I met at El Rosario who are working to protect these precious sites.

This site is one of several that have become ecotourist attractions, bringing much-needed foreign currency to this poor part of Mexico. In return, the guides and wardens work hard to stop logging interests from felling the precious trees that the monarchs depend on, or encroachments from agriculture such as avocado farming. It's not without risk. I was shocked to learn that, at the start of 2020, one of the guardians of this place, Homero Gomez Gonzalez, was found murdered. Three days later, Raúl Hernández Romero, a tourist guide, was also found dead. These people belonged to a land collective that seeks to find sustainable uses for these mountain forests, against the interests of those who would make a short-term profit from their exploitation and in the process not only destroy a natural wonder of the world, but also threaten the very existence of this familiar butterfly across most of North America.

I have nothing but admiration for the people of El Rosario. As we packed our kit for the last time and prepared to leave, I offered to leave the bird field guide that had so engrossed our local team. I suggested this to our fixer, who passed on the offer to a village elder. A simple gift, I thought – and a simple transaction. However, at the end of the day, as I emerged under the archway with its dazzling monarch paintings, I was greeted by a crowd of villagers. This was to be a formal handover ceremony, with speeches on both sides. I praised the work the community was doing not just in protecting these winter sites, but in so doing, ensuring that monarchs can fly back to North America to give delight to people four and a half thousand kilometres, and a whole world, away. In return they told me that this single book would form the start of the El Rosario nature library to help teach local children about all the treasures on their doorsteps. I wish them every success in the hard task they face.

(*overleaf*) The monarch groves of Mexico's Sierra Madre Mountains are one of the natural wonders of the world, but they're under severe threats from climate change and agricultural encroachments.

6

Flower Power

The complex relationships between insects and plants

Natural selection cannot possibly produce any modification in a species exclusively for the good of another species; though throughout nature one species incessantly takes advantage of, and profits by, the structures of others. If it could be proved that any part of the structure of any one species had been formed for the exclusive good of another species, it would annihilate my theory, for such could not have been produced through natural selection.

Charles Darwin, *The Origin of Species*, 1859

Darwin had a real problem with selfless relationships. His theory of evolution by natural selection demanded that each and every individual behave entirely in its own interests, to maximize its own chances of surviving and breeding at the expense of any and all of its competitors. Taking up Darwin's baton, Richard Dawkins has recast this fundamental concept of biology into the selfish behaviour of genes, whose only role in life is to make more copies of themselves, which they do with callous indifference, always assuming, of course, that a molecule can be callous.

Yet cooperation is just as much at the heart of nature as selfishness. For Darwin, as for a great many biologists since his time, understanding how such partnerships form and how they are maintained despite the stark precepts of natural selection, is a cornerstone of evolutionary theory. And there is no more fertile field for such studies than the complex and varied partnerships between insects and plants.[1] But these intimate associations also go a long way towards explaining why insects, along with flowering plants, have been so successful.

Darwin was well aware that the entwined lives of insects and flowering plants may have played a part in the rise to dominance of both groups. Yet, the apparent alacrity of that rise, as flowering plants seemingly appeared from nowhere and in great diversity during the middle of the Cretaceous period, troubled him greatly. For Darwin, evolution proceeded at an imperceptibly slow pace, and flowering plants had no business just popping into existence like that. The French palaeobotanist Gaston de Saporta suggested to Darwin that the driving force behind this proliferation of flowering plants was the development of a close relationship with insects, especially as pollinators. The idea impressed Darwin, who wrote back to Saporta, 'your idea that dicotyledonous plants were not developed in force until sucking insects had been evolved seems to me a splendid one'.[2] Yet still he fretted. Late in his life he wrote a letter to his friend and confidant, the botanist Joseph Hooker, in which he described this apparent explosion of life as an 'abominable mystery'.

> The rapid development as far as we can judge of all the higher plants within recent geological times is an abominable mystery... Saporta believes that there was an astonishingly rapid development of the high plants, as soon [as] flower-frequenting insects were developed and favoured intercrossing... I sh'd like to see this whole problem solved.[3]

A century and a half later we've solved some of those problems, but found a few new ones. The blossoming of the flowering plants was not

(previous page) A red-tailed bumblebee, *Bombus lapidarius*, harvests pollen from a dandelion. Flowering plants and insects have evolved hand in hand for tens of millions of years, enhacing the diversity of both groups.

as abominably quick as Darwin thought. New fossil finds have pushed back the origin of flowering plants to the early Cretaceous period, while molecular dating techniques suggest an even earlier origin, in the Jurassic or Triassic periods, allowing time for a more sedate rise to power.[4] But what of the idea that insects and flowering plants enhanced each other's success?

COEVOLUTION

A belief widely held since Darwin's time is that the burst of evolution of flowering plants during the Cretaceous period coincided with a rapid diversification of insects – a smoking gun suggesting that the fate of these two groups was closely tied together through a process that eventually came to be called coevolution. This is a reciprocal process in which changes in a population of one organism spur changes in a second organism, which in turn spur further changes in the first.[5] As this evolutionary tit for tat is played out through many generations, the two partners rapidly evolve in step.

If the subtleties of this relationship are different across the geographical range of either partner, perhaps due to differences in local ecology, or if one of the partners differs over the range of the other, then

Spring in Oklahoma's Wichita Mountains. Carpets of flowering plants, such as tickseed, *Coreopsis lanceolata*, provide endless opportunities for insects.

coevolution could drive rapid speciation.* Coevolution could be a driving force creating diversity through many different kinds of relationships, for example between predator and prey or between host and parasite. However, it's often been cited as a major factor in the rapid radiations† of insects and flowering plants, whose varied relationships include those between flowers and their pollinators or between herbivores and the plants they eat.

Have the destinies of both insects and flowering plants really been so intimately bound by such relationships? We've just seen that flowering plants evolved much earlier than originally thought, but so too did many groups of pollinating insects. Molecular dating techniques are not without controversy, but suggest that moths (Lepidoptera), for example, may have evolved as early as the Carboniferous period, and bees and wasps (Hymenoptera) in the Permian period, long before there were any flowering plants. Subsequent bursts of speciation among flowering plants and insects sometimes appear synchronous, but just as often don't bear much relationship to each other.[6] For example, a new analysis of changes in insect diversity through time has revealed a rise in diversity, at least among insect families, in the early Cretaceous period, pre-dating the expansion of flowering plants by many millions of years.[7] Moreover,

* This is usually called the Geographical Mosaic Theory and is just one way that coevolution could drive speciation.

† An evolutionary radiation is the rapid increase in the number of species or of higher taxa such as genera or families. A well-known example on a small scale is among Darwin's finches on the Galapagos Islands, where a single colonizing species of finch rapidly radiated on the islands to create the current thirteen species.

a new analysis of the radiation of flowering plants has shown that, while they did increase in diversity during the Cretaceous, they didn't come to dominate terrestrial ecosystems until the Palaeogene (66 to 55 million years ago), after the end of Cretaceous period.[8] All of this points to a much more complicated story than the one frequently told, and one that happened over a much longer period of time than originally supposed.

To further complicate the picture, as more data is gathered on a wide variety of partnerships throughout the living world, coevolution, as strictly defined, is – in many cases – proving a tricky concept to prove beyond doubt, and some now consider that true coevolution wasn't the major force driving the diversity of both flowering plants and insects.[9] If we look closely enough, the relationships between plants and insect are not at all how they appear at first sight. As we're about to find out, some of the classic examples of coevolution may have other, equally intriguing, explanations. The fates of insects and flowering plants do seem to be bound together, although in many different and often unexpected ways.

Many insects eat plants, and the bigger the variety of plants that exists, the bigger the variety of insects that can feed on them. Some groups of beetles, such as leaf beetles and weevils, feed exclusively on flowering plants, and following the proliferation of such plants during the Cretaceous and Palaeogene, these beetles found themselves with boundless new opportunities. Leaf beetles and weevils are now two of the most diverse groups of any insects, their success underpinned by the increasing success of flowering plants.[10]

Since plants don't want to be eaten, they've come up with an armoury of physical or chemical defences, which plant-feeding insects promptly find ways to circumvent. Some can detoxify a plant's chemical defences, others, like monarch butterfly caterpillars that feed on milkweeds stuffed with noxious cardiac glycosides,* store the plant's toxins and use them for their own defence. In addition to cardiac glycosides, milkweeds have sticky latex running through their veins, which gums up the mouthparts of leaf-eating insects – unless they know what they're doing. Some beetles that, like monarchs, specialize on milkweeds, cut through the leaf veins below where they want to feed, draining out the latex before it reaches their chosen dining area.[11]

To get round the wide variety of ways in which plants defend themselves, many insects must specialize – to focus on circumventing

* This is usually called the Geographical Mosaic Theory and is just one way that coevolution could drive speciation.

Flower-rich grasslands on the steep slopes of the Cotswolds in Britain's south-west are alive with insects, such as marbled white butterflies, *Melanargia galathea*.

just one specific set of plant defences. In doing so, they end up locked in an arms race with their lunch, a process of measure and countermeasure that gradually generates an increasing diversity of both plants and insects. Butterflies and moths which – as we've already discovered – are one of the mega-diverse orders of insects, are good illustrations of this. As caterpillars, many of them feed on just one particular plant (or perhaps a few different but related plants) whose defences they have managed to thwart – as is the case with those milkweed-devouring monarchs.

These tight relationships between butterflies and their food plants led ecologists Paul Ehrlich and Peter Raven to coin the term 'coevolution' back in the 1960s, and to suggest that this had been a driver of species diversity in both groups.[12] This seminal paper influenced a generation of biologists, although – as we saw earlier – Darwin and some of his contemporaries were already thinking along these lines, albeit without using the word 'coevolution'. Darwin was fascinated by the relationships between flowers and their pollinators, and this might seem like an obvious place to begin exploring how partnerships between flowering plants and insects work. We'll get to that later, but there's another group of plants that has evolved much more unexpected liaisons with insects – plants that have taken to eating insects.

A large milkweed bug, *Oncopeltus fasciatus*, is just one of a whole suite of insects that have circumvented the milkweed's toxic chemical defence.

CARNIVOROUS PLANTS

Many plants have turned the tables on the leaf-munching hordes of insects and evolved ingenious and often macabre ways of trapping and digesting insects for their own nutrition. On the face of it, this would seem like a one-way relationship, and no more a partnership than the relationship between a gazelle and a hungry cheetah – but let's look again at the exploits of carnivorous plants.

Early botanists, such as Carl Linnaeus, refused to believe that such plants really ate animals. A generation before Darwin, this went against the natural order of things as ordained by God. Animals ate plants – not the other way round. However, with their fascination for Gothic horror, the Victorians enthusiastically embraced carnivorous plants, although they often let their imagination run wild. Fictitious accounts of man-eating plants were widely believed. These stories were lapped up by an audience eager to hear of trees whose branches came to life 'with the fury of starved serpents' to devour sacrificial maidens.[13] Charles Darwin himself was captivated by carnivorous plants, although his interest lay more in how such relationships evolved than in lurid tales of human sacrifice.

Darwin spent many years studying the details of how these insect traps worked and how the plants digested their prey. Sundews, *Drosera* spp., were one of his favourites and he became so obsessed that, according to his wife, Emma, he treated them more like animals than plants. Perfectly understandable. I talk to my sundews. These plants, of which three species live in Britain's bogs, have one of the simplest traps – sticky blobs of mucilage perched on hairs that cover the plant and glue small insects to its leaves. Many sundews, including the ones that Darwin cultivated, react to a trapped insect by curling the leaf around the prey, not perhaps with the fury of a starved serpent, but still very quickly for a plant. When I replicated Darwin's experiments a few years ago for a film on carnivorous plants, it took only twenty minutes or so for a sundew to ensnare its victim, though with time-lapse photography we could speed up the process to just a few seconds, which would doubtless have enthralled a Victorian audience.

Many plants are covered in sticky hairs that serve to dissuade insects from nibbling on their leaves, and it's only a small step from this to deliberately trapping insects, then evolving enzymes that can digest their prey, along with leaf glands that absorb the resulting nutrients, as Darwin observed in his sundews. However, these last steps have also been taken by a surprising number of plants, including some that we

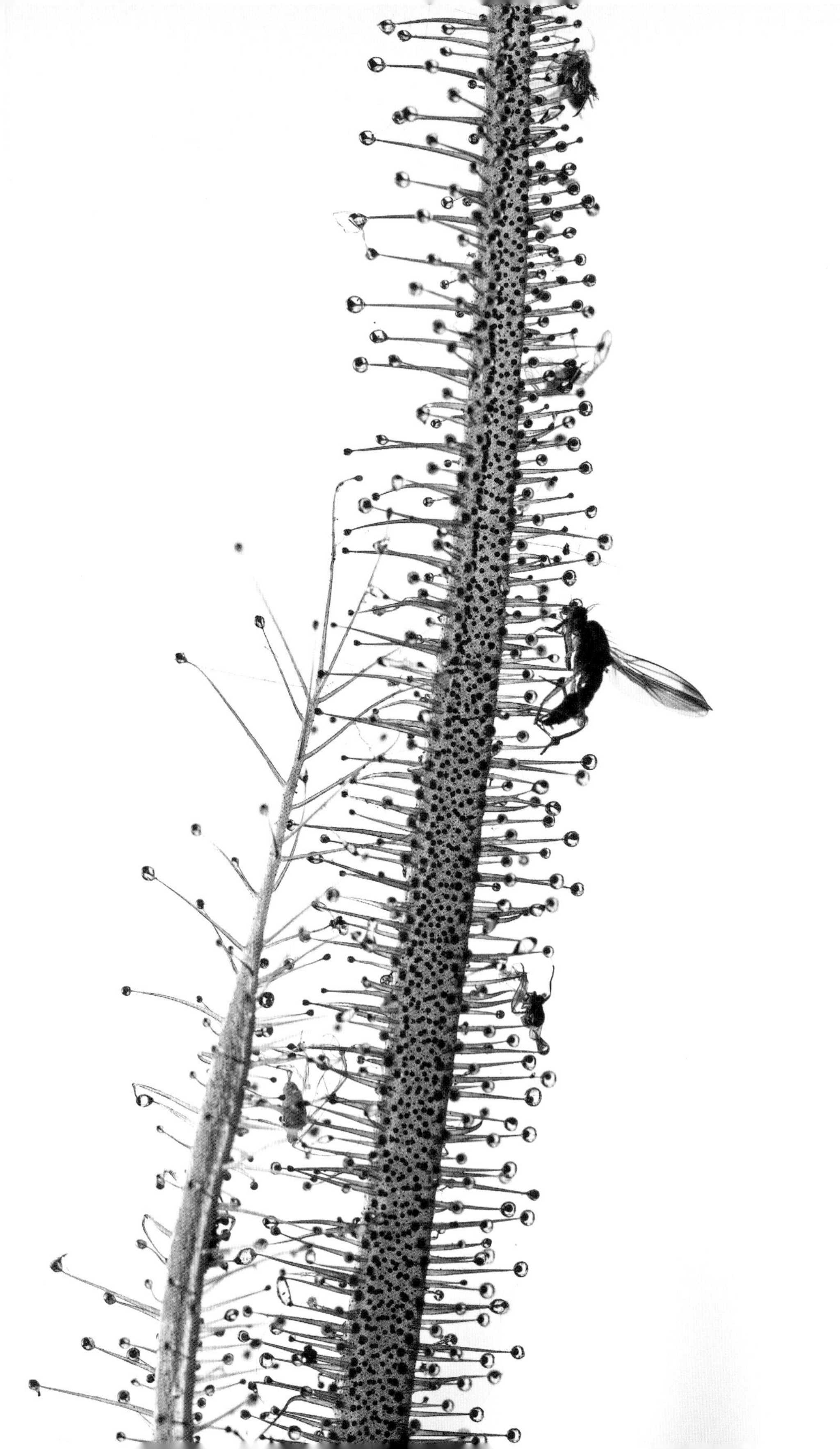

certainly wouldn't think of as carnivores, such as potentillas, petunias and geraniums (which should give you pause for thought when you next do a bit of gardening). In many ways these familiar garden plants are more truly carnivores than some of the more exotic plants that are commonly classed as carnivorous.

One such is a sticky shrub from southern Africa, often called the fly bush (*vlieëbos* or *vlieëbossie* in Afrikaans). Fly bushes are not related to sundews and there are just two species, *Roridula gorgonias* and *R. dentata*, both confined to the floristically spectacular Cape Province of South Africa. Their lance-shaped leaves grow in clusters at the ends of long, woody branches and each leaf is covered in stout, spiky hairs, all of which bear sticky blobs at their tips. These blobs are made of resin that is a much stronger glue than the mucilage of sundews. *Roridula* can therefore snare much larger insects and occasionally even small birds. It's such a good flytrap (reportedly four times stronger than some commercial flypapers)[14] that South Africans sometimes hang bunches of *Roridula* in their houses to trap flies. The problem is that once *Roridula* has trapped an insect it can't digest it. It has none of the proteolytic* enzymes that sundews use to break down their insect prey. It needs help – from another insect.

A trick I learnt when finding a *Roridula* bush is to blow gently across the plant, which – rather surprisingly – causes some of the 'trapped' insects to jump up and start running around. These are mirid bugs, *Pameridea* spp., tiny insects (just a few millimetres in size) that live their whole lives on a *Roridula* bush. Other mirid bugs are widespread and common on many plants, including those that have sticky hairs for defence. They pick their way around a plant's defences and use their hypodermic mouthparts to suck the plant's sap. To cope with the superglue world of *Roridula*, *Pameridea* (a different species lives on each of the two kinds of *Roridula*) have become more muscular and developed a greasy cuticle so that they can push their way through the tangle of sticky traps with impunity,[15] and they've switched from a vegetarian to a carnivorous diet.

When I gummed a large insect onto *Roridula*'s sticky leaf, nearby *Pameridea* bugs took an immediate interest. They slowly began to creep out of the forest of hairs, circling around the struggling insect, probing it with their stiletto mouthparts, then darting back under cover as the

* Proteolytic enzymes are chemicals that break down proteins, often into their constituent amino acids, which can then be reused for building a plant or animal's own proteins.

The leaves of the forked sundew, *Drosera binata*, are covered in hairs that produce sticky blobs of mucus at their tips. These trap small insects, which are then digested by the plant.

insect twitched. It reminded me strongly of a pack of wolves carefully sizing up their prey before attacking. Since *Roridula* can trap large and powerful insects, their caution makes sense. Wait until the prey is exhausted by its struggles, then move in for the kill.

Eventually, the bravest Pameridea touched the prey with the tip of its proboscis and when this was met with no final convulsive struggles, the bug stabbed the trapped insect and began to suck up its liquid lunch. Other bugs, large and small, emboldened by their meal's final demise, crowded in to find their own place at the feast. It was a real family affair, with tiny juveniles jostling for space alongside fully grown adults. But there's no such thing as a free lunch. Now it's time for Pameridea to repay *Roridula* for providing such a fine meal, and Pameridea pays off its debt by defecating all over the leaves.

Pameridea poo is the ideal foliar fertilizer, pre-digested into a form that's easy for *Roridula* to absorb through its leaves.[16] The bug can't live without the plant and the plant can't live without the bug – an example of an obligate partnership* and an illustration of how an arms race can turn into an intimate relationship. Probably, like many plants the world over, *Roridula*'s ancestors evolved sticky hairs to deter plant-feeding bugs. Mirid bugs the world over have developed countermeasures to such deterrents, and still abound on sticky but non-carnivorous plants. However, as *Roridula* became ever stickier and began to trap big insects, *Pameridea*'s ancestors spotted an opportunity to switch to a more nutritious diet. After this, *Roridula* could avail itself of bug excrement to obtain nutrients missing from the local soils.

Thus did enemies become friends and each now depends on the other. It's a close partnership, despite the fact that each partner got there

* *Obligate partnerships* are those on which both partners depend – one can't survive without the other – as opposed to *facultative partnerships*, which are more optional for each partner. Each can survive on its own, although the partnership may bring additional benefits.

through entirely selfish motivations, in keeping with the Darwinian world view. Darwin himself knew of *Roridula* and considered it a bona fide carnivorous plant, although later researchers, realizing that it lacked digestive enzymes, assumed it was merely using its sticky leaves to protect itself. The ever-present Pameridea bugs, they reasoned, were the plant's pollinators – but the truth is far more fascinating. Through its partnership with an insect, *Roridula* is functionally carnivorous. Moreover, it's not the only carnivorous plant that needs a little help from its friends.

The rainforests of Borneo hide an astounding range of pitcher plants, *Nepenthes* spp., which grow as vines, sometimes for tens of metres through the trees. They produce elaborate pitchers from the tips of their leaves that often take on very different forms on the upper and lower parts of the vine. Each is partially filled with a digestive liquid, the plant's version of a stomach, and each has a different strategy for filling its belly, often reflected in the shapes of the pitchers.

One of the great privileges of thirty-five years of making wildlife films is the time spent with experts who know their own patches with such a complete intimacy that I'm frequently left spellbound. Exploring the forests of Borneo with American ecologist and photographer Chien Lee was just such an experience. Chien moved from California to Borneo in the mid-1990s and spent much of his time thereafter studying *Nepenthes* pitcher plants. He had discovered some intriguing pitcher plants in the lowland forests of Sarawak, on the island of Borneo, which had such unusual relationships with insects that I was eager to capture them on camera. Even before we found any of these oddities, my attention was grabbed by a huge pink and green pitcher dangling just above the forest floor, which turned out to belong to a plant called *N. rafflesiana*.

Conventional *N. rafflesiana* certainly is not, but – in this remarkable family of plants – it is a typical enough representative of its kind to illustrate the sci-fi lifestyle adopted by pitcher plants and to show the next of many unexpected twists in the tale of interaction between plants and insects. The pitcher itself is a complex structure that grows from a tendril at the tip of a leaf, with a ridged rim running around the edge of the flask and a lid held over the open mouth to prevent the all-too frequent torrential showers from flooding the pitcher and washing away valuable nutrients. This species specializes in catching ants, which are drawn to the brink of the trap by sweet nectar produced in nectaries around the rim. Yet, for all its elaborate design, the plant doesn't seem particularly efficient at catching ants. As we filmed the pitcher, we

The only home of the *Pameridea* bug is among the sticky leaves of the fly bush, *Roridula*. It feeds on insects trapped by the plant – and returns the favour by fertilizing the plant with its droppings.

watched ants sauntering around the rim without a care in the world and helping themselves to a free lunch.

It turns out that those downpours are a critical part of the pitcher plant's strategy. The ridged rim is a super-wettable surface and once covered by a thin film of water it becomes super-slippery. Now the feeding ants soon lose their footing and tumble into the pitcher, where they are digested.[17] Unlike the *Pameridea* bugs scuttling around on *Roridula* bushes, there seems to be no sense of any partnership in this relationship – it looks like a straightforward case of a predator catching its prey. Such an interpretation, however, may be a little too short-sighted. Some scientists see this more as an alliance between ant and plant. There are times, during dry spells, when the ants can feed in safety and take their sugary booty back to the colony. At other times, the plant benefits by digesting some of the ants. As we'll see in a later chapter, workers of social insects are more or less expendable as long as the colony as a whole benefits. In essence the colony is more like a single superorganism. So, both ant colony and plant gain from this partnership, even if quite a few ants are sacrificed along the way.[18] There are, however, less brutal partnerships with ants in these forests.

The fanged pitcher, *N. bicalcarata*, looks like it might have stepped straight out of *The Little Shop of Horrors*. The two long 'fangs' projecting down from the lid house some of the biggest nectaries of any pitcher plant. Insects tempted by this treat find themselves dangling precariously over the pitcher on a finely pointed, slippery surface and soon drop to their doom. However, we also saw swarms of ants clambering around the fangs and the rim of the pitcher with apparent impunity. These ants all belonged to one species, *Camponotus schmitzi*, and – extraordinary to relate – it turned out that these deadly traps were their home.

The pitcher even provides the ants with living quarters in a twisted and swollen section of the trap's stem, and with ample food. Those huge nectaries provide all the sugar an ant could desire, but getting protein

(above) The fanged pitcher, *Nepenthes bicalcarata*, houses symbiotic colonies of ants in the thickened stems that support the pitcher.

(left) The spectacular pitcher of *Nepenthes rafflesiana*, growing in Kubah National Park in Sarawak.

needs a bit more work. We watched in fascination as ants descended into the pitcher, nimbly climbing down to the surface of the water inside and hauling out drowned insects. These feisty little ants can also swim and think nothing of plunging into the digestive fluid to haul out a particularly tasty morsel.[19] The ants are adept at clambering around the pitcher's walls and rarely spend more than thirty seconds immersed in the pitcher's fluid, so they remain unaffected by the concoction of digestive enzymes.

In return for the pitcher's generous hospitality, the ants offer a full valet service. Firstly, being ants, they are aggressive predators and deter plant-eating weevils from settling on the pitcher. They also clean the rim, removing fungal growth and keeping it clean and slippery – all the better to catch food for both the plant and themselves.[20] Finally, even having its prey stolen helps the pitcher plant. The ants remove large insects that, if left in the pitcher, would putrefy and damage or kill it. So the relationship benefits both parties. The ants are supplied with food and a nest, while the fanged pitchers benefit from being kept clean. Without their partners, these pitchers grow poorly compared to those harbouring active colonies of ants.

Clambering through the steep forests of Kubah National Park, just outside of Kuching, Chien finally introduced us to one of the pitcher plants that had originally drawn us to Borneo. This one was the white-band pitcher, *N. albomarginata*, named for a conspicuous white band running just below the rim. He explained that this one, rather than feeding on ants like *N. rafflesiana*, was a termite specialist, and we soon found several hundred thousand potential victims marching across the forest floor. The white-band pitcher favours termites of the genus

Hospitalitermes, which form long, broad columns of foraging workers snaking out from their nest to food sources.[21]

The white-band pitcher doesn't produce much nectar but the termites aren't after nectar – they're fungus eaters. The white fuzzy band around the pitcher's lip does look a bit like fungal growth, but it's probably not enough on its own to fool a termite. However, the felty texture of the band serves as an ideal bed for the spores of real fungi to settle and grow, so the pitcher provides a genuine source of termite food.

Life for a white-band pitcher is all or nothing. Most pitchers that we found were empty, but they live longer than most other pitchers and can afford to wait. When a termite eventually discovers the pitcher, it will recruit others from the nest, and soon one of those busy trails will wind its way towards the plant. Hundreds of termites crowd onto the pitcher, chewing at the white band; inevitably, eager new arrivals shoving into the crowd from below push their nest mates above into the pitcher. After an hour or two of searching, Chien found one of these lucky pitchers, crammed to the brim with dead termites. Once it has digested its feast, this individual pitcher will die, but the plant will produce others to patiently wait their turn.

This looks more convincingly like a simple, one-way, predator–prey relationship, but here too there may be a partnership lurking beneath the surface, much like that of *N. rafflesiana* with ants. The termite super-organism may benefit from the high-quality fungus growing on the pitcher, while the plant certainly benefits when one of its pitchers is stuffed with dead termites. This relationship is not so well known among botanists and entomologists, so it made a great new story to film. The problem was how to capture it when so few of the pitchers catch termites and all the action happens after dark. After a couple of fruitless nights under Borneo's spectacular starry skies, I realized that another plan was needed. We decided to collect part of a termite colony and some pitchers, and stage an encounter under more controlled conditions. Without access to a studio, the controlled conditions turned out to be my room back at the hotel in Kuching, the state capital of Sarawak.

Unusually, our production office had booked us into a rather luxurious hotel in the centre of town, which had the advantage of spacious rooms – plenty of space to build an escape-proof set, covered in polythene to keep the humidity high, just as termites like it. We then placed some pitchers on the set and in a short time the first termites found them and started to recruit more workers to harvest the white band. That day happened to be my birthday and the cameraman, Kevin Flay, thoughtfully surprised

Termites, *Hospitalitermes* sp., feed on the lip of a white-band pitcher, *Nepenthes albomarginata*. Inevitably, as numbers build up, many will fall into the pitcher to be digested by the plant.

me with a bottle of my favourite malt whisky, so I was able to lie on my bed, sipping Scotch and watching an extraordinary insect–plant interaction that hadn't been filmed before unfolding on a monitor. It was everything we could have wanted, but then I noticed some unexpected movement out of the corner of my eye. No, it wasn't the Scotch. There was a solid line of termites heading purposefully up the curtains and across the ceiling. So much for our escape-proof set. Fortunately, we were scheduled to leave the hotel the next morning. I haven't been back since.

The forests of Kubah National Park are home to yet another curious pitcher plant. This one is a vegetarian carnivorous plant or, to be more accurate, a detritivore, though it too depends on help from insects. Patches of the flask pitcher, *N. ampullaria*, growing on the forest floor, look like Roman amphorae neatly stacked in the hold of a merchant ship. Most kinds of *Nepenthes* have a large lid held over the pitcher, but the flask pitcher's lid is tiny and held well away from the rim of the pitcher. The flask, open to the rainforest canopy above, catches leaves as they fall. Unfortunately though, leaves are tough to break down so, like *Roridula*, it needs help.

Lying on my belly in the damp gloom of the rainforest floor and illuminating the pitchers with my torch, I could see that each flask is a miniature pond, packed full of life. Some were home to dozens of tiny tadpoles of the black-spotted sticky frog, *Kalophrynus pleurostigma*, swimming around their private pool with just the tiniest twitches of the tips of their tails. Curiously, these tadpoles don't feed – they have enough yolk reserves to see them through from hatching to metamorphosis but the pitchers provide them with a secluded pond in which to grow. Other pitchers swarmed with mosquitoes, and all were packed with protozoans and bacteria. These microbes digest the leaves collected in the pitcher and, in turn, mosquito larvae feast on the burgeoning populations of microscopic organisms. The larvae excrete their nitrogen-rich waste into the water, which is then absorbed by the pitcher. By processing the leaves that drop into the pitcher, the microbes and mosquitoes provide up to half of the nitrogen the plant needs.[22]

On the other side of the world, in North America, there are yet more insect-trapping pitcher plants. These New World plants, *Sarracenia* spp., are not related in any way to the Old World *Nepenthes* pitcher plants of south-east Asia and Madagascar. Instead of dangling from a leaf tip, *Sarracenia* pitchers grow up from the ground, though they too have a lid over the pitcher mouth that both shields the pitcher from rain and secretes nectar to draw in victims. Some species grow in vast drifts,

Flask pitchers, *Nepenthes ampullaria*, catch leaves that fall from the canopy, but they can't digest their vegetarian diet without the help of insects.

creating eye-catching displays that no doubt serve to attract even more insects, just like fields full of flowers.

The white-topped pitcher, *S. leucophylla*, is one of the most dramatic, growing to nearly a metre in height and carpeting bogs in the south-eastern United States for hundreds of metres. In one bog I watched dozens of giant swallowtails, *Papillo cresphontes*, fluttering through these displays, just as they would over a flower meadow. They paused frequently to suck nectar from the underside of the lid, but being far too big to fall into the tall, narrow pitcher, they were not part of the plant's plan. Smaller insects alighting on the lid were in much greater danger. Downwards-pointing hairs on the underside of the lid make keeping a firm foothold very difficult and, preoccupied with feeding, many insects fall into the soup of digestive juices at the bottom of the pitcher. If that didn't make life precarious enough for a small insect, other dangers lurk in these traps. green lynx spiders, *Peucetia viridans*, concealed in green-striped camouflage, pick off insects that escape the deadly drop.

I'd noticed crab spiders (Thomisidae) hiding under the rims of *Nepenthes* pitchers in Borneo, also taking advantage of insects drawn to the conspicuous pitchers, but recent studies suggest that this too may be more of a two-way relationship. These spiders capture large flying insects that may have the aerial skills to avoid falling into the pitcher. Once they've had their fill, they drop the carcass into the pitcher, where what remains is digested and absorbed by the plant.[23] I wonder whether green lynx spiders, which hide on many of the pitchers of North American

plants, perform the same duties, and whether the large butterflies were more at risk than I surmised. In any case, the crab spiders in Borneo provide another example of an apparently one-sided relationship that turns out to be a true partnership.

Lurking spiders are not the only similarity between Old and New World pitcher plants. Empty the contents of a large *Nepenthes* pitcher into a collecting tray and you'll find a wriggling assortment of creatures, such as the mosquito larvae that help break down leaves in the pitchers of *N. ampullaria*. Fifty-nine different animals have been recorded in pitchers of this species, the highest recorded for any *Nepenthes*,[24] perhaps reflecting the plant's need for assistance in breaking down indigestible leaves – but the pitchers of other species all contain their own rich fauna.

There is a whole ecosystem in these miniature ponds, including predators. Many of the pitchers I examined contained the fat, sausage-like larvae of a predatory mosquito, *Toxorhynchites* sp., which specializes in devouring other mosquito larvae. By inserting a filming window into the sides of such pitchers we could watch the voracious behaviour of these larvae as they grabbed hold of other larvae nearly their own size and quickly chewed their way from one end to the other. It might seem that, by reducing the number of larvae helping to digest leaves, these predators are depriving the pitcher plant of nutrients. In fact, the opposite is true. *Toxorhynchites* digests other insect larvae that would not normally be eaten by the pitcher plant, so its excretions add yet more nutrients to the broth in the pitcher.[25]

While exploring the bogs of North America, I armed myself with a turkey baster to suck up the liquids inside *Sarracenia* pitchers, since it's not so easy to pour out the contents of a rooted pitcher plant. My

(*above*) The large predatory larvae of *Toxorhynchites* mosquitoes eat other larvae in the pitcher's pond – they'll even attack each other.

White-topped pitchers, *Sarracenia leucophylla*, form impressive stands in bogs across the southern United States.

improvised method worked well, and on closer examination the New World pitchers, like their Old World counterparts, turned out to contain thriving communities of insects, crustaceans and protozoans, though in even greater diversity than the *Nepenthes* communities. In a comparative study, some *Sarracenia* pitchers, such as the purple pitcher, *S. purpurea*, had well over twice as many species as *N. ampullaria*.[26] This may be because *Sarracenia* pitchers have less powerful enzyme systems and may therefore depend even more on the community of creatures that they shelter.

In both the Old and New World pitcher plants, the structure of these tiny aquatic communities is remarkably similar, although different species play the roles of filter feeders or predators in each case,[27] and some types of pitcher plant house species that are found nowhere else. Two mosquito larvae, *Wyeomyia smithii* and *Metriocnemus knabi*, known colloquially as the pitcher plant mosquito and the pitcher plant midge, only live in the ponds of the purple pitcher. The larger mosquito lives nearer the surface and tackles large prey as it falls in. The bits floating down are processed further by the midge, and the excretions of both feed the plant. It's just as well these insects work together in such an efficient way – the purple pitcher has hardly any enzymes of its own. The pitcher is mostly filled with pure rainwater, making it the perfect home for the wide range of creatures found there.

So, far from being simple predators, carnivorous plants have formed all kinds of intimate partnerships with insects, providing them with both board and lodging. One additional partnership that I haven't mentioned yet is one that is common to many other flowering plants. Carnivorous plants also need insects to pollinate their flowers, although, given their lifestyle, they must be careful not to eat their pollinators. Among other flowering plants, their relationships with pollinators provide limitless opportunities to study the many ways in which plants and insects work together.

PLANTS AND POLLINATORS

Although angiosperms possess many other evolutionary and ecological innovations, flowers are their most obvious attribute, and the intimate relationship between flowers and their pollinators is widely regarded as a major factor in the success of both groups. However, as we saw at the start of this chapter, the story is not this simple.

Before flowering plants arrived on the scene, the world was dominated by gymnosperms, which survive today as conifers and cycads, along with

The purple pitcher, *Sarracenia purpurea*, produces few digestive enzymes, so is dependent on insects living in the pitcher to break down drowned insects into a form the plant can absorb.

the peculiar gingko tree and the even stranger *Welwitschia* and its relatives (Gnetales). It has long been assumed that ancient gymnosperms, which – like flowering plants – produced copious quantities of pollen, relied on the wind for pollination. Then, so the story goes, the world blossomed for the first time. Insects seized on these new opportunities and, driven by coevolution, both plants and insects rapidly diversified.

However, long before flowering plants appeared on the scene, insects were being drawn to the nutritious pollen of gymnosperms.[28 29] Inevitably, a feeding insect would become dusted with pollen and so carry it to the next plant it visited, where some may be dusted off, so pollinating the plant. Pollen is precious since it's part of the plant's reproductive process, but ancient gymnosperms probably produced excess pollen as a reward for their pollination partners. Many modern cycads are still insect pollinated, so the rise to dominance of both insects and flowering plants can't simply be down to the forging of a successful new partnership. Such a relationship already existed among ancient gymnosperms and insects – indeed one extinct group of gymnosperms (Bennettitales) even evolved structures that bore some resemblance to flowers.

However, real flowers are different. It's likely that insects located gymnosperm pollen by scent, whereas flowers, which do also use scent, rely much more on conspicuous visual signals to draw pollinators. Flowers also reward their insect partners with nectar, a high-energy

flight fuel. So, all in all, it probably made sense for insects to switch their allegiance from gymnosperms to angiosperms. The structure of their pollen suggests that almost 90 per cent of the earliest fossil flowers were adapted to animal pollination, of which 40 per cent were general insect pollinators and nearly 30 per cent had already evolved a more specialized relationship with a narrower range of insects.[30]

The evolution of the angiosperms was undoubtedly a game changer for all terrestrial life, but – like so many pivotal moments in evolution – this new era began with what seemed like an inconsequential chance event. We can find the evidence of this ancient accident in the most primitive flowering plants alive today – a shrub growing on the remote island of New Caledonia to the east of Australia. *Amborella* is not going to win any prizes at a flower show. Its flowers are small and inconspicuous, but a detailed analysis of its genome has revealed something far more spectacular. Before they diversified, the earliest flowering plants doubled up their entire genome.[31] This meant that they had a whole load of spare genes, since the original set was more than enough to cope with normal life. These additional genes were therefore free to evolve and to create novel structures, of which flowers would eventually become the most obvious.

Darwin realized that the way in which plants and their pollinators have adapted to each other could provide strong support for the theory of evolution that he would eventually lay out in *The Origin of Species*. So, two decades before its publication in 1859, he began a detailed study of a wide range of flowers, both exotic specimens sent by collectors around the world and those commonly found around his home in Downe in Kent.

> During the summer of 1839... I was led to attend to the cross fertilisation of flowers by the aid of insects, from having come to the conclusion in my speculations on the origin of species, that crossing played an important part in keeping specific forms constant. I attended to the subject more or less during every subsequent summer.

In fact, by 1839 Darwin had largely figured out his ideas on natural selection, the mechanism by which evolution occurs, but was reluctant to publish such revolutionary ideas for fear of a backlash from the religious Victorian establishment. Instead, he spent decades seeking irrefutable confirmation, especially among plant and insect partnerships. These extensive studies eventually resulted in several books dealing with pollination, including a book on orchids published just three years after *The Origin of Species*,[32] and containing such meticulous studies that it

Peacock butterflies, *Aglais io*, are drawn to the copious nectar produced by hemp agrimony flowers.

drew yet more people to share his views on the mechanisms underlying evolutionary change.

One of the most extreme examples that Darwin examined concerned a beautiful orchid from Madagascar, the Madagascan comet orchid or Malagasy star orchid, *Angraecum sesquipedale*. The comet's tail is a spur some 30 centimetres long, running from the back of the flower and producing nectar only at its very far tip. No one had seen any insect capable of reaching this nectar, and therefore pollinating the plant in the process, but based on observations of butterfly orchids, with much shorter spurs, growing near his home, Darwin was confident enough to predict that there must be a moth, probably a hawkmoth, with a 30-centimetre tongue, flying around the Madagascan forest.

This mystery was picked up by Alfred Russel Wallace, who is something of an unsung hero in the story of our understanding of evolutionary processes. Unlike Darwin, a man of Victorian privilege, Wallace came from a working-class background and had to fund his travels around the world by gathering specimens and sending them back to England to be sold to collectors. He endured extraordinary hardships, not least when the ship on which he was travelling home after many years in South America caught fire in the middle of the Atlantic and sank, taking with it almost all of his specimens and notes – and forcing Wallace and his fellow voyagers to abandon ship. Undaunted, he then headed on to the Malay Archipelago, where, delirious from a malarial fever, he came up with the exact same mechanism of natural selection that Darwin had been toying with for many years.

The two men knew each other and corresponded, but the letter that landed on Darwin's desk in 1858 outlining Wallace's ideas shocked Darwin out of his decades-long prevarication. He quickly put together an outline of his ideas in what became *On the Origin of Species by Means of Natural Selection, or the Preservation of Favoured Races in the Struggle for Life*[33] (mercifully shortened to *The Origin of Species* in recent times). However, Darwin did the gentlemanly thing: his and Wallace's discoveries were jointly presented to the Linnean Society of London on 1 July 1858.

Wallace later followed up the comet orchid story in an essay called 'Creation by Law', in which he confidently illustrated the still undiscovered hawkmoth. Although Darwin felt that the moth looked more like a flying termite, he was impressed by Wallace's ideas. Wallace had already measured the tongue of a large hawkmoth from southern Africa called Morgan's sphinx moth, *Xanthopan morganii*, but this moth was unknown in Madagascar, and furthermore its tongue was a mere 18

The Madagascan comet orchid, *Angraecum sesquipedale*, only produces nectar at the very tip of its long spur. For a long time, its pollinator, with a tongue to match the spur, was unknown.

centimetres in length – nowhere near long enough to tackle the spur of a comet orchid. Despite this, Wallace and Darwin remained convinced that a similar moth, with an even longer tongue, must be the pollinator of the comet orchid. 'That such a moth exists in Madagascar may be safely predicted; and naturalists who visit that island should search for it with as much confidence as astronomers searched for the planet Neptune,—and they will be equally successful!'[34]

In 1903, their prediction was proved correct when a subspecies of *X. morgani* with a much longer tongue was found in Madagascar. It was named *X. m. praedicta* to celebrate the prescience of the two great Victorian naturalists. The adaptations of moth and orchid are each uniquely adapted to the other in such an extreme way that this example has often been used as the perfect illustration of coevolution. To effectively pick up or deposit pollen, the moth needs to push its head deep into the flower. So, from the orchid's point of view, it's best if the moth has to stretch into the flower as far as it can to collect the nectar. This favours plants with a spur slightly longer than the moth's tongue. The moth, though, soon adapts and evolves a longer tongue, forcing the orchid to evolve an even longer spur, and then the moth an even longer tongue. This arms race seems a good fit for the strict definition of coevolution.

Other factors, however, might be at play. One suggestion is that hawkmoths originally evolved long tongues so that they could feed from flowers without getting too close to them. Those crab spiders and lynx spiders that I found lurking in pitchers in south-east Asia and North America also hide in flowers, where some crab spiders are so well camouflaged that they are almost invisible until they pounce on a nectar-feeding insect. When moths turned up in the Madagascan forests with super-long tongues, evolved for feeding at a safe distance from a flower, the orchid may have switched to these strong flyers as new pollinators. But, for the moths to be effective, the orchid had to evolve a super-long spur to match the moth's tongue, in order to force the moth to insert its head deep enough into the flower to pick up pollen.[35] The process is one-sided, the orchid adapting to the moth but not the other way round. In other words, there is none of the tit for tat that marks true coevolution. This mechanism is now called the Pollinator Shift Model. Whether or not this is valid in the case of the comet orchid–hawkmoth partnership, it does seem to explain a number of other close relationships between insects and plants. Comet orchids perched high in rainforest trees and fast-flying moths are not easy to study, so the exact nature of their relationship remains a mystery, but elsewhere there are partnerships that

Spring flowers carpet the ground in Namaqualand in southern Africa. Several intriguing insect–plant relationships have evolved here.

are much more amenable to close scrutiny.

In South Africa there's another example of well-matched partners that is much less known, but is no less spectacular than that between the comet orchid and the hawkmoth – and even more informative. This region is blessed with a uniquely colourful variety of plants, from dazzling displays of desert annuals in areas such as Namaqualand in the Northern Cape, through a whole spectrum of spectacular bulbs, to the strangest of succulents. Not surprisingly, there is an equally wide assortment of insects buzzing around all these flowers. One group of flies (Nemestriidae) has evolved tongues almost as long in proportion to the body as that of Morgan's sphinx moth but, unlike the hawkmoth, these flies can't neatly roll their tongues under their heads when they're not in use. So, these one-and-a-half-centimetre-long flies are forced to fly with 8 centimetres of thin proboscis trailing behind them. They feed on a variety of flowers, some with short tubes, others with long ones, and both flowers and flies are much more common than the scarce comet orchid, which means that it's been possible to make more exhaustive observations on this partnership.

Flowers with shorter tubes, whose nectar is easily reached by a fly without it needing to insert its tongue fully, aren't pollinated so effectively. Long-tubed flowers force the fly to push right into the flower and in doing so it picks up much more pollen.[36] Under such pressure, it makes sense that the plant evolves a longer tube. Conversely, the flies drink

more nectar when they can reach it easily, which should put evolutionary pressure on the flies to evolve longer tongues. Each partner, in acting in its own interests, creates pressure on the other to make changes – exactly the scenario envisioned long ago by Darwin and Wallace, and that today we call coevolution.

As mentioned at the start of this chapter, if the specific partners vary across their range or if the selective pressures differ, then coevolution like this can produce new species and drive a growing diversity. That looks like a distinct possibility in the case of another bizarre South African insect and its plant partners.

The two dozen or so kinds of *Rediviva* bee are only found in this part of the world, and the females are instantly recognizable by their greatly elongated front legs, giving them their rather unimaginative common name of long-legged bees. They use their long legs to probe into the spurs behind the flowers of certain orchids and other kinds of flowering plants. However, they're most commonly seen buzzing around the flowers of a plant called twinspur, *Diascia* spp., equally unimaginatively named for two long spurs that arch from the back of the flower. With their two long front legs, *Rediviva* bees can insert a leg into each spur. This time

A long-legged bee, *Rediviva* sp., uses its long front legs to reach into the spurs of *Diascia* flowers to extract oils.

they're not after nectar or pollen but oil, which they mix with pollen as food for their larvae or use to line the walls of their burrows.

The mix of flower species varies across the bees' range, with the result that the bees encounter different lengths of flower spurs in different areas. When leg length in two species of bee was measured in many different parts of their range, it matched the lengths of the local flower spurs.[37] Further analysis showed that most of the variation in leg length was caused by selective pressures imposed by the plants, so fitting at least part of the definition for true coevolution.[38] It's easy to see how these geographical differences, if maintained for long enough, could result in the formation of new species and an increase in the diversity of both bees and flowers.

However, the natural world rarely makes life simple. The pattern seen in *Rediviva* bees and *Diascia* plants could also be explained by pollinator shifts, with the plant evolving to fit the pollinator but not vice versa. A complex statistical analysis of the interactions between fourteen bee species and thirty-four twinspurs (from records held at the South African Museum) showed that this might be exactly what is happening. Only around ten of the relationships seemed to be true co-speciation, with bee and plant evolving in step, whereas sixty to eighty could be accounted for by pollinator shifts.[39] In other words, the diversity of bees and twinflowers has arisen in several different ways. In most cases the diversity of twinspurs has been driven by the bees, but the flowers have had much less influence on bee diversity. Natural selection operates in complex environments and responds to a baffling array of factors. The lesson we are drawing from many of these plant–insect partnerships is that they do generate diversity, but in a whole series of different and often unexpected ways.

In all these cases, pollination is an accidental by-product of the pollinator's behaviour. As Darwin observed in the quotation at the head of this chapter, no organism can evolve behaviour or structures solely for the benefit of another organism. Even so, we now know of insects that do deliberately pollinate plants – a phenomenon known as active pollination. Darwin must be spinning in his grave. How could such a thing ever happen? One answer is revealed by a tiny wasp, without which we'd have no figgy pudding to overindulge in at Christmas.

MUTUALLY ASSURED SURVIVAL – THE CASE OF FIGS AND YUCCAS

Fig wasps (Agaonidae) are the only pollinators of figs. Insect and plant are entirely dependent on each other for their continued survival – no wasps, no figs, and no figs, no wasps. Figs have a peculiar, enclosed inflorescence called a syconium with a narrow entrance hole just big enough for a tiny female fig wasp to enter. A few female wasps may enter the syconium, but soon the hole closes over and seals the wasps inside. At this stage in the syconium's life only the female flowers are mature and the wasp's first job on entering the inflorescence is to deliberately pollinate the female flowers with pollen that she's carrying, which allows the seeds and fruits to start developing. Next, the female wasp lays eggs deep in the ovaries of the female flowers. The wasps hijack the development of some of the seeds and induce the fig to form a gall of nutritious tissue instead of a seed. This will sustain the larvae through their own development until they emerge as a new generation of wasps.

Wingless males emerge first and mate with females while they are still embedded in their gall. That's virtually all they do – though they do have one other critical task to perform. Before they die, they chew holes in the wall of the syconium through which the female wasps escape to find new figs. When the females emerge, the male flowers of the fig are ready and waiting so, before leaving, each female picks up a batch of pollen that she will use to stimulate fruit development when she enters a new

fig. The fig sacrifices a few seeds to feed its pollinators, and in return the wasps deliberately pollinate figs. There has to be a balance since the fig doesn't want to sacrifice too many seeds, but each syconium soon seals its entrance hole after wasps begin to enter, so the plant restricts the number of females allowed in and, in this way, is able to limit the damage.

The highly specific behaviour of fig wasps seems to have evolved just once, perhaps around 87 million years ago, and now there are 750 species of fig, many with their own dedicated species of agaonid wasp. It's very tempting to see this as another case in which two partners have speciated together, dancing in step to create a bewildering diversity of matched pairs – although, again, it may be more complicated than that. It's often assumed that each fig has its own highly tuned species of pollinator, but a closer look reveals that about a third of fig species employ several different types of wasp for the job of pollination. Sometimes a fig may have several different pollinators across its range, suggesting that the wasps are speciating faster than the figs, which is hardly surprising since the life cycle of a fig is a hundred times longer than that of the wasp.[40] Whatever the fine details, this special relationship has clearly spawned a profusion of both figs and wasps.

Walk through the fantastical landscapes of Joshua Tree National Park in California's Mojave Desert and you're wandering through a world created by a partnership with close similarities to that of figs and wasps. The crooked branches of the Joshua trees end in a spiky tuft of bayonet-like leaves, revealing them to be a species of yucca, *Yucca brevifolia*. Other yuccas, such as the Mojave yucca, *Y. schidigera*, form the shrub layer of this strange desert forest. In all there are around fifty yucca species growing across large parts of North and Central America, and each is dependent on a moth to perform the same task that wasps carry out for figs. These moths belong to one family (Prodoxidae), and the females have specially adapted mouth parts to carry a ball of yucca pollen with which they actively pollinate the female flowers of their particular yucca. Like the fig wasps, yucca moth larvae then develop in the flower, in this case feeding on the growing seeds.

In most cases one kind of yucca works with just one species of pollinating moth and the two are entirely dependent on each other, though again like the fig wasps, this rule is not always strictly followed. A few types of moth feed on up to six different yucca species, although it's thought that around 70 per cent are faithful to just one.[41]

In both cases a few renegade kinds of yucca moths and fig wasps cheat on the system. After all, these selfish partners are in it for their

The Mojave Desert in California, a landscape created by intimate partnerships between yuccas, such as Joshua trees, *Yucca brevifolia*, and yucca moths.

own gain, and if there's a way to secure the benefit without incurring the cost, then natural selection will find a way. Several species of both wasp and moth lay their eggs in the flowers of figs or yuccas without going to the bother of pollinating them, relying instead on the more diligent species to do that job. Their larvae then feed on the developing seeds. Any relationship, even those as close as wasps and figs, or moths and yuccas, is open to cheating by either side and, in fact, it's often the plants that have come up with the most ingenious ways of deceiving their partners.

SEX AND LIES

The simplest scam is not to bother producing expensive nectar, relying on the fact that other kinds of flowers in the neighbourhood are all offering sweet rewards. These deceitful plants look much the same as all the others, with colourful flowers, so insects visit them, probing deeply, trying to find nectar before they eventually leave disappointed. However, by then they will have probably picked up the plant's pollen. Orchids are the real masters of this form of trickery. It's thought that a third of all species of orchid don't produce any nectar at all, although a recent examination of some duplicitous Australian orchids discovered minute traces of nectar in species previously thought to offer no reward,[42] so they may be not so much cheats as teasers.

Food and sex are the two strongest drives in life, and if a plant can't fool its pollinators over food, then perhaps it can over sex. In the profusion of flowers colouring the South African countryside there's a kind of daisy that has evolved an interesting twist for its pollinators. With bright orange flowers, *Gorteria diffusa* grows in eye-catching displays often covering large areas, an unmissable signal to the insects that pollinate it. One of these is a kind of bee-fly, *Megapalpus capensis*, of which both sexes visit the flowers to sup nectar with their long probosces. Some forms of *Gorteria* flowers carry dark marks on some of their petals, which at first glance could be mistaken for a visiting insect. It's certainly enough to fool some of the male bee-flies, which stop drinking and become decidedly aroused. They attempt to mate with the flower spots and, in their frenzy, pick up more pollen than if they were simply feeding from the flower.[43] By adding a touch of sexual frisson to its flowers, *Gorteria* has enhanced its chances of pollination.

Again, however, it's the orchids that have taken this form of deceit to its highest (or perhaps lowest) levels. As far as we know, *Gorteria* is the only plant other than orchids that has stooped to such underhand measures, but sexually deceitful orchids are widespread in Australia and Europe, with a few examples in Central and South America and in South Africa.[44] However, given the prevalence of this trick among orchids, I imagine it's very likely that other examples will turn up elsewhere. *Gorteria* daisies show how this form of deceit may have first evolved, but orchids have turned it into a master class.

In Europe, almost all the sexually deceitful orchids belong to one genus, *Ophrys*, the bee orchids, named from the resemblance of the lower lip of the flower to an insect. That's no coincidence. They lure male wasps or bees and, depending on the species, the shape and texture of the lip helps the duped male to get a grip as he tries to mate with the flower. This orientates the male in such a way that he picks up the orchid's pollen.[45] But it's scent that really convinces a male that he's found a partner.

Fly orchids, *Ophrys insectifera*, are pollinated by small wasps that are fooled into thinking the insect-like flowers are female wasps.

Female bees and wasps release chemical signals to attract mates. These pheromones are usually complex mixtures of different chemicals, a unique blend for each species, so a male can be certain that he's tracking down a female of the right species. Well, almost certain. *Ophrys* orchids manufacture nearly identical chemical blends and each kind attracts a particular species of bee or wasp.[46] Quite a few Australian orchids play the same trick but use completely different sets of chemicals to draw in their victims, in this case thynnid wasps.

Among the 350 species of Australian spider orchids, *Caladenia* spp., there are three different ways of attracting pollinators – nectar rewards, food deception and sexual deceit. Just as for *Ophrys* orchids, different kinds of spider orchid dupe different wasp species, and a male trying to mate with the orchid flower ends up with the orchid's pollen mass glued to his body, in just the right position to brush the female part of the flower when the unfortunate insect is fooled again. However, a few

Australian orchids make absolutely certain that pollen is picked up by exploiting a peculiarity of the biology of their pollinators.

Male thynnid wasps are fully winged, but females are wingless and look more like ants, so much so that the common name of one widespread species, *Diamma bicolor*, is blue ant. To get around, a female blue ant depends on a male to give her a lift. Drawn to an unmated female by her scent, the male lands on top of her, then, like a heavy-duty helicopter, lifts off and carries her to a flower to feed. Several orchids have exploited this behaviour in a most unexpected way. They have evolved complex hinge joints as part of their flowers, below which the flower both looks and smells like a female thynnid wasp. When a male obligingly lands on the flower and tries to lift off with the 'female', the flower simply rotates around the hinge, causing the male to crash into the upper part of the flower where the pollen is stored. Undeterred, the male tries again and again, hammering into the orchid's anthers until he finally gives up and flies off carrying the orchid's pollen.

These orchids, among them the flying duck, *Paracaleana nigrita*, hammer orchid, *Drakaea glyptodon*, large duck orchid, *Caleana major*, and the fifteen or so species of elbow orchid, *Arthrochilus* spp., have the weirdest pollination mechanism that I've ever seen; it's impossible not to smile as a poor male insect is repeatedly flung against the orchid's anthers and stigma, though, of course, for the wasp it's no laughing matter.

The evolution of these close associations, of one orchid with just one species of insect, has led some to conclude that sexual deceit has driven speciation and an increase in diversity in these orchids and perhaps in their pollinators. Yet again, though, it probably isn't that simple. Closer observations reveal that this neat one-to-one relationship doesn't always hold up. Despite using species-specific scent lures, a single kind of sexually deceptive orchid may attract several kinds of wasp or bee. Moreover, the diversity of *Ophrys* orchids may not be quite as impressive as it appears at first sight. Although some botanists recognize several hundred species of these orchids, others see most of this apparent diversity as variations in a much smaller number of species. Even so, the impressive number of sexually deceitful orchids around the world suggests that their rather offbeat approach to pollination has played a part in their success.

These deceitful relationships are often with wasps or bees, but they're not the only insects to be frustrated by orchids. There are some 400 species of *Pterostylis* orchids scattered across the Australasian region that have touch-sensitive flowers. They are pollinated by flies, which trigger a floral trap when they land on the lower lip of the orchid. The

Cobra lilies, such as *Arisaema consanguineum*, smell of decomposing corpses, which draws scavenging beetles and similar insects that are busy searching for such places to lay their eggs.

imprisoned flies pick up the orchid's pollen as they struggle through the single escape route provided by the flower. The flowers offer no nectar reward, and it's not clear what draws the flies to this humiliating encounter. However, in one species, *P. sanguinea*, which attracts tiny fungus gnats (Mycetophilidae) as pollinators, the gnats have been seen trying to copulate with the flower, which suggests that here again, sexual deceit is at work.[47]

Many other orchids deceive male flies, some with both scent and tantalizing images of female flies. The flowers of one South American orchid, *Telipogon peruvianus*, are bordered in bright yellow, while the centre is dark red. In between is a white area crossed by narrow striations of red. To our eyes, it's a flower of exquisite beauty, but to a male tachinid fly's eyes it looks more like a female tachinid feeding on a yellow flower – and is therefore well worth investigating. The orchid's flower is all the more convincing since it releases the same scent that female tachinids use to attract males.[48] Completely irresistible.

If bees, wasps and flies all fall for this deception, why not other insects? In fact, most of these poor hoodwinked males are either bees and wasps (Hymenoptera) or flies (Diptera), although recently an orchid from South Africa, *Disa forficaria*, has been caught in flagrante delicto with a longhorn beetle. It's clearly a case of sexual deceit since the beetle was seen to ejaculate onto the flower as well as picking up the orchid's pollen in its moment of bliss.[49] It's not surprising that this story had

been missed until now. The orchid is vanishingly rare; only eleven plants have been found in the past two centuries.[50] Until a plant turned up in 2018, it hadn't been seen since 1966, and that plant promptly disappeared again in 2019. But its devious relationship with longhorn beetles gave scientists an unusual tool to search for other specimens of this orchid. They had isolated and identified the floral scent that proved so seductive to male beetles and in testing various versions of this in 2020, they managed to attract three beetles, all of which were carrying pollen from *D. forficaria*. Obviously, the plant still existed somewhere close by, although the beetles were far better equipped to find it than botanists.[51]

The chemical mimicry of orchids knows no bounds. In South Africa one species, *Satyrium pumilum*, due to a set of chemical compounds identical to those produced by rotting flesh, smells an awful lot like the decomposing body of an animal. The smell is enough to fool flesh flies, which take it to be a source of rotting food for their larvae.[52] Although sexual deceit seems rare outside of the orchids, this kind of subterfuge – the promise of somewhere that an insect can lay its eggs – is commonly used among a much wider variety of plants.

The tallest inflorescence (group of flowers) in the world does exactly this. The flower spike of *Amorphophallus gigas*, sometimes called the giant voodoo lily, grows to the astonishing height of 4 metres. Perched on top is a fleshy vase from which protrudes the structure that gives the genus its name. *Amorphophallus* means 'misshapen penis' and, without going into unnecessary detail, it's a pretty accurate description. Apart from its off-putting appearance it also produces an off-putting smell of rotting flesh. An even bulkier inflorescence sprouts from the titan arum, *A. titanopsis*. It may reach over 3 metres in height but grows directly out of the ground, making it the biggest inflorescence in the world. It too produces nauseating waves of scent that waft through the night-time rainforest. For good reason, the locals call it *bunga bangkai*, the corpse flower. I grow much smaller kinds of *Amorphophallus* in my garden, along with the related *Arisaema* and *Arum*, and all produce unpleasant smells when they flower – so much so that when other spring gardens are heavy with sweet scents, mine smells as though several neighbourhood cats have died in the flower beds.

All of these plants belong to the aroid family (Araceae), and all follow a similar pattern in their pollination. *Amorphophallus* species largely attract carrion beetles, though other aroids draw in flies or even bees. The flowers are arranged on the lower part of a central spike and enclosed by the vase-like structure. The female flowers mature first, when the

plant is at its smelliest, and the beetles crawl down into the vase. They squeeze past the female flowers and, if the beetles are carrying pollen, this gets brushed off and pollinates the flower. At this point, the plant seals the top of the vase and imprisons the beetles. Now, the male flowers mature and release their pollen, and late the next day, when the beetles are released, they must brush past the male flowers to escape, so picking up more pollen.

The world's biggest single flower, *Rafflesia arnoldii*, also stinks. This huge red flower erupts from an underground stem, the only part of the plant to break cover. Otherwise, it grows as a parasite on the roots of a rainforest vine. The flower, maybe a metre across, opens from an enormous bud, like a giant cabbage, which erupts onto the forest floor and – once mature – draws in hundreds of flies. It's a curious fact that the world's largest inflorescence and the world's largest flower are both carrion mimics. Although not world beaters, the flowers of many other carrion mimics are also huge, which raises the intriguing possibility that being big somehow helps to convince pollinators that the flower really is a rotting corpse.

One possibility is that such flowers are attempting to draw in only those insects that prefer big carcasses. Attracting a smaller range of

Rafflesia keithii growing in a rainforest in Sabah. Rafflesias are the world's largest flowers, and both look and smell like a lump of rotting meat.

pollinators that are more likely to visit only other giant flowers may give the plants better odds that their pollen will end up on a plant of the same species. Many carrion-mimicking flowers also generate heat. This strange phenomenon serves several purposes. It helps to vaporize the foul scent of the plant and make it smellier, and it fools insects into thinking they have found a nice, warm rotting carcass that will make a cosy home for their larvae. The temperature that a flower can reach depends on its size, so giant flowers can heat up more than smaller ones.[53]

As much as my garden smells pretty rank in spring, so too does the house. The culprits this time are an entirely unrelated group of plants, the stapeliads. These are fleshy succulents from South Africa and they produce a huge bloom compared to their size, a fleshy red thing that bears a close resemblance to a chunk of meat – and it certainly smells like rotting meat. Colour, texture and smell are all enough to fool our local flies, and I often see them depositing eggs on the petals of the flower.

Once pollination is complete, plants even persuade insects to disperse and plant their seeds. Plants from a wide range of families produce seeds with a fleshy cap, an elaiosome, which contains a whole mixture of valuable nutrients. When ants find these seeds, they carry them back to their nest, bite off the elaiosome and feed it to their larvae. The seeds are discarded but now lie safely planted in the ant nest. Ants are everywhere, so they've become frequent partners with plants, often with dramatic effects.

ANTS AND PLANTS

We've already encountered the fruitful association between ants and the fanged pitcher. One of the many advantages of these ant guests in this relationship is that they prevent leaf-eating beetles from nibbling on the plant. Ants are generally aggressive and armed with either stings or powerful jaws, and often both, so if a plant needs a defence force, there's nothing better than recruiting the neighbourhood ants. Some plants grow extrafloral nectaries to feed their army but others, like the fanged pitcher, go to the trouble of providing homes as well.

In both Africa and Central America, bull's horn acacias develop bulbous bases to the many spines that cover them. Ants can chew holes in these and inside each spine there's enough space to house part of their colony. In return for food and lodging the ants will take on any comers. The long, sharp spines of acacias alone deter many herbivores but, in Africa, giraffes seem oblivious to this threat. Their tough, leathery

The flowers of stapeliads are such convincing mimics of rotting flesh that flies often lay their eggs on the petals.

tongues can strip bushels of leaves from a tree – unless it's defended by ants. When such a tree is shaken, the ants stream out of every spine and swarm over the offending giraffe's tongue and face. Never mind a size difference that amounts to several hundred thousand times, the giraffe soon gives up and heads off to find a less well-defended meal.

Sometimes this relationship grows even closer. Many Central American acacias have teamed up with ants of the genus *Pseudomyrmex* and not only provide their ant infantry with thorns in which to live and nectaries on the leaves, but also detachable structures, Beltian bodies, on the leaf tips that contain high-quality ant food. Some of these partners are now entirely dependent on each other. The ants need the acacias for food and shelter, and without ants an acacia is soon stripped bare and dies.[54]

Cecropia trees are also very common in Central and South America, growing along riverbanks or sprouting up in areas where the vegetation has been cleared. As pioneer species, *Cecropia* trees are involved in a mad scramble for light in a newly opened patch of forest, competing with other small trees and shrubs as well as vines. In this battle, they are aided by ants of the genus *Azteca*. *Cecropia* stems are partly hollow to provide living space for the ants, and new stems also grow a spongy tissue that the ants can either eat or use to build structures within the hollow stems. Where a leaf meets a stem, another specialized tissue produces small white Mullerian bodies that resemble ant eggs and fit neatly into ants' jaws. These bodies are packed with nutritious glycogen. This is often called animal starch since plants normally make a different form of starch, but *Cecropia* makes glycogen because the ants find this easier to digest.

When a queen ant first bites through into the hollow stem to start a colony, she also inoculates the stem walls with a special fungus she has carried with her from her own nest. Once this fungus is established, she'll use it to feed her first larvae; later, her workers will use it to feed the ever-expanding family. The plant offers the ants everything they might need, and in return the ants defend the plant from attack. If any leaves are damaged, the plant releases an alarm substance to which a passing ant reacts. It runs around, quickly checking the area for the culprit, then lays a trail back to the main stem, to the nearest entrance hole. Dozens of workers stream out and follow the trail to scour the area and deal with anything that has had the temerity to munch on a *Cecropia* leaf.

If it's a big intruder, like a giant katydid, the tiny ants still have the upper hand. The plant provides a carpet of hairs matched to the size of the ants' claws, which gives them a firm foothold and enough leverage to

Many acacias have bulbous bases to their spines, which can be hollowed out by ants to create homes all over the tree. In return for lodgings, the ants defend the tree against even the biggest grazers.

tackle the biggest leaf nibblers. The *Azteca* army tackles plant invaders too. Any vines that try to scramble over *Cecropia* in the race for light are snipped off by the ants. A proportion of the food eaten by the ants also ends up as nitrogen in the plant tissues, presumably from the refuse heaps that accumulate inside the stem, so *Cecropia* can grow quickly and free from competition until it prevails in the race for light. As with some of the acacias, this relationship has grown so tight that some kinds of *Azteca* ant are found nowhere else but on *Cecropia* trees.

Azteca ant colonies help *Cecropia* succeed in the dynamic world of forest gaps, but another kind of ant shapes the whole structure of large patches of forest. Walking through the Amazon rainforest takes you through a bewildering variety of trees, up to a hundred different kinds in each hectare. However, occasionally you'll come across a patch, maybe hundreds of square metres in extent, dominated by just a single species of tree. Just which species depends on exactly where you are in this vast rainforest, but across a large area it's a tree called *Duroia hirsuta*. Locals believe these patches are cultivated by an evil forest spirit and they were long known as *Devil's gardens*. It wasn't until 2005 that scientists found the true cultivator, an ant called *Myrmelachista schumanni*.

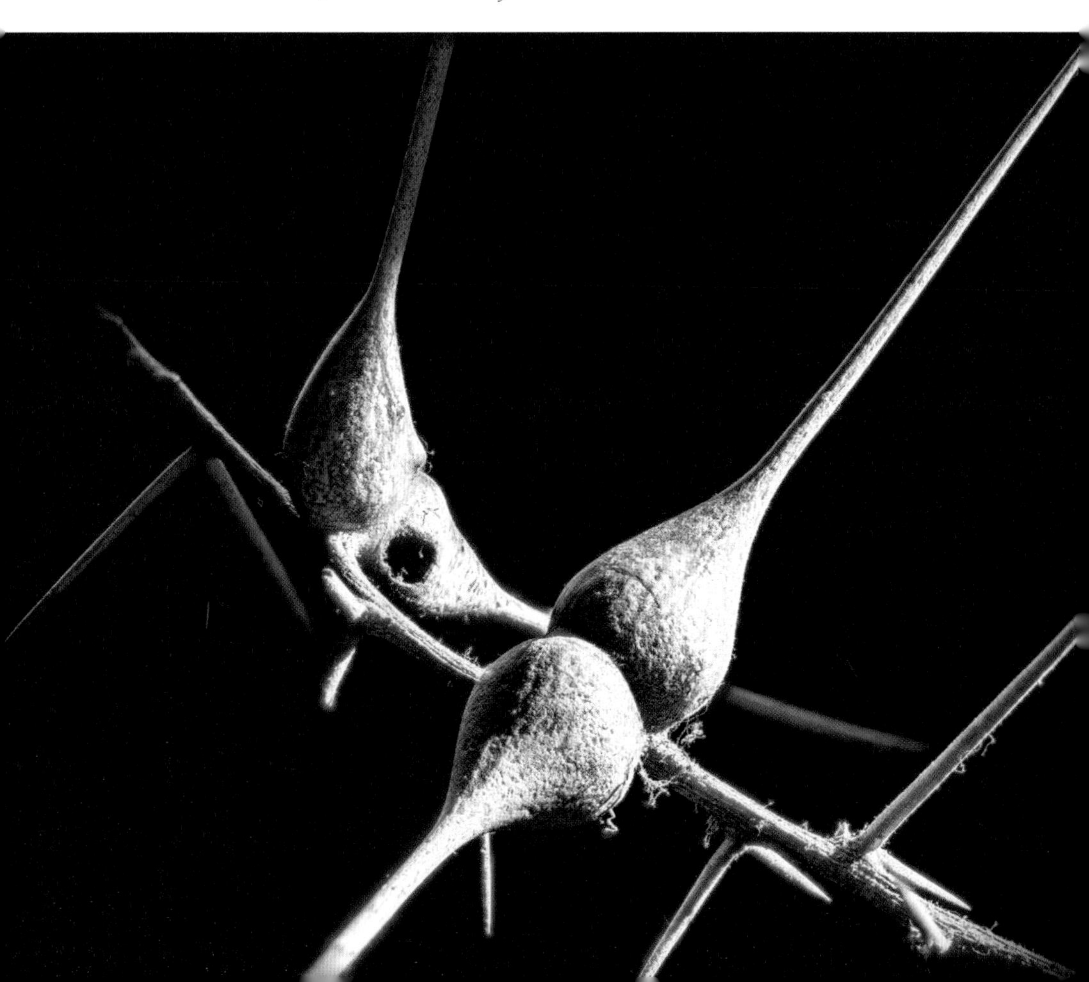

These ants live in the stems of their host, which – like many of the other examples we've explored – provides the ants with bed and breakfast. In the Amazon, it's often rare to find two specimens of the same kind of tree close to each other, but that's of no help to the ants. They prefer to live in *Duroia* trees, so the only way to expand their colony is to encourage more *Duroia* trees to grow close together. The ants do this by poisoning every other kind of plant that tries to grow here – spraying or injecting formic acid* into seedlings. This action constitutes the first known instance of ants using their chemical weapons as a herbicide. The result is a monoculture of their favoured tree that might extend to a patch of 600 trees, allowing the ants to build a vast metropolis. Some colonies comprise as many as 3 million workers and 15,000 queens. With a workforce of that size the ants can maintain their gardens for long periods of time; some of these ant colonies and their glades may be at least 800 years old.[55]

The success of the ants' strategy of forest management has recently been underlined by the discovery that Devil's gardens may be more widespread than was previously thought. Two thousand kilometres from the nearest examples in the western Amazon, in the cloud forests of French Guiana, *Myrmelachista* ants cultivate gardens of a different kind of tree, *Cordia nodosa*. Curiously, they only create gardens above 500 metres, even though the tree occurs below this altitude as well. Clearly, there is still a great deal we don't understand about how this relationship works.[56] That's equally true of the countless other partnerships that exist between flowering plants and insects. Only a few such relationships have been studied in real detail; it is certain that there are many more intriguing partnerships hidden out there, in places that we have yet to even see.

It would be hard to overestimate the importance of these relationships in the success of both insects and flowering plants. In the past, when it was believed that these relationships grew and blossomed in the Cretaceous period, the process became known as the Cretaceous Terrestrial Revolution. Now that we know the process began earlier than the Cretaceous and probably peaked in the Palaeogene period, after that giant asteroid brought the Cretaceous to a dramatic close, it is perhaps better termed the Angiosperm Terrestrial Revolution (ATR),[57] and a real revolution it was.

It might seem reasonable to assume that Earth's biodiversity is equally

* The name derives from *formica*, the Latin word for 'ant'.

distributed between the oceans and the land (including freshwater bodies), or perhaps even biased in favour of the marine world since life began in the sea, and oceans now cover 70 per cent of the planet. In the past the split did seem more equitable – until flowers appeared. Now terrestrial biodiversity is ten times greater than that in the oceans, in large part the result of all the partnerships that became possible with these new types of plants.[58] Of course, many different animals are involved in relationships with flowering plants, but insects predominate. Their partnerships with flowering plants have transformed the ecology of the planet in fundamental ways. As a child of the Sixties, it's gratifying to realize that flower power really did revolutionize the world.

Coevolution between insects and flowering plants is often cited as the reason for the success of both groups and it's likely that coevolution, as strictly defined, has been at play in some of the cases we've explored. However, this generalization hides a whole suite of partnerships that work in very different ways. In the course of their evolution, some insects and flowers have even become embroiled in such tight relationships that now neither partner can survive without the other. Yet in all these cases, each partner is working solely out of self-interest. These are all selfish partners – and Darwin can rest easy in his grave.

7

The Mating Game

How insects make more insects... and lots more insects

Then there is a fly, which looks something like our house fly. These settle on the beach to eat the worms that wash ashore—and any time, you can see there a belt of flies an inch deep and six feet wide, and this belt extends clear around the lake—a belt of flies one hundred miles long. If you throw a stone among them, they swarm up so thick that they look dense, like a cloud.

Mark Twain, *Roughing It*, 1872[1]

There's no doubt that insects are very good at making more insects. If it weren't for predators, parasites and parasitoids, many of which are themselves insects, we'd soon be knee-deep in them. A common housefly, for example, laying a couple of hundred eggs could, in theory, leave 2 million descendants in just two generations, and in not many more generations could have produced enough offspring to cover the surface of the Earth in a solid layer of flies. Luckily, insects are as good at dying as they are at reproducing, but even so there are times and places where circumstances allow insects to demonstrate their immense powers of multiplication.

SWARMS

One such place is Mono Lake, perched high in California's Sierra Nevada Mountains. Very roughly circular and about 20 kilometres across, it is fed by streams from the mountains that surround it. However, the water has no outlet other than by evaporation under the hot Californian sun, so over time mineral salts in the water have become steadily more concentrated. Mono Lake is now a soda lake, with such high concentrations of sodium carbonate that it is three times saltier than the Pacific and, at pH10, dangerously alkaline.[2] Calcium flushed into the lake from the surrounding rocks reacts with dissolved minerals to produce calcium carbonate that grows as tufa into towering sculptures around the lake, creating a strangely alien landscape.

Not much can live in these alkaline waters apart from brine shrimps and the larvae of a kind of brine fly called the alkali fly, *Ephydra hians*. With

(*previous page*) A pair of large red damselflies, *Pyrrhosoma nymphula*, lay their eggs. The female carefully inserts each egg into plant tissue, where it'll be safe until it hatches.

few competitors and predators, both species exist in truly astronomical numbers. As adults, the flies emerge from the lake and carpet the shoreline in the black slicks encountered by Mark Twain in the quote at the start of this chapter, taken from his book *Roughing It*, which describes his time spent travelling around post-gold rush California in the 1860s. Even today, walk along the shores of Mono Lake and the flies rise in thick black clouds to knee height. It's like wading through a dark swirling mist. The adaptability of insects allows them to make homes in the harshest of places, such as soda lakes. It's a tough life here but worth it since they can build up such extraordinary numbers. At the height of summer, perhaps 100 million flies buzz around the lake's shore.

There's not much for all those flies to eat around these barren shores, though beneath the surface, algae also thrive in the alkaline water. These algae fed them as aquatic larvae and now the adult flies have no option but to crawl back below the surface to rasp at algae-coated rocks. Ever observant, Mark Twain described the incongruous sight of flies happily submerging themselves.

> You can hold them under water as long as you please—they do not mind it—they are only proud of it. When you let them go, they pop up to the surface as dry as a patent office report, and walk off as unconcernedly as if they had been educated especially with a view to affording instructive entertainment to man in that particular way.[3]

The flies are equipped with tiny hairs that trap a bubble of air around their bodies as they submerge. Beneath the water, they look like little silver beads as they potter over the rocks – amazing to watch, not least

Brine flies, *Ephydra hians*, swarm around the shores of Mono Lake in California.

because it is such an impressive feat. The high concentrations of sodium carbonate in the lake make objects submerged in it more 'wettable' than would be the case in water without such high levels of salinity; any ordinary fly would quickly drown if it tried to swim in these waters. However, brine flies are much hairier than everyday flies – a third more hairy according to the scientists who have bothered to count their hairs.[4] When they have finished feeding, they crawl back out of the water and turn from silver back to black as the bubble bursts, leaving them, as Mark Twain so wittily observed, bone dry.

Yet even here, the flies don't have it all their own way. California gulls nest on islands in the lake and they have an enormous appetite for brine flies. They trawl for the flies by running through the swarms, beak agape, snapping wildly to the left and right. The flies are so dense that the gulls are barely visible through the cloud, and they can't fail to grab a beakful every time they snap their bills shut. In late summer, the flies also face migrants arriving from the north. Easy and abundant protein on tap makes Mono Lake a key refuelling stop for birds heading south for the winter. Beginning in late summer, around 80,000 Wilson's phalaropes, delicate wading birds, call in at Mono Lake to moult, and feast on flies and shrimps. By October, three-quarters of a million eared grebes have also joined the banquet. Non-breeding grebes hang around Mono Lake all year, feeding largely on brine flies in the early summer, but switching to brine shrimps as birds that have finished breeding pile on to the lake in the autumn. These huge flocks of birds devour 60–100 tons of brine shrimps every day at this all-you-can-eat buffet.[5] Despite such gargantuan feasting, both flies and shrimps still swarm in their millions.

There are similar hypersaline lakes elsewhere, colonized by different brine flies, though in equally impressive numbers. The tiny island of Laysan lies about halfway along the Hawaiian archipelago, a world away from the incessant throng of Honolulu, more than 1,000 kilometres away to the east. At its centre is a large oval lagoon made hypersaline by high levels of nitrates and phosphates. These have been washed in from the guano produced by generations of Laysan albatrosses that nest in this remote spot. It couldn't be more different from Mono Lake in its lofty mountain valley. Laysan rises to a maximum height of around 12 metres, yet its lagoon is also teeming with brine shrimps and another kind of brine fly, *Neoscatella sexnotata*.[6]

They too attract the attention of the local birds, in this case a vanishingly rare duck, the Laysan duck, found nowhere else but on this atoll. The ducks catch flies in the same way as the gulls, running through

Cluster flies, *Pollenia* sp., get their name from their habit of hibernating together in large numbers, but they sometimes also congregate on flowers in the spring – a good chance to find a mate.

the dense swarms, head low, snapping in every direction – but being a duck, the sight is a lot more comical. Yet in neither case do the predators seem to make much of an impact, so astronomical are the numbers of flies and so good are they at maintaining these impressive populations.

Hypersaline lakes lie scattered across the western half of the United States – places such as Lake Abert in Oregon and the Great Salt Lake in Utah, which have their own blankets of brine flies. Elsewhere, other types of flies can sometimes reach equally unimaginable numbers. In some years, the towns around Lake Winnebago in Wisconsin are invaded by lake flies. These are mosquitoes that belong to the family Chironomidae, which, luckily for the good folk of Oshkosh or Lake View, are a family of non-biters, feeding instead on nectar. They spend their early lives on the lake bed as worm-like larvae, often called bloodworms or redworms. They get their common names from their colour, created by high concentrations of haemoglobin in their bodies, which helps them extract vital oxygen from oxygen-poor mud. They swarm through the deep lake sediments in extraordinary numbers. One survey found that nearly half of all the creatures down there were redworms.[7]

Lake Winnebago has a long tradition of winter spearfishing for sturgeon, a major source of income for towns around the lake, and lake fly larvae are a major source of food for these impressive fish. The more redworms there are, the fatter the sturgeon.[8] Even so, due to the

prodigious ability of insects to make more insects, there are enough larvae left over to form swarms of adults so thick that, at the start of May, it looks as though a mist has descended on the local towns. Just which towns are affected and how badly depends on which way the wind blows the swarms, but the flies are sometimes dense enough to drive residents away for the ten days or so of an adult lake fly's life. The reality is, however, that their local winter economy depends on these flies, and they should also be grateful that they don't live on the shores of Lake Victoria in Africa, where lake flies put on an even more remarkable show.

Swarms here drift over the water like towering plumes of smoke visible for miles across the lake. There are some chironomids among these multitudes. although the majority are midges from a different family, Chaoboridae. These are often called phantom midges since their larvae are as transparent as glass, apart from silvery air sacs that they use to adjust their buoyancy so they can float effortlessly, suspended in the water column above the legions of bloodworms below. Similar swarms fly above Lake Malawi,* and smaller numbers fly over Lake Tanganyika. When one of these clouds floats across the shoreline of Lake Malawi, the locals don't flee in terror; they run for their baskets, attached to long poles, which they sweep through the swarms to collect as many flies as they can. Crushed, boiled and shaped into patties, they are dried in the sun before being eaten. Around Lake Malawi these fly burgers are called *kungu* and for some reason they have a hint of caviar in their flavour.

Plenty of animals around these African lakes also relish lake flies. As in Lake Winnebago, fish devour the larvae, while birds pour in to feast on the aerial swarms. Yet, the fact that there are still enough flies to form choking clouds is testament to the extraordinary reproductive powers of insects. For most insects, that process of creating the next generation begins when two insects manage to find each other. Often that's no easy job. Insects are tiny creatures living in a very big world, so even just finding a partner can be fraught with difficulties. However, insects have come up with all kinds of ingenious ways to overcome these problems.

One simple way is to swarm in such abundance that finding a partner is all but inevitable. That's what lake flies around the world are doing. They synchronize their emergence as adults so that they are all on the wing at the same time. It's a strategy also used to spectacular effect among other river and lake insects. Each species of mayfly often emerges

* This huge lake in Africa's Great Rift Valley is bordered by several countries and is known as Lake Malawi in Malawi, Lake Nyasa in Tanzania and Lago Niassa in Mozambique.

in close synchrony, known to fly-fishermen as a hatch. It's important for these fishermen to know which species emerge when, since trout will focus on whichever species is abundant at the time with such intensity that they won't feed on other species. So, in the often hopeless task of trying to out-think a fish, a fly-fisherman's flies have to be close mimics of the mayflies of the moment. Driven by the fickle minds of fish, fly-tying has become something of an art form. However, on a few rivers, even the best fisherman's flies would go unnoticed – so abundant are the mayflies.

Ephoron virgo occurs in rivers across Europe and North Africa and, for a few days during August, it emerges in vast clouds. From the Danube in Hungary to the Ebro in Spain, mayfly blizzards rise from the rivers at dusk. On the Tisza river in Hungary the long-tailed mayfly, *Palingenia longicauda*, Europe's largest species, measuring 10 centimetres from its head to the end of its long tail filaments, also forms vast if short-lived swarms that rise from the river in the afternoon. Across the Atlantic yet another species, *Hexagenia bilineata*, swarms over the Upper Mississippi river, while clouds of the closely related *H. limbata* fill the air over the western Great Lakes. Some of these swarms have been estimated to contain nearly 90 billion insects.

Adult mayflies live for only a day, some for just a matter of hours, and their only job is to find a mate and, in the case of females, to lay eggs. Male mayflies are equipped with long front legs that they deploy to hook a female from the swarm. Their task is aided by peculiar upwards pointing eyes that detect any females entering the swarm above them. After mating, a male's purpose in life is complete and he dies. The female has still to lay her eggs, but before she does so, she makes a flight upstream. The larvae live for several years in the river, during which time they are likely to have drifted downstream, so a female must fly upstream to compensate, at the end of which she will simply crash onto the water's surface, where her last act is to release her eggs. Even as darkness falls, the river guides the mayflies on their journey. They can see polarized light reflected from the water's surface, so the river stands out as a bright ribbon snaking across the land. Unfortunately, there are other beacons of polarized brightness in the modern landscape.

Bridges cross all these rivers at frequent intervals, obscuring the mayflies' view upstream. Worse, many of the bridges are lit at night and the glare draws the mayflies from their vital task as they form towering, disorientated columns above the bright lights. If this isn't enough of a distraction, the asphalt road surface reflects polarized light in exactly the same way as the river surface, causing the females to crash onto the road and waste all their eggs.[9] After such a swarm, the road surface is slick with dead and dying mayflies. In Lacrosse, Wisconsin, on the Upper Mississippi, I've seen mayfly bodies piled so high along riverside roads and around nearby gas stations and parking lots that town officials had to mobilize the snow plough fleet to clear them. Uncountable numbers of mayflies die around these light traps and not one of them has been able to contribute to future generations. Not surprisingly, the size of these spectacular mating swarms is now declining.

I was lucky enough to witness these swarms several decades ago, but times have changed. A study published in 2020 by scientists from the University of Notre Dame in Indiana has shown that the swarms on the Upper Mississippi have declined by 50 per cent since 2012, while those around Lake Erie have collapsed by more than 80 per cent.[10] Artificial illumination is not the only culprit. Pesticides from surrounding farmland and climate change all have their effects on mayfly populations, but scientists from the Danube Research Institute in Hungary have found a way of at least mitigating the effects of artificial lights. By attaching bright LED lights to bridges and pointing them downwards towards the river, they were able to guide the females back to the water's surface. This simple trick was so effective that many fewer females left the river.[11]

Swarming is a common way to ensure that males and females find each other. Often the swarms are composed of male flies that hold station in a towering cloud over a landscape feature, such as a tree or building, known as a swarm marker. These swarms are conspicuous and easy for females to locate. The swarms of St Mark's flies, *Bibio marci*, are so obvious on early spring days in Britain that they often draw our attention as well. These flies get their name from the fact that their brief appearance in the sky falls around St Mark's Day on 25 April. They belong to a family of flies (Bibionidae) famous for its mating swarms.

In the same family, lovebugs, *Plecia nearctica*, make more of an impact on human lives than St Mark's flies. For a few weeks in spring and autumn they can become a nuisance in the states bordering the Gulf of Mexico. Males dance in enormous swarms and females drawn to them have their choice of partner.[12] Once paired, the aptly named lovebugs remain in an extended copulation for most of the rest of their adult lives. They fly like this, coupled tail to tail, still in huge swarms. If they cross highways, they spatter windscreens in such numbers that they create a driving hazard. In some years these swarms of mating insects, sometimes also called honeymoon bugs, have been recorded flying across a quarter of the state of Florida, and airline pilots reported them reaching heights of 500 metres.

Plecia nearctica is common in Central America but was only recorded in the United States in 1940, in Texas and Louisiana. However, in the last few decades of the twentieth century, lovebugs spread rapidly around the Gulf and became startlingly abundant. This sudden appearance of swarms of paired insects spurred an interesting conspiracy theory – that lovebugs were created by genetic engineers at the University of Florida.[13] The truth is less likely to make an episode of the *X-Files*. Swarms are often recorded out over the open ocean, blown by the wind, so their spread is probably a natural migration.

The vast majority of insects lead more hidden lives than these swarming fly species, and for them finding a mate is much more difficult. They need to find ways of broadcasting their location, of evolving long-distance signals that unambiguously declare their species and their intent. There are many ways of doing this – using sound, scent or light – and different insects have exploited all of these, often in ingenious ways.

LOVE SONGS

A peaceful summer meadow, filled with the relaxing chirps of grasshoppers and crickets, is anything but if you're a grasshopper. All these songs are urgent pleas by males for a mate before the summer ends. Each species sings a different tune, so it's possible to lie on your back in the grass and do a species survey with your eyes closed. At least, that's true in Britain, which has only around thirty species. Cross the Channel to Europe and you will find close to 700 species. Here, some songs are so similar that it takes a sonogram, a visual representation of the song, to tell them apart, although apparently the females' ears are good enough to do the job without the aid of technology.

Grasshoppers, crickets and katydids (bush crickets) are all related as members of the order Orthoptera and most of them sing, although their method of producing songs is different in each group. Grasshoppers have a series of pegs along the inner face of the enlarged femur of each rear leg. These powerful legs evolved for jumping, but with typical insect economy have also been co-opted to make music. The grasshopper scrapes its legs over a rigid area of its forewing, which at rest lies folded along its back, like scraping a bow over a violin string, an action known as stridulation. Different arrangements of pegs and different choreographies produce different songs.

In full song, a grasshopper's legs often move so fast that they are just a blur, and it's hard to appreciate how precise these movements have to be to generate the distinct song recognizable by a female of the same species. So, recently I collected a range of species and filmed them with a high-speed camera to slow down their movements forty times. I found that different species moved their legs at different speeds, and some moved both legs in synchrony, creating pulsing songs, while others used their legs out of phase to produce a more continuous song. The degree of phase difference between the legs was also different for different species. A female is drawn to a male by his distinctive song, and, in some species, she'll sing back to signal her consent. For others the arrival of a female is just the start of a much more elaborate ritual, which I'll return to later in the chapter.

Crickets and bush crickets sing by rubbing one forewing over the other. Their wings have a file-and-scraper system, with hardened pegs on the underside of a vein on one wing that are drawn across a hardened ridge on the other wing, but as usual it's a little more complicated than that. The wings also have enlarged areas, clear of major veins, that are free to vibrate like the membrane of a loudspeaker and which amplify the song. These structures are referred to rather poetically as the harp and the mirror. Among the crickets (Grylloidea) the two wings are quite similar, with both left and right wings having pegs, and a harp and mirror, but the true crickets (Gryllidae) always sing with their right wing over the left, using the pegs on the right wing. Mole crickets (Gryllotalpidae) are more ambidextrous and can alternate the positions and roles of the two wings.

Bush crickets (Tettigoniidae) reverse this arrangement and sing left over right. They also differ in how their songs are amplified. In true crickets, the main resonator is the harp, and the mirror seems of little importance. In bush crickets the mirror is critical. This thin area of membrane is supported by a frame of thickened veins, and as each peg scrapes over the file the frame is distorted, then flips back to its original shape, so vibrating the membrane of the mirror. The left and right wings of bush crickets are more differentiated than among the true crickets, and in some the stridulatory file has been lost from the right wing.

The most extreme differentiation so far discovered is in *Ectomoptera nepicauda*, a bush cricket from East Africa. The left wing has just the file and the right wing only the stiffened ridge and the soundbox of the mirror. This bush cricket uses its specialized wings to produce an unusual song of frequency-modulated pure tones high up in the ultrasound range.[14]

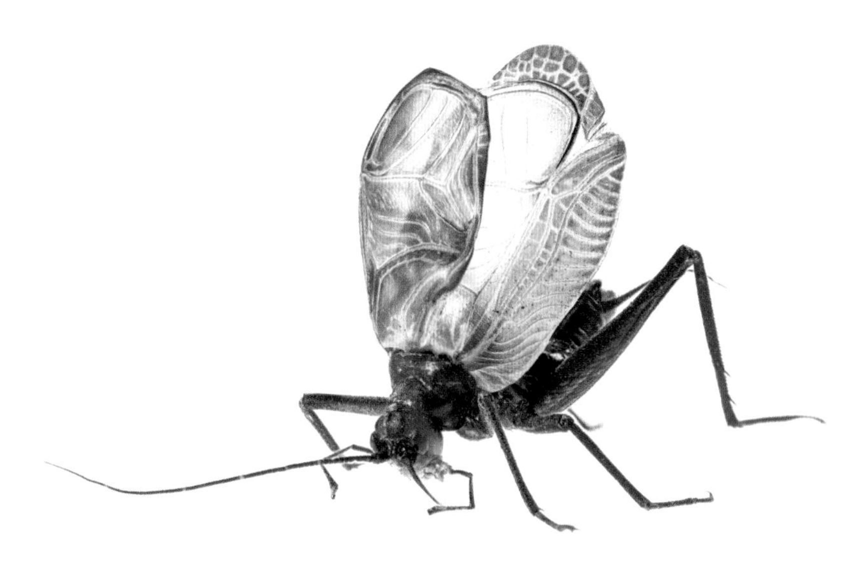

To create such high frequencies, *E. nepicauda* moves its specialized wings at extremely high speed, so that the file rubs over the scraper as quickly as possible. In the same way, you can get a higher tone from running your finger along a comb at a higher speed. Other bush crickets, though, pitch their songs even higher.

A South American bush cricket, as yet with no scientific name, produces pure tone pulses at an astonishing 129 kHz. For comparison, humans can hear up to 20 kHz, at least when young, and bats mostly operate in the range 15–90 kHz – although some can produce sounds up to 200 kHz. The South American bush cricket produces its high-pitched song with relatively slow wing movements, so the large amount of energy needed to oscillate its soundbox powerfully enough to produce ultrasound must therefore come from energy stored in the file-and-scraper system itself, almost certainly with the aid of the insect super-rubber, resilin. These songs can also be loud.

A more recently studied species from South America, a strange spider-like bush cricket called *Arachnoscelis arachnoides*, produces an ultrasonic song that reached 110 decibels, as loud as a chainsaw.[15] These little bush crickets are not uncommon in parts of Colombia, so I suppose we should be grateful that they sing well above the threshold of our hearing, otherwise we'd be deafened by their chorus. In fact, the songs of many species of orthopteran are well beyond the range of human ears, so we are oblivious to the true symphonies produced by this insect orchestra.

As is usual with insects, startling new discoveries are being made all the time. In 2014, more South American bush crickets were discovered

The Ethiopian bell cricket, *Homoeogryllus xanthographus*, sings by rubbing one wing over the other. Large areas of thin membrane and the box shape of the raised wings both amplify the song.

that could sing at an even higher pitch. There are three species in the aptly named genus *Supersonus*, and their songs have been recorded at 150 kHz, and even louder than those of *Arachnoscelis*. It's been claimed that they make the highest-pitched sounds of any animal, although a species of bat, *Cloeotis percivali*, has been recorded reaching 210 kHz.[16] Nevertheless, it's a remarkable output for a tiny insect. The tiny size of their wings helps them to achieve this since they can move them quickly. In addition, the right wing is concave, which creates a closed space between the wing and the body of the bush cricket. This is analogous to what happens with the box of a loudspeaker, where a big box enables deep bass notes to be reproduced faithfully. In this case the tiny box resonates at a super-high frequency, but amplifies the sound to a deafening level – as long as you have the ears to hear it.[17]

Amplification is the goal for all these creatures – to broadcast their love songs as far as possible and to draw in females from far and wide. Most grasshoppers lack major adaptations to amplify their songs and their calls are audible over a few tens of metres at most. However, in Africa, male bladder grasshoppers, *Bullacris membracioides*, develop huge, bloated bodies housing large air sacs that act as resonating chambers. Unlike in other grasshoppers, the file is on the side of the enlarged abdomen and the scraper on the inside of the femur. By scraping their legs over their resonant bodies they can produce a song loud enough to be heard nearly 2 kilometres away, the longest distance recorded for any insect. Nevertheless, their extreme body adaptations are only part of the explanation for their penetrating songs.

Bladder grasshoppers usually sing at night, when temperature inversions in the atmosphere often form. These inversions create a layer of cool air lying low over the ground capped by a layer of warmer air above, the reverse of the normal condition during the day, when the air gets progressively cooler with increasing height. An inversion creates a distinct boundary in the atmosphere, which has the property of reflecting sound waves. In effect, an inversion creates a 'sound channel' that continuously bounces sound waves back to the ground.* This allows a bladder grasshopper to broadcast its song much further. If recordings of their songs are played during the day, when this night-time sound channel is absent, they travel for just 150 metres.[18] Bladder grasshoppers have harnessed atmospheric physics in their efforts to be heard as far away as possible.

Crickets and bush crickets, as we've seen, have resonators built into their wings that create louder songs. They also lift their wings slightly from their bodies when they sing, to create a resonant space, taken to an extreme in *Supersonus*. In addition, some make use of external amplifiers. Mole crickets live in tunnels beneath the ground and the male carefully shapes the entrance to his burrow to increase the volume of his song. The burrow has two openings, shaped like the horns of old-fashioned record players. Just inside the entrance, the burrow narrows briefly, then flares into a more bulbous structure. This is where the singing male sits, facing into the tunnel. The horns efficiently convert the energy in his song into vibrations of a much larger volume of air that broadcast the song for nearly half a kilometre.[19]

Some crickets are even more sophisticated. Tree crickets, *Oecanthus* spp., have large, racquet-shaped wings with extensive areas of thin, transparent membrane that serve as resonators. They hold their wings vertically, well away from their bodies, as they sing, and in this way can produce quite loud sounds. However, a problem facing all these creatures is that sound radiating from the rear surface of the wings interferes with sound emanating from the front, which reduces the intensity of the song. Tree crickets have come up with an ingenious way of solving this problem.

A male tree cricket chews a cricket-wing-sized hole in the centre of a suitable large leaf. He then sits with his head poking through the hole, so that when he lifts his wings, they fill the space he has carved out. In

* A similar phenomenon is responsible for the fact that radio stations broadcasting on medium wave can be picked up from much further away at night.

The song of the great green bush-cricket, *Tettigonia viridissima*, is so loud, males singing in roadside vegetation can be heard from inside a passing car.

this position, as he sings, the leaf acts as a baffle, preventing sound from one surface of the wings interfering with sound from the other, which dramatically increases the volume of the song.[20] More remarkable still, this construction fulfils most modern definitions of animal tool use; by modifying and using an external object (the leaf) to amplify its song, the tree cricket deserves its admission, along with chimpanzees, dolphins, parrots and crows, to the elite club of animal tool users.[21]

Grasshoppers and crickets are not the only songsters of the insect world. Many of the true bugs (Hemiptera) stridulate, although the most remarkable must be one of the lesser water boatmen, *Micronecta scholtzi*, which sings with its penis. *M. scholtzi* is tiny, a fraction over 2 millimetres long, yet its song is so loud that it can be heard by a person sitting by the side of a pond in which these creatures are living. It creates this intense sound by rubbing a file on one of its genital segments across a ridge on the abdomen. Scaled for its tiny size, *M. scholtzi* might be the loudest of all insects, although that honour is usually accorded to cicadas.

Cicadas are also true bugs, but they produce their deafening songs in a different way from lesser water boatmen. Males have two large, circular organs – known as tymbals – on either side of the abdomen. Tymbals are slightly domed structures, supported by a rigid frame. When muscles attached to the frame contract, it is distorted and the tymbal flips from convex to concave with an audible click. Resilin incorporated into these structures then causes the tymbal to flip back to its original shape as soon as the muscles relax, emitting another audible click. A cicada can contract and relax its tymbal muscles so quickly that the clicks merge into one continuous sound. Air spaces inside the male cicada's body amplify this sound to painful levels, loud enough to cause hearing loss in an entomologist foolish enough to spend too long in close proximity to a singing cicada.

The loudest cicada on record is *Brevisana brevis*, a species from Africa whose song reaches nearly 107 decibels at a distance of 50 centimetres, although several other species come very close to this intensity. However, the volume of a cicada's song is directly related to its body size: the larger the cicada, the louder the song – and there are some very large cicadas living in the forests of south-east Asia. The empress cicada, *Pomponia imperatoria*, with a body length of 7 centimetres, is the largest cicada in the world, and although the volume of its song hasn't been accurately measured, it's likely to be louder than that of the current record holder.[22] Its song is a peculiar and very un-insect-like bugling sound that heralds nightfall in Asian rainforests.

A lesser marsh grasshopper, *Chorthippus albomarginatus*, like most grasshoppers, sings by rubbing pegs on the insides of its broad femurs against ridges on its forewings.

A cicada's song, like that of grasshoppers and crickets, is intended to draw females, but some cicadas have adopted a two-pronged approach to courtship by singing and swarming at the same time. Periodical cicadas, *Magicicada* spp., spend a long time as nymphs underground, then all emerge together over the space of a few days. Marching hordes of nymphs erupt from the soil like a scene from a horror film before swarming up tree trunks, where they moult into their adult forms. Then the males begin to sing together – an ear-splitting chorus that has to be heard to be believed.

Periodical cicadas have a peculiarly mathematical life cycle. Some spend seventeen years underground, others thirteen years – both prime numbers. One theory suggests that this long life cycle, based on a prime number, makes it impossible for predators or parasites to track the cicadas and predict when there will be a glut of prey. As important as reducing predation, the enormous numbers of cicada nymphs that emerge all but guarantee that males and females will meet up. As a female approaches a singing male, she calls back. The females of a few species also have tymbals, but most announce their presence by flicking their wings to make a sharp click. Males react instantly to this and follow the sound until they find the female. Male cicadas are so tuned in to this sound that you can persuade a male to follow your hand by clicking your fingers – providing hours of amusement while you slowly go deaf.

Predators and parasites are a real problem for all these songsters. Loud sounds that precisely pinpoint your location can be used just as easily by enemies as by potential mates. Periodical cicadas have found one way around this by swamping forests and parks at unpredictable intervals with so many individuals that predators or parasites can't react or make any inroads into their numbers – but many other singing insects aren't so lucky. Australian mole crickets, with their loudspeaker burrows, attract bright blue parasitic wasps, blue ants, so called because the female is wingless and resembles an ant. We met these creatures in the last chapter, where the males were drawn to female-mimicking orchid flowers. Those that manage to find real females start a process that ends badly for mole crickets. After tracking a mole cricket by its song, the female blue ant paralyses the cricket, then lays an egg on it. When its larva hatches it has a mountain of fresh cricket meat to eat.

Singing crickets also attract their own problem parasites in the form of a fly, *Ormia ochracea*. Ormia belongs to the fly family Tachinidae, which – like many wasps – are also parasitoids. This fly is a cricket specialist and can home in on cricket songs with incredible accuracy, using 'ears' located in a cavity underneath its thorax that consist of two membranes separated by a stiff ridge of cuticle down the centre. Whenever the sound of cricket song reaches Ormia's ears, the membrane on the side nearest to the song starts vibrating first, before the membrane on the far side. The two membranes therefore end up vibrating out of phase. Depending on the phase difference – which may be only a thousandth of a second and therefore difficult for the fly to perceive directly – the two vibrating membranes distort the central strut of cuticle in different ways. It's the way in which this central ridge is distorted that allows the fly to pinpoint a singing cricket.[23]

Ormia is so effective at tracking down crickets that on the Hawaiian Islands some crickets have stopped singing altogether. On Kauai and Oahu, among populations of *Teleogryllus oceanicus*, a mutation has spread that reduces the sound production structures on the wing, making the male's wings more like those of females.[24] This adaptation arose on Oahu, then a few years later appeared on Kauai, but genetic studies have revealed that this wasn't through mutant Oahu crickets somehow colonizing Kauai. Instead, silent crickets arose on Kauai independently where a different genetic change produced the same end result – wings without any files or scrapers.[25] This is an example of convergent evolution – when natural selection produces similar results in response to similar environmental pressures. There are countless examples of this throughout nature, but it's much rarer to have such a clear picture of the genetics behind this phenomenon.

As a means of avoiding parasitoids, silence is a somewhat extreme measure. Far better to develop a more secure channel. Many leafhoppers and planthoppers (Hemiptera) also sing by stridulation, but instead of broadcasting their songs through the air, they transmit their songs through their bodies or feet into the plants on which they are sitting. Their songs are then transmitted as vibrations through stems and leaves, to be picked up by any females within earshot – a private communication network that excludes distant eavesdroppers. Often a female will call back and the pair perform an elaborate duet as they track each other down. The plants in your garden are all resonating to the complex love songs of bugs when all we gardeners can hear is silence. Your garden is, quite literally, alive with the sound of music. Although we are not exactly sure how, some leafhoppers and planthoppers can also transmit their songs across gaps between plants, so partners can find each other over considerable distances.[26]

The ground also serves as a good conduit for songs of courtship. Well

known in South Africa, a group of darkling beetles (Tenebrionidae) uses this method to good effect. There are many different kinds of tok tok (or tok-tokkie) beetles that attract females by banging their hard bottoms on the ground in an insect version of Morse code. Each beetle uses a different pattern that unambiguously declares which species it is. Some keep up a steady rhythm but change the intensity of the taps, others accelerate their taps until they reach a buzzing crescendo. When the female hears the right code, she responds with her own taps and, like planthoppers, the two duet until they finally find each other.

A WHIFF OF ROMANCE

Another way to avoid unwelcome eavesdroppers is to use scent signals. Many insects use chemicals, called pheromones, which are both silent and invisible, to advertise their whereabouts. Another advantage of pheromones is that they can drift on the wind and could, in theory, carry the signal over much longer distances than even the loudest insect song. However, the further it travels from the source, the more the chemical trail breaks up, making it harder to track to its source. Many moths use this form of signalling, although it is often only effective over a hundred

A female emperor moth, *Saturnia pavonia*, releases a pheromone from her abdomen, a scent trail that drifts over open heathland to draw males from far and wide.

metres or so. Even so, there is some evidence that silkmoths can track chemical trails over a kilometre or more,[27] and there are anecdotal reports of moths homing in on females from 11 kilometres away.[28]

There's no better demonstration of the calling power of pheromones than the emperor moth, *Saturnia pavonia*, a species of silkmoth from Europe in which the female releases a pheromone to attract males. Emperor moths fly over open heaths during the day, but even in areas where they're abundant, such as the New Forest in Hampshire, they're not always easy to find. All that changes if you make the effort to collect late-stage caterpillars the previous year and keep the resulting pupae over winter. It's possible to sex the pupae and segregate males from females. Then, in April, when the adults emerge, you can take unmated females contained in a small net cage onto the heath – and be amazed.

The female adopts a characteristic posture as she exposes glands in her abdomen that release a pheromone. In no time at all, the first male is fluttering eagerly around the cage, soon followed by others. Where were all these males hiding when I walked over this same patch of heath this time last year without a virgin female in tow? Moth enthusiasts call this procedure 'assembling' and it doesn't take long to assemble half a dozen males. If I've handled a female carelessly and contaminated my hands with her pheromone, I too will soon be pursued by ardent males wherever I go.

This astonishing ability of female moths to conjure up males has been known for a long time. By the late eighteenth century, moth hunters were taking their caged virgins onto the heaths around London:

> Upon this is founded the practice of 'sembling', as it is called by the London collectors, among whom… it has long been in use, for entrapping the males of the fox-moth, the grass-egger and others.
>
> It is a frequent practice… with the London Aurelians,* when they breed a female of the lappit-moth, and some other day-flying species, to take her in a box with a gauze lid into the vicinity of the woods where, if the weather is favourable, she never fails to attract a numerous train of males, whose only business appears to be an incessant, rapid, and undulating flight in search of the females. One of these is no sooner descried, than they become so enamoured with their fair kinswoman, as absolutely to lose all fear for their own personal safety.[29]

* *Aurelian* is an old term for a lepidopterist – a collector or breeder of butterflies and moths. It has nothing to do with the Roman emperor of that name. It is derived from *aurelia* – an old word for the pupa or chrysalis. Chrysalis comes from the Greek word *chrisos* (χρῡσός), for gold. *Aurelia* is the Latin equivalent, from *aureum*.

(overleaf) A male Chinese oak silk moth, *Antheraea pernyi*, has huge antennae covered in tiny sensory hairs, that detect the slightest whiff of the female moth's pheromone.

Since this time, it was assumed that it was smell that somehow drew the males, although no one knew how the males could detect a female's scent over such distances. The French entomologist Jean-Henri Fabre studied this phenomenon at the start of the twentieth century by experimenting with a relative of the emperor moth, the great peacock moth, *S. pyri*. He was able to lure 150 males to his house in one evening. He too suspected that the males were drawn by scent, although he also wondered whether the females were creating some kind of mysterious electrical field.[30]

We now know not only that it is smell that males are responding to, but also the exact chemical formulae of many of these sex pheromones. Researchers have even synthesized artificial pheromones, including one for the emperor moth that works just as well as the real thing and which I've used when filming these insects. A few dabs around the edge of the lens and males cavort in front of the camera for as long as the scent lasts. These sex pheromones consist of molecules made up of chains of carbon atoms, usually between ten and eighteen atoms long, depending on species. To this central chain, side branches consisting of alcohol groups, or sometimes aldehydes or acetates, may be added. The chains also vary in the number of double bonds between the carbon atoms.[31] If

your organic chemistry is getting a bit hazy, don't worry. The bottom line is that there is enough variation in this basic molecular structure to allow each species to have its own distinct scent. In addition, the scent often also consists of several different chemicals, and the blend of these components can also be varied to make sure the perfume is unique to each species. This is important, since males don't want to waste their time tracking an alluring scent to its source only to find it's brought them to the wrong species.

We also now know how the males home in on females with such accuracy, and again this is exemplified by the silkmoths. Male silkmoths have particularly extravagant and feathery antennae that sit on their heads like two enormous radar dishes, while females possess much more modest antennae. A male's antennae are covered in microscopic hairs that capture the specific pheromones of his species. A tiny nerve fibre runs down the centre of each hair, which fires an electrical signal in response to pheromone molecules attaching to the hair. The large surface area of these antennae ensures that a large volume of air is sampled and that even a small whiff of a female pheromone is not missed. A single molecule of the pheromone binding to a sensory hair is enough to trigger a nerve impulse, although a male doesn't react until many of the hairs are being triggered and he's sure he's picking up a genuine signal.

Once a male has detected the scent of a female, he sets off upwind, but his task is far from easy. The pheromone plume will have been broken up into wisps and strands by the wind. So the male zigzags across the plume, at a slight angle to the prevailing wind. When he flies out of the plume and his antennae fall silent, he turns and flies back across the plume. This flight pattern, called counterturning, is hardwired into his brain, and keeps him broadly within the pheromone trail and moving towards its source. However, if the wind has recently changed direction, the plume will have a sharp bend in it. Tacking upwind will now eventually lead the male out of the plume for good. At this point, a new flight pattern, called casting, kicks in. He now flies at right angles to the wind, first one way, then the other, slowly increasing the length of each leg until he either bumps into the plume again or simply gives up.[32]

In most cases, females release the pheromone and males do the tracking – but a few moths like to do things differently. Males of the gangis moth, *Creatonotus gangis*, an Asian and tropical Australian species, inflate two enormous appendages, branched into two at the base, from their rear ends. Called coremata, these structures are longer than the moth's entire body and covered in long hairs. It appears that though the

A female promethea moth, *Callosamia promethea*, releases her pheromone after nightfall – but scent works just as well in the dark as it does in daylight.

moth is being attacked by four nasty-looking centipedes. The coremata release a pheromone that attracts females and judging by their size, they must release an awful lot. They're also responsible for a lot of panic phone calls when one of these is discovered on the wall of a house in Queensland or the Northern Territory. With his coremata fully inflated, the male gangis moth is a terrifying sight. Gangis moths are members of the tiger moth family (Arctiidae), and several other family members are similarly equipped, if not quite as well endowed as the gangis moth.

In North American salt marsh moths, *Estigmene acrea*, a male makes up for his smaller unbranched coremata by gathering with several other males to release pheromones. On average, around nine or so males occupy a small patch of vegetation early in the evening. This has been described as a lek, a mating strategy more familiar among birds such as the black grouse, a European bird of upland grassland and heath edges. On the grassy moors of the Yorkshire Dales in northern England, I've watched male black grouse gather in groups of around ten to twenty just before first light to parade and strut. They each defend a tiny territory, not much bigger than they are. It contains nothing of value to the female, but the rules of the game state that the territories at the centre of the lek are the most valuable and therefore the most fought over. In the end only the strongest males can hold on to central territories, so when a female turns up she knows straightaway who will make her best partner. Males on the central territories get to mate with many more females than males around the edges.

There's no evidence that female salt marsh moths choose one male over another at moth leks. All the males stand an equal chance so it's not quite the same strategy as that pursued by black grouse and many other birds. Perhaps by gathering together like this, salt marsh moths are simply increasing the power of their pheromone signal, a brighter beacon for females to find. Later in the evening, the males retract their coremata and it's the females' turn to release pheromones. Now, the situation reverts to the more familiar story among moths of males tracking down females.[33]

Curiously, gangis moths, and a closely related species, *C. transiens*, do the same thing. After a few hours of signalling in the early evening, males deflate their coremata and the females take on the task of releasing a pheromone.[34] No one really knows why these moths alternate between two different mate-finding strategies each night.

Many insects manufacture their pheromones from scratch, but some need precursors from their food. This is true of many male-released

Female ivy bees, *Colletes hederae*, mate only once so, when a female first emerges from her burrow, she is pounced on by dozens of eager males forming a 'mating ball', somewhere in the middle of which is the female.

pheromones, including those of the gangis moth. The male needs to store pyrrolizidine alkaloids from his larval diet to manufacture his pheromone, and the eventual size of his coremata will be determined by how rich in alkaloids his diet as a caterpillar was. More alkaloids equates to bigger coremata.[35] His limited supply of chemicals from which to manufacture pheromones may explain why a male gangis moth only calls for a limited time each night.

A few insects use chemicals that can be found in abundance in their environment to attract mates, like wearing an alluring perfume, which gets around supply problems and saves on the biochemical expense of making their own pheromones. The best-known examples are the orchid bees (Euglossini), some 200 kinds of stunningly iridescent bees that buzz like flying jewels through the rainforests of Central and South America. They are important pollinators of many tropical plants but in particular orchids, of which 700 species are dependent on these bees.[36] Male orchid bees seek out flowers with particular fragrances, and the right plants can attract dozens of bees at the same time. The males will use these scents to entice females, but first they have to collect and store them.

They do this by landing on the flower and spitting on it. The chemicals responsible for the fragrances are not water soluble but do dissolve in fatty solvents, a process only too familiar to perfumiers. They call this 'greasy extraction' or 'enfleurage', and orchid bees use exactly the same trick. The bees secrete a fatty solution from glands near their mouth and spit it onto a petal where it dissolves the chemicals that make up the perfume. The bee then soaks up this scented mixture from the flower with his hairy front legs and lifts off. Hovering in front of the flower, the bee deftly transfers the perfume from his front legs to his hind legs, which have greatly swollen tibias that act as storage containers. He'll return to the flower to repeat the process, but eventually he'll seek out other flowers, in the process transferring pollen between plants, until he has plenty of perfume stored in his hind legs. Now it's time to put it to good use.

He finds a perch in a forest clearing and transfers some of the perfume from one of the storage chambers to a tuft of hairs on the opposite rear leg. These exposed hairs allow the perfume to evaporate and drift downwind. Often the male hovers close to his perch once he has soaked these tufts of hairs with scent, so the air currents from his beating wings also help vaporize the scent and waft it on its way.[37] Each species makes its own blend of perfume, so with any luck the male will attract a female of the right species to his perch. However, to avoid any confusion, different species display at different times or choose perches at different heights in the forest. Once mated, a female heads off to build a nest and stock it with pollen, so she too visits flowers and, as she does, also cross-pollinates them. Some females travel tens of kilometres to collect pollen, so both male and female orchid bees are vital strands in the complex web of rainforest life.

A great many insects rely on scent to find each other, including the enigmatically named 'dark fireflies'. This group of little-known diurnal beetles are the most primitive members of the family Lampyridae, the bulk of which are more famous for using glowing lanterns in the night to signal between the sexes, and in the process creating spectacular natural light shows that can't fail to stir the soul. The details of how these beetles use light to bring males and females of the right species together in the dark forest are intriguing and have kept evolutionary biologists busy for many decades.

BRIGHT SPARKS

Lampyrids are broadly divided into fireflies or glow-worms, a distinction based on how they use their lights as signals. Female glow-worms are wingless and each evening clamber to the top of a tuft of grasses, where they light up their glowing tails. They burn their lanterns continuously, lighting up grassland with eerie green pinpricks of light. Male glow-worms don't glow, but they do have wings and fly nightly patrols in their search of these glowing females. The females of some glow-worms also release pheromones, like the ancestral dark fireflies, to draw in males from further afield. The males then switch to the light signal to guide them on their final approach.

Glow-worms are the common lampyrids in Europe, and the European glow-worm, *Lampyris noctiluca*, was once a familiar part of the countryside. Sadly, destruction of its species-rich grassland home has meant that its ghostly green glows are no longer common. Glow-worms also occur in Asia but are surprisingly scarce in North America. Here, the fireflies predominate. Instead of a continuous glow, fireflies use intermittent, flashing signals. In these species both males and females produce light and it's the males that, at least initially, signal to the females.[38] They do this in flight, often in large numbers, which creates a mesmerizing effect as tiny fairy lights drift through the forest, winking on and off.

There are many species of fireflies across North America and often several kinds live together in the same glade or patch of forest, so the

trick is to make each signal distinctive, otherwise all the fireflies would waste a lot of time courting unsuitable partners. Grasshoppers and bush crickets have evolved distinct songs, while moths have evolved different chemicals, used in different combinations, to create unique scent signals. With light signals, it's all about evolving codes.

Back in the 1960s, the entomologist Jim Lloyd was the man who cracked the firefly code. Over his long career, he spent thousands of nights driving the back roads of the United States, headlights off, staring off into the darkness for the flashes of fireflies. Whenever he found a colony, he stopped and set up camp and meticulously recorded the flash patterns of the fireflies, before collecting them to ascertain their identity. It gradually became clear that each firefly species has a unique signal, differing in number of flashes and flash duration. Males also have different flight patterns, so they write a different light signature in the air. The big dipper firefly, *Photinus pyralis*, for example, flies a descending loop during each flash, so creating a 'J'-shaped light trail.

When a female sees the right signal for her species, she flashes back and again her message must be unambiguous – I'm the same species as you. She too has a distinct pattern of flashing, but her timing is also critical. She must wait for just the right amount of time, different in different species, before signalling back, otherwise a male will ignore her signals. How the males keep track of such precise details of timing in the disco world of a firefly forest is beyond belief. As if this isn't complicated enough, the males of some fireflies tightly coordinate their individual performances with their neighbours, all lighting their lanterns together to become an orchestra of light or, in the words of Tuft's University entomologist and firefly enthusiast, Sara Lewis, a 'synchronous symphony'.[39]

In the Smoky Mountains of Tennessee, *P. carolinus* lights up the forest with such impressive, constantly shifting, luminescent symphonies that they draw hordes of tourists. There are now so many that numbers visiting a site near Elkmont in the Great Smoky Mountains National Park must be strictly regulated, with a fleet of shuttle buses to ferry people from parking areas to the forest.

In south-east Asia, *Pteropteryx tener* also puts on phenomenal displays, although this species doesn't flash in flight. Instead, thousands and thousands of males perch on the leaves of *Sonneratia* mangrove trees along tidal creeks and light up in unison. The effect is both hypnotic and strangely disorientating as you drift past the display in a boat. Known as *kelip-kelip* in Malaysia, *Pteropteryx tener* – like *P. carolinus* in Tennessee

A wingless female common glow-worm, *Lampyris noctiluca*, lights her green lantern after dusk as a signal to attract winged males.

– now draws firefly tourists in large numbers and provides a valuable source of income to the region's economy.

Conspicuous firefly displays among the mangroves had always been familiar to local people, but it wasn't until the American ichthyologist Hugh Smith witnessed these sights in Thailand in the 1920s that they became much more widely known. In the journal *Science*, he described the extraordinary spectacles along the Chao Phraya river near Bangkok.

Similar spectacles elsewhere in the world are still being discovered by firefly tourists. Near the town of Nanacamilpa in Tlaxcala in southern Mexico, another species, *Photinus palaciosi*, lights up the summer forest with displays that surpass even those of the Smoky Mountains. Organized tours to this firefly hotspot began as recently as 2011, and in the last decade tourism here has sky-rocketed, bringing much-needed cash to a chronically poor part of the country.

From Mexico and the United States, to Taiwan and Malaysia, around a million people now travel to see firefly displays, but as welcome as tourist dollars are, there's rising concern for the welfare of the colonies. Uncontrolled access means thousands of trampling feet – and a death sentence for those species with flightless females. As infrastructure grows, light pollution from buildings swamps precisely coded signals and prevents the beetles from breeding. Even light spilling from tourists' mobile phones as they eagerly video the light show can cause problems. The International Union for the Conservation of Nature (IUCN) has a firefly specialist group that has recently surveyed all the firefly tourist hotspots and has come up with recommendations that will allow us to continue to marvel at one of the natural world's most extraordinary spectacles in a way that doesn't threaten the fireflies' habitats.[40] It is to be hoped that those who manage these sites will have the foresight to act on these suggestions.

Using these different kinds of long-range signals, tiny insects manage to locate each other over surprisingly long distances. But this initial contact is only the beginning of the story. After all, a bit like internet dating, there's no telling what kind of potential partner will respond to your signals. Females make a much bigger investment in the next generation than males, since eggs must be packed with energy-rich yolks, whereas sperm are nothing more than swimming packets of DNA. Eggs are expensive to make whereas sperm are cheap. So, females need to be choosy. They want to know that their precious eggs are being fertilized by the best sperm around. Males, on the other hand, will try it on whenever they can.

After using claspers at the tip of his abdomen to grab a female, the male large red damselfly, *Pyrrhosoma nymphula*, first fills his secondary sex organs with sperm from the primary organs at the tip of his tail. Then the female must reach up to connect to these secondary organs.

BATTLE OF THE SEXES

While making a film on dragonflies in 2018, I spent many weeks over the summer recording their behaviour around a lake in Wiltshire, in south-west England. The shores of this lake were alive with some twenty different species of dragonflies and damselflies, most of which were males. Females were busy feeding over meadows and woodland, well away from the throng of males, to build up energy reserves for those expensive eggs. When a female is ready to find a partner, she ventures on to the lake and is almost immediately pounced on by a male. However, I soon discovered that the males are far from discriminating. I watched one poor female common blue damselfly, *Enallagma cyathigerum*, alight on a reed stem, only to be grabbed by a male red-eyed damselfly, *Erythromma naias*. Red-eyes are much bigger and look nothing like common blues, but it would appear that, for a male damselfly, any female damselfly is worth a shot.

Luckily for species integrity, the mechanics of this union won't work. Damselfly mating requires the male to latch two hooks at the tip of his abdomen onto the front of the female's thorax to begin the process. Mating between different species of damselfly is prevented by a 'lock-and-key' mechanism: in this case successful docking is prevented by the fact that the hooks of a red-eyed damselfly do not fit the thorax of a common blue. The male red-eyed damselfly left frustrated after a few minutes of trying.

Next, the long-suffering female was visited by a male azure damselfly, *Coenagrion puella*. Common blue and azure damselflies do, at least, look very similar, so this mistake is more understandable, but again the precision mechanics prevented him from locking on. Finally, a male common blue showed up and immediately made a connection. Once hooked up, the pair took to the air in tandem and disappeared into the bushes to consummate a long-awaited union.

Many years ago, I came across an even more graphic (and amusing) demonstration of a total lack of discrimination by males in Australia. *Julodimorpha bakewelli* is a kind of jewel beetle (Buprestidae), as stunningly

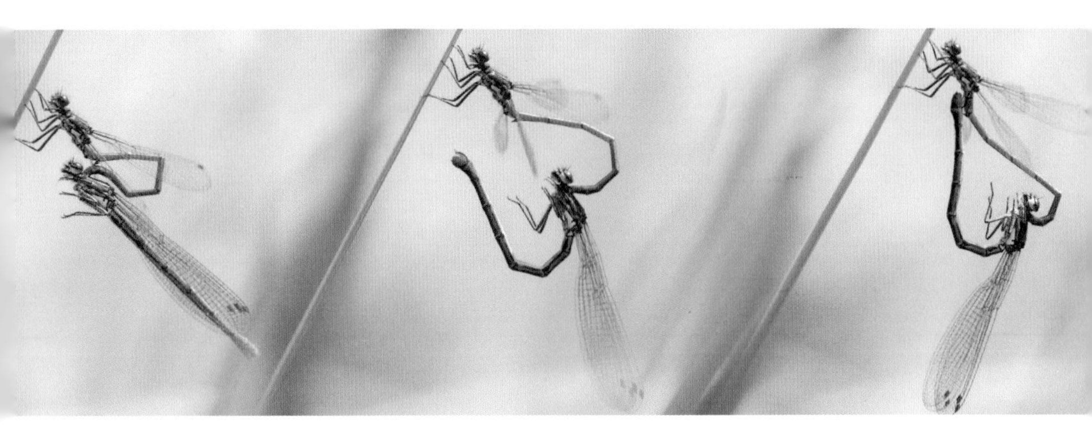

beautiful as its name suggests. The beetles are an iridescent golden-brown and covered in fine dimples – uncannily like certain Australian beer bottles. In fact, they're so similar that empty 'stubbies' launched from passing cars and landing in prime jewel beetle habitat cause a real problem for the beetles. Male beetles are tuned in to the dimpled amber texture of females, which they use to locate their partners, but when a beer bottle lands, they can't believe their eyes. It's like a giant female. Bigger females lay more eggs, so males like to mate with the biggest girls around and a beer bottle trumps any female in the neighbourhood.

Whole gangs of males descend on the bottle and extend their genitalia, pushing and probing, trying to find a way to mate with this super-female. They're so engaged that they seem oblivious, even when local *Iridomyrmex* ants turn up and start chewing on their exposed organs, often killing the beetles in the process.[41] We filmed this bizarre behaviour for the *Alien Empire* series in the early 1990s and it became one of the most widely shared and longest-remembered sequences, no doubt in part because it sparked a whole series of jokes stereotyping Australian males. It did, however, have a serious side. The fact that female beetles were coming a poor second to beer bottles prompted concern that populations of *J. bakewelli* could die out. The solution turned out to be very simple. The beer company that made the bottles heeded the concerns of the conservationists and changed the design of its bottles. Once the dimples were eliminated and the bottles made from smooth glass, they lost all their appeal to male beetles.

In most cases, females are far choosier than males, so they need ways of discriminating between males and selecting the fittest. Returning to the night-time forests of North America, illuminated by the coded signals of fireflies, it turns out that even if a male does flash out the correct sequence he may not get a reply. Females respond to fewer than half the males they see.[42] Although the flashes of males of the same species look identical to our eyes, clearly the females are seeing something different.

It's possible to mimic the flash pattern of a male firefly with LEDs, then subtly change flash intensity, duration and frequency to see what catches a female's attention. In species such as *P. consimilis*, whose code consists of a brief sequence of flashes, females respond to 'males' that flash the fastest.[43] In species that use a single pulse, such as *P. pyralis*, it's the intensity of the pulse that makes a female sit up and take notice.[44]

Do these flashier males really make the best partners? *P. ignitus* is another single-flash species, but in this case females find longer pulses more attractive. Detailed studies of this species showed that males with longer

A male damselfly's secondary sex organs contain structures that scoop out any sperm in the female from previous matings. This male azure damselfly, *Coenagrion puella*, remains attached to the female while she lays her eggs, to stop other males from latching on and removing *his* sperm.

pulses also produced larger spermatophores, the packets that contains the male's sperm, but also a gift of proteins and amino acids for the female.[45] Adult *Photinus* fireflies don't eat so the bigger this food parcel, the better. The amino acids are quickly incorporated into the eggs, and these extra resources help the offspring get off to a good start in life.

Male crickets and katydids are also well known for providing nutritious packages to their partners. They transfer sperm along with a nutritious gelatinous blob, known as a spermatophylax,* which in some cases may be a third of the male's body weight – a huge investment in his partner and in the next generation. The spermatophylax protrudes from the female after mating so all she needs to do is bend round to eat it, incorporating the nutrients into her own body as well as into her growing eggs. However, the gift may not be as generous as it looks at first sight.

Males and females want different things from their relationship. It is often beneficial for a female to mate many times. This provides a final test of male fitness as sperm from different partners compete to fertilize her eggs. A male, on the other hand, wants to be the father of all her eggs, leaving as many copies of his genes as possible in the next generation. Some studies point to evidence that his nuptial gift also contains substances that reduce the female's desire to mate again.[46]

A female is on safer ground when her nuptial gift is less easy to tamper with. Hanging flies, *Harpobittacus*, are predators that get their name from a habit of hanging beneath branches on long legs and grabbing passing prey. When it's time for a male to go courting, he catches a tasty insect as a succulent offering to a female. She gratefully accepts the gift and while she's preoccupied with eating, the male mates with her – the insect version of a dinner date. Males that can catch bigger items of prey are more attractive and get to mate more often, but they can also be cagey with their gifts. If a male has been lucky enough to catch something really big, he grabs it back from the female as soon as he's finished mating and uses it to entice another female. A female choosing a male with the biggest offering may not even be picking the most skilled predator. Male hanging flies are not above piracy and they frequently steal prey from other males to save the bother of hunting themselves.[47] What's a girl to do?

In many animals, females judge males on their courtship displays, which often involve extravagant adornments and energetic dance moves.

* The spermatophores of many crickets and bush crickets are made up of a flask-shaped structure (ampulla) containing the sperm and a sperm-free gelatinous structure, the spermatophylax, which contains high concentrations of nutrients.

Female giant Asian mantids, *Hierodula membranacea*, are known to practice 'sexual cannibalism'. She may eat her partner after mating if he can't make a quick getaway, although this behaviour is less common among mantids than is generally believed.

There are no better examples than the birds-of-paradise from New Guinea and northern Australia. Males are decked out in the most flamboyant plumes imaginable and cavort ostentatiously in front of females. Many insects also go in for elaborate courtship routines.

SONG AND DANCE ROUTINES

One of the great pleasures of a warm late-summer day is to lie on a grassy hillside and watch the antics of grasshoppers. We've already discovered that males use their distinctive songs to attract females, but when a female does turn up some species then switch to a song-and-dance routine. The calling songs of closely related species are often quite hard to tell apart, at least to human ears, but the short-range courtship songs are more specific and serve to make sure both partners are of the same species. To make doubly sure, some grasshoppers also combine their courtship songs with a unique visual display that can only be described as charming.

It's not hard to devise a way of watching these antics. Simply house a collection of grasshoppers of one species in an aquarium planted with a turf of rough grasses. An incandescent light over the tank is usually

enough to start them singing, and once a male encounters a female he'll perform his whole courtship display. In the UK, the rufous grasshopper, *Gomphocerippus rufus*, is found on dry grassland in southern England and is not too hard to identify. Although variable in colour, with differing patterns of grey and warm brown set against a reddish abdomen, both sexes have distinctive antennae. Females have antennae that are thickened and darkened towards the end, tipped with a light patch. Male antennae have enlarged and flattened tips, coloured in contrasting black and white.

The males sing a dry, rattling calling song that lasts around 3–5 seconds. Once a female approaches, the male begins his display. He starts by standing tall, tilting upwards and rhythmically waving his four white palps, the segmented appendages that surround his mouth. At the same time, he swings his head from side to side, with his conspicuous antennae held out rigidly to the sides to emphasize his movements, gradually accelerating in intensity. Viewed from his level and head on, the effect is always comical to my eyes, although deadly serious for the male. He finishes this part of his dance by flicking his antennae quickly backwards then all the way forwards and downwards with a deliberate movement that seems like a courteous bow. Following this, he sings a brief snatch of courtship song while waving his palps wildly. He repeats the whole process for up to fifteen minutes. If the female is impressed enough, she sings a short song back to him and almost immediately the male leaps onto her back to mate.

The courtship rituals of mottled grasshoppers, *Myrmeleotettix maculatus*, are even more energetic. Their calling songs are soft and pulsing, lasting for 10–12 seconds. In the presence of a female, a male seems to pulsate with energy. His legs scraping over his wings move so fast that they are just a blur. His antennae are held stiffly erect and his body bounces slightly to the rhythm of his song. Then he flicks his back legs forwards over his head, and as they move back to their singing position he lays his antennae flat along his thorax. When he starts his song again, he jerks his abdomen up and down in time to the music, and tilts his head manically from side to side. As I watch his performance culminate in a blissful union, I can only feel that he's well and truly earned it.

Is the female grasshopper judging the male's performance to assess his fitness to father her offspring? The answer in some cases is that she probably is, but it's not as clear as it is for birds-of-paradise. I've watched rufous grasshopper males barely begin their elaborate routine before they are rudely interrupted by an overeager female singing her acceptance. She's had no chance to judge the male. Others have witnessed

A male giant swallowtail, *Papilio cresphontes*, flutters close to a female. Many butterflies pack pheromones into specialized wing scales which, when dusted on to the female, help persuade her to mate.

mottled grasshoppers behaving in the same way. Perhaps the courtship serves more to switch the mood of the female. She spends most of her time feeding to nourish her eggs and it has been suggested that courtship serves simply to switch the female's programme from 'eat' to 'mate'.[48]

However, I doubt whether it's quite that straightforward. The sex life of grasshoppers is remarkably complex. Rufous grasshopper females, for example, behave differently as they age. Young females (1–4 days old) are entirely focused on eating and refuse the advances of any male, no matter how stellar his dance routine. Older females (5–6 days old) enter a state of passive acceptance and will allow mating, sometimes with courtship, sometimes without. If a female is still a spinster after six days, she actively seeks out a male, singing herself to attract a partner, or responding with unseemly haste to his calling song by promptly singing back, so the duetting pair can find each other without delay. After each mating the cycle is reset. Males court females in the passive state far more than those in any other state. Long courtship at this stage seems to calm the female and stops her from wandering away, so the male can mate as soon as she lets him.[49]

Courtship may also play a role in species recognition, since the choreography is distinctly different in each species. However, not all grasshoppers have elaborate courtship routines and species-specific songs

The conspicuous, flattened tips to the antennae of a male rufous grasshopper, *Gomphocerippus rufus*, are part of the regalia for his courtship dance.

are probably the most important way in which grasshoppers recognize their own kind.[50] So, small changes in these songs can isolate populations and create new species. This is exactly what has been happening on the Hawaiian archipelago, where there are at least thirty-eight different kinds of native cricket, *Laupala* spp., all with different songs.

The most likely explanation for this rash of speciation is that some females showed a preference for a slightly different song, which makes males that sing this variant more successful in mating and favours a new generation with more singers of the new song. Eventually, a new population is formed, in which males all sing a different tune, effectively isolating them from the original population as a new species. This is a process called sexual selection, which stands alongside natural selection as a powerful force driving evolutionary change. In fact, these obscure little crickets have the highest rate of speciation of any arthropod yet studied.[51]

Courtship, along with courting gifts, provides one way for a female to judge the fitness of a male. There are other, more subtle cues such as the intensity of a male's colours. Only males in the peak of condition can afford to splash out on flashy pigments. The simplest way, though, is to encourage males to slug it out, and the one left standing must be the fittest.

TO THE VICTOR GO THE SPOILS

Fighting over females has driven the evolution of all kinds of weapons among male insects, but perhaps the most spectacularly armed are the giant scarab beetles (Scarabeidae). The Atlas beetle, *Chalcosoma atlas*, has three long horns, one on its head and two projecting from its thorax, which make it look like a miniature *Triceratops* dinosaur as it trundles through the rainforests of south-east Asia. In the New World, Hercules beetles, *Dynastes* spp., have two long horns, sometimes as long as the rest of their bodies, one on the thorax and one on the head.

By moving the head relative to the thorax, all these beetles can use their horns as callipers to grasp a rival. Like tiny sumo wrestlers, each tries to lift its opponent off the ground and throw him onto his back. The outcome of the struggle is eventually decided when one beetle has had enough and retreats or is hurled out of a tree.

Frog-legged beetles, *Sagra* spp., also live in the rainforests of south-east Asia. They get their name from the male's greatly enlarged and curved hind femurs, armed with stout spines on their inside edges. Frog-legged beetles live in large aggregations on rainforest vines so it's not too hard for a male to meet a female. The problem is hanging on to her. If a single male spots a mating couple, he tries to muscle in. Initially, the incumbent male wraps his curved back legs around the intruder, and he responds in like manner as the two indulge in a squeezing match. If this doesn't settle the issue, the intruding male then attaches himself to the thorax of the mating male with his front and middle legs, and wraps his powerful back legs around the vine. Now, by squeezing hard, the intruder can pivot and, with a bit of luck, lift the mating male off the female and move in himself.

The bigger and more powerful your legs, the more males you can lever off their partners, so there's strong evolutionary pressure to evolve the biggest rear legs possible. They have already become so big that they're useless for walking. Male frog-legged beetles drag their hind legs behind them when they move. This is another example of sexual selection, this time between members of the same sex.[52]

Some true bugs (Hemiptera) also fight with their rear legs; the most striking of these are flag-legged bugs, *Anisoscelis affinis*. The males have developed large, colourful flanges on their back legs that they wave at rivals to warn them off. However, if two bugs have equally bright and extensive flanges, these displays escalate into fights in which the bugs

The males of many giant scarab beetles are armed with horns for combat with rival males. With two thoracic horns and one growing from his head, this Atlas beetle, *Chalcosoma atlas*, looks like a miniature *Triceratops* dinosaur.

go rear end to rear end – the exact opposite of a face-off. Backing into position, they try to beat each other into submission with their legs.

Many crickets and bush crickets already have powerful back legs for jumping. These have been co-opted as instruments for singing, but in some species they're also used for fighting. Cave crickets or camel crickets (Rhaphidophoridae) are peculiar-looking crickets with extremely long antennae and equally long rear legs. Those of the males of so-called 'nutcracker' camel crickets are not only long but curved, allowing them to grasp a rival and squeeze him into submission. The wētās of New Zealand and Australia also belong to this family (see page 49). Some reach enormous sizes, while others live in dense aggregations inside caves. Fights are frequent among males in these colonies, again using their long back legs as weapons. The conflicts of one species, *Pachyrhamma waitomoensis*, have been carefully studied and show that, despite a wētā's bulk, it's not body weight that counts among these insects, but the length of their legs. The longest legs always triumph.[53]

Male combat leaves the fittest males to mate, which is a benefit for females, but well-armed, aggressive males may also pose a threat to the females. The males of one North American camel cricket, *Pristoceuthophilus marmoratus*, use their powerful back legs both for combat and to coerce

Male yellow dung flies, *Scatophaga stercoraria*, are much larger than females. They hang out on piles of dung and capture any females that visit.

unreceptive females into mating. Virgin females are usually happy to mate with victorious males, but once mated they are more concerned with egg-laying and so rebuff any further male advances. It's these females that male camel crickets often grab and force into mating.[54]

This seems to be an unusual case of male weapons being used for an alternative mating strategy – but many other insects are also less than gallant in their attempts to win a female. Most male katydids provide an expensive food gift as part of their sperm package, and this largesse gives them ample time to mate while the female is occupied with her meal. However, some males no longer go to all this expense and effort. Instead, they have turned their cerci (the appendages on the tip of the abdomen) into strong grappling hooks that firmly hold the female in place until the male has mated.[55]

In the case of a pond skater, *Rheumatobates rileyi*, the males have evolved similar grapples on their antennae to subdue females,[56] while male sagebush crickets, *Cyphoderris strepitans*, have developed what can only be described as a gin trap on their abdomens to trap females.[57] In this last case, mating is brutal on both partners. While the female is secured, she chews on the male's fleshy hindwings and feeds on the haemolymph that oozes out. So the male ends up providing a nuptial food gift even though he has the means to trap females against their will.

All of this seems a long way from the courtly songs and dances of grasshoppers or perfume-wafting orchid bees. But needs must. Finding a partner and leaving the legacy of a new generation is the culmination of the often brief life of an adult insect. Yet the story goes on. Their offspring must now survive in a hostile world and in most cases they must do so on their own. However, a surprising number of insects don't leave this to chance. Instead, they take on parental duties with as much commitment as any bird or mammal.

A male banded demoiselle, *Calopteryx splendens*, uses his coloured wings to display to rival males and to impress females.

8

The Next Generation

Insects as caring parents

You are not tired, I hope, of hearing about the Scavenger Beetles with a talent for making balls. ...now I wish to say a few words of yet another of these creatures. In the insect world we meet with a great many model mothers: it is only fair, for once to draw attention to a good father.

Jean-Henri Fabre, *Fabre's Book of Insects*, 1921[1]

Driving from Boston to Durham, New Hampshire, is like driving ever deeper into a Stephen King novel; white picket fences, white clapboard churches, immaculate small towns, and a primness and properness that must make the good people of the Granite State very happy. Set, as many of King's novels are, against the stunning natural and architectural beauty of the New England states, their macabre side is all the more terrifying. This link with the modern master of horror is particularly apt given my reason for making this journey. We were here to visit the labs of Michelle Scott at the University of New Hampshire, to film an insect story every bit as gruesome as any of King's imaginings.

We arrived to discover Michelle's lab lined with shelves stacked high with plastic boxes, each containing a dead mouse buried in dirt. This wasn't some twisted allusion to *Pet Semetary*, a King novel set in neighbouring Maine, but board and lodging for beetles with a truly extraordinary family life. The high societies of the insect world – those of the ants, bees and wasps (Hymenoptera) and the termites (Blattodea) – are based on extended families, albeit sometimes extended to tens of millions.[1] We'll explore these bustling societies in later chapters, but the first step in understanding how social insects arose is to take a close look at insects that exhibit less elaborate family lives.

Insects have the option of one of two broad strategies to ensure the success of future generations. Many simply play the numbers game. Pump out as many eggs as you can (and insects excel at this) and if the statistics are on your side, enough will survive to make sure your genes survive into the future. Alternatively, you can invest time and energy into making sure that your offspring have a better than average chance of surviving. The downside to such parental care is that it's only possible to look after a limited number of eggs or larvae. Depending on how protracted or involved parental care is, this limits the number of young that can be produced by each female, although the smaller numbers should be offset by a greater chance of survival. Just which strategy makes evolutionary sense depends on many and various ecological factors, which means parental care is found across a wide range of different insects.

Although insects are not widely seen as doting parents, a surprising number do have a more caring side. Compared to birds and mammals, most of which invest considerable time and energy into rearing their young, it's been estimated that only around 1 per cent of insects have any contact at all with their offspring.[2] Given the sheer number of insects, that's still a lot of species and some way of helping their young get off to the best start in life has been discovered in more than 2,000 different

(previous page) An emperor dragonfly, *Anax imperator*, goes to a lot of effort to find just the right plant stem into which to insert her eggs. Here they'll be safe until they hatch.

A burying beetle, *Nicrophorus vespilloides*, tends its larvae, nestled inside a buried mouse.

kinds of insect, across 19 orders and 164 families.[3] It's among these varied behaviours that entomologists and sociobiologists are seeking the origins of the great insect super-societies. The beetles in Michelle Scott's lab have particularly complex and flexible family lives, which has made them good experimental animals for scientists trying to understand the factors that drive the evolution of both parental care and social behaviour across the whole of the animal kingdom. They also make great television.

The seventy-five or so species of burying (or sexton) beetle, *Nicrophorus* spp., live right around the northern hemisphere, where they spend much of their adult lives hunting for the mortal remains of small birds or mammals. They track them down by scent and, if luck is on their side, a male and a female will bump into each other on the carcass. If not, the male releases a chemical signal that guides females to him and his trophy. Once paired up, the beetles set about burying the body. With a fitting sense of the macabre, I planned to film the start of this process in a classic New England cemetery, and the beetles duly obliged by interring a mouse I provided. They began by excavating soil from under the carcass, slowly lowering it into its final resting place. Between bouts of digging, the pair wandered over the carcass, clipping its fur, cleaning and preparing it for their future family. If the soil is hard or compacted, the beetles might only dig a shallow grave, but in the right conditions they can bury their prize nearly half a metre down, where it

will be much safer from the prying noses of skunks and racoons – and from other burying beetles.[4]

Once the beetles have buried the carcass, it's impossible to film the rest of their story in the wild. Instead, in a corner of Michelle's lab, we built small filming sets to house her experimental beetles and since she had pairs at different stages of rearing their larvae, we could quickly build up a picture of the extraordinary events that unfold in their burial chamber.

Below the ground, both parents continue to shave the carcass. Like miniature embalmers, they also smear it with anti-microbial secretions from their rear ends to prevent the carcass from decomposing too quickly. Once finished, they spit fluid from their mouths onto the top of the neatly prepared body, which partially digests the flesh and creates a small hollow. This will become a grisly nest for their larvae when they hatch. In some species the larvae can feed themselves on the mouse consommé accumulating in the bottom of their nest, but in others, including the one we were filming, they're also fed by their parents. The adult beetles make a rasping noise to call their larvae, who rear out of their nest to be fed – a gruesome parody of sweetly begging chicks. In such species, the larvae sicken and die if not fed by their parents, even though they're surrounded by food, which suggests that the adult beetles provide some additional supplements vital to larval health.[5] But in case I paint too charming a picture of the family life of the burying beetle, these beetles also practise brood reduction. They are smart enough to assess the size of the body they have buried in relation to the size of their family; if they discover that they have too many young to feed, they kill the excess.[6]

Burying beetles can make other, even more sophisticated decisions about their family life, which is why they have become such ideal animals for the study of evolution of parental care. They have advanced parenting skills but, before we explore these in more detail, we should look at some simpler examples to get an idea of the factors that drive some insects to invest more heavily than others in their young.

GIVING THE NEXT GENERATION A HEAD START

I can witness these first glimmerings of parental concern without ever leaving my own garden in Bristol in the south-west of England. In the back corner stands a rank and scruffy patch of garlic mustard, *Alliaria petiolata*, one of the food plants of the orange-tip butterfly, *Anthocharis cardamines*. These butterflies overwinter as pupae so are on the wing early

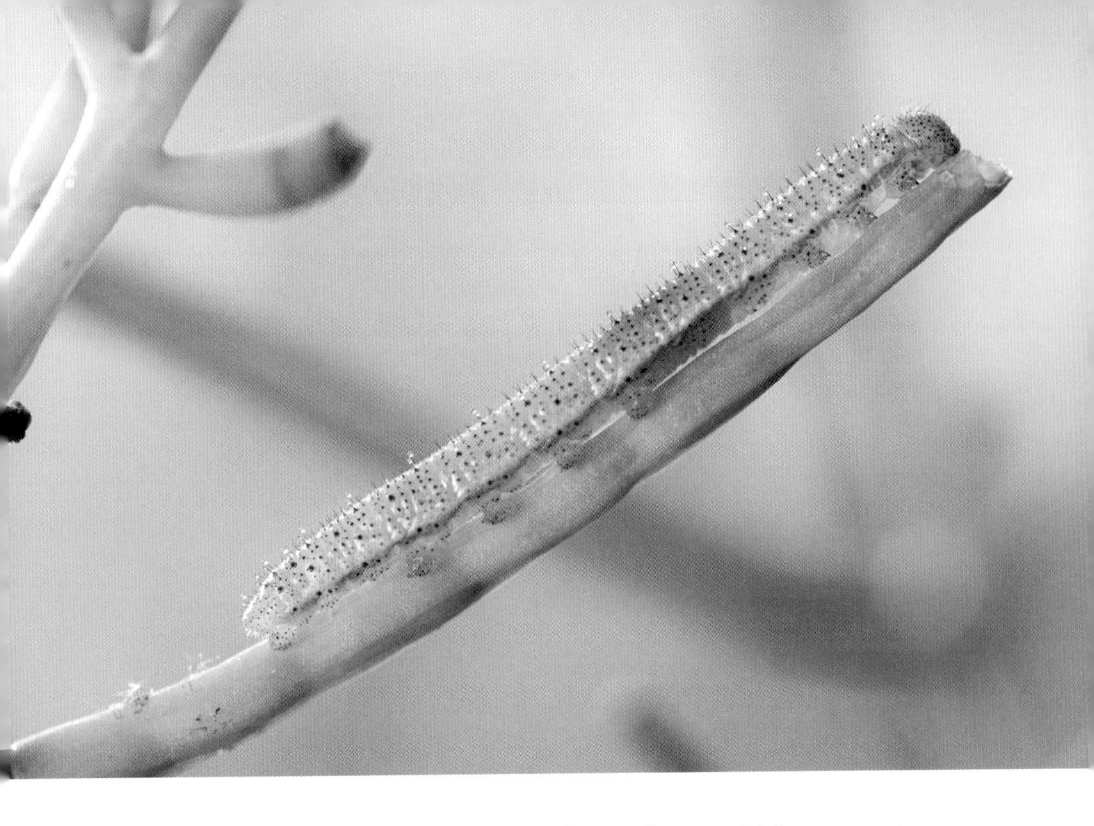

in the year, a welcome sign that winter has finally passed. The orange tips of the males' wings are as bright and uplifting as the spring sunshine. But I'm watching a female, lacking the orange tips, flit over my carefully nurtured weeds, giving them the once-over. And yes, she's impressed enough to select one of the long seed pods on which to lay her tiny egg. The caterpillar, which bears more than a passing resemblance to those seed pods, will munch its way down each one, like eating a giant cucumber. If it encounters another orange-tip caterpillar, it will eat that too. For vegetarians, they have a very cannibalistic streak. A female orange-tip takes care to avoid this and thoroughly checks each plant for eggs laid by other orange-tips. If she finds one, she'll move on and look for an unoccupied plant.

Many other butterflies behave in a similar fashion, but this parental diligence opens them up to some clever countermeasures from the plants they target as food. After all, as much as a caterpillar doesn't want to get eaten, neither does a plant. In Central America, there's a dazzling diversity of butterflies belonging to the genus *Heliconius*, all of which feed on passionflower vines. The vines try to protect themselves from hungry caterpillars with a wide range of toxic chemicals in their leaves, but also employ a couple of rather underhand tricks to dissuade

The caterpillar of the orange-tip butterfly, *Anthocharis cardamines*, feeds on the seed pods of garlic mustard, although it will eat another orange-tip caterpillar if it bumps into one.

Heliconius butterflies from laying eggs on them in the first place. They grow extrafloral nectaries at the bases of their leaves to encourage sweet-toothed ants that might also relish a few newly hatched caterpillars. On top of that, they grow little yellow blobs on their leaves and tendrils that bear an uncanny resemblance to *Heliconius* eggs. Experiments have shown that, like orange-tips, female *Heliconius* butterflies avoid plants that already have eggs from the same species laid on them. They also avoid plants that have grown egg-mimicking blobs.[7] It's not easy trying to be a caring parent.

Even if an insect never sees its young, placing them on the right food, free from competition, is at least a rudimentary form of care. Those unwelcome large and small white butterflies, *Pieris brassicae* and *P. rapi*, which are busy smothering your carefully tended cabbages with eggs that hatch into voracious caterpillars, have homed in on the distinctly 'cabbagy' smell of cabbages, so their young are born into a world of limitless food. A great many other insects follow the same pattern.

Sometimes it's more important for an insect to find a secure site for her eggs, but she still needs to make sure her larvae will be close enough to a food source that they won't starve while searching for it.

The silver-washed fritillary, *Argynnis paphia*, is a spectacular European butterfly, flitting and gliding through forest glades on black-chequered orange wings. The silver-washed fritillary is a butterfly of high summer, which is when it lays its eggs. Although the caterpillars hatch in late summer, they won't feed or grow until the following spring, after they've hibernated. So, it's critical for this butterfly to provide a safe and snug site for them to pass the winter. A silver-washed fritillary places her eggs in crevices in tree bark, where the newly hatched caterpillar can shelter from the winter, but being a good mother, she still makes sure that the caterpillar's food plant, the common dog violet, is growing in profusion nearby. She examines patches of violet plants to assess their quality, and if satisfied that there will be enough high-quality food next spring, she finds a tree close by on which to conceal her eggs.[8]

Hiding eggs is a simple but effective way of avoiding losses to predators or parasitoids, so many insects adopt this tactic. The female pink-winged stick onsect, *Necroscia annulipes*, provides one of the most remarkable examples of this method of egg protection that I've ever seen. As each egg emerges from her body, she holds it at the tip of her abdomen, so it's clearly visible. The egg is enormous, and one end of it is shaped like a chisel blade – and indeed, that's exactly how she uses it. She positions herself head down on the trunk of a tree and begins hammering her egg into the wood. Once she has made an impression, she stops hammering and begins twisting her abdomen back and forth, screwing the egg further into the wood. She only releases it once it is completely buried, then, finally, uses the cerci at the tip of her abdomen to pack bits of loose wood into the hole so that her egg is totally hidden.

An insect's egg can equally be seen as a form of investment in care; a female insect ensures the embryo has a place to grow that meets all its needs by producing an intricately sculpted survival capsule. Insect eggs are impressive structures that are impermeable enough to prevent the growing embryo from drying out, yet permeable enough for it to breathe. They are so sophisticated that my old professor and mentor, Howard Hinton, saw fit to write three very large volumes on the subject.[9]

More recently, I worked with a colleague and dear friend, Alfred Vendl, a professor at the University of Applied Arts in Vienna, Austria, to visualize the elaborate architecture of an insect egg in a remarkable new way. Alfred heads up the Science Visualization Lab Angewandte, set up to develop new ways of visualizing microscopic structures, and since he also runs a company that makes science documentary films, we have worked together on many projects over the last two decades.

A female short-winged conehead, *Conocephalus dorsalis*, has a huge, curved, blade-like ovipositor which allows her to lay her eggs deep inside plant stems.

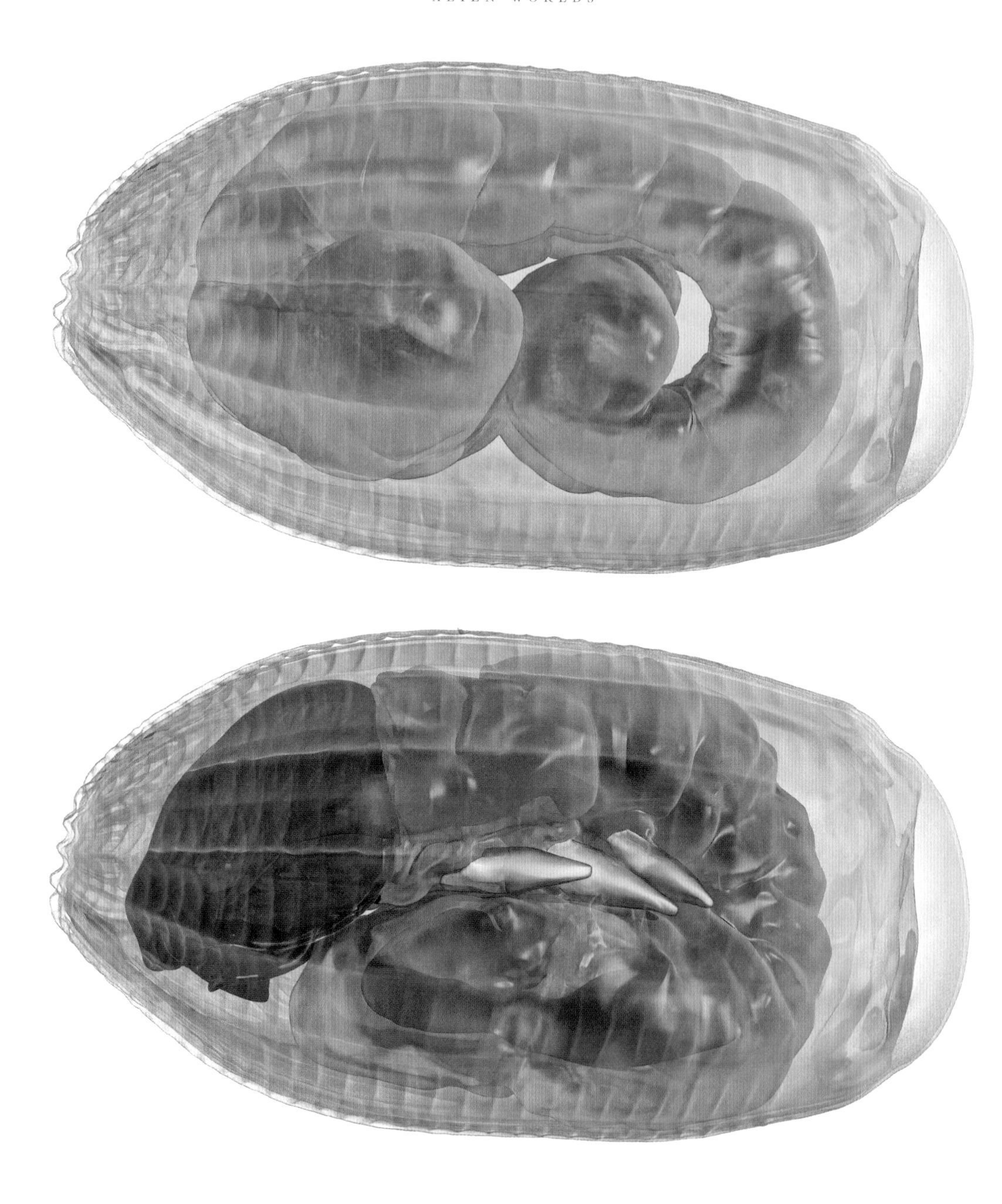

Micro-CT scans of large white butterfly, *Pieris brassicae*, eggs reveal the caterpillars developing inside. The top egg is one day old, the bottom six days old, and almost ready to hatch.

I chose the eggs of the large white butterfly as our latest subject because it's all too easy to persuade this species to lay a large number of eggs on cabbages and, looking down my binocular microscope, I could see that the eggs have an intriguing pattern of ridges and cross-ridges on their surfaces. Once my butterflies began to lay, I carefully detached the eggs from the cabbage leaf and fixed them in a chemical solution that prevents any degradation in their structure, even at a microscopic level. I then sent them to Vienna, to Stephan Handschuh, a colleague of Alfred's at the University of Veterinary Medicine. He has perfected the art of micro-CT scanning even the smallest specimens – a technique that uses X-rays to produce a 3D image of the internal structure of an object.

Next, the egg was placed in a scanning electron microscope, which uses an electron beam to scan the surface of an object in incredible detail. Now we had data on both the internal structure of the egg and its surface texture. All we needed to do was to combine this to produce a revelatory image of an insect egg. That task fell to Martina R. Fröschl in the Science Visualization Lab Angewandte, who produces astonishing computer graphic images from data such as we'd obtained from our egg. When the result popped up on the screen, I was left open-mouthed. We could see the structure of the egg in stunning detail, but the shell was transparent so we could see the tiny developing caterpillar curled up snug and safe inside, almost ready to hatch and demolish more cabbage leaves. More than that, we could even see the caterpillar's insides – its gut and excretory organs, and even its tiny brain. Howard Hinton would doubtless have been as stunned as I was to see such an image.

Eggs obviously provide protection for the growing embryo, but often in unusual ways. Owlflies are large, colourful insects that are found in the warmer regions of the world and can be quite abundant in places such as the south-western United States. They look a lot like dragonflies, although in reality they are related to lacewings and ant lions (Neuroptera). They lay their eggs on twigs but in two distinct clumps, one mass containing the growing embryos and a second one arranged as a palisade below the first. These palisade eggs, called repagula, contain

Newly hatched caterpillars of the large white butterfly, *Pieris brassicae*. Their mother carefully placed her eggs on a succulent food plant although the caterpillars' first meal is their own eggs shells.

no embryos and are purely for defence. Ants certainly seem reluctant to cross this barrier. The fertile eggs hatch into fearsome-looking predators, armed with huge, scimitar-shaped mandibles. Even so, while they are still small, they stick together on the branch, remaining behind the protection of the repagula. There are ways, however, to protect a clutch of eggs without having to sacrifice any.

Assassin bugs (Reduviidae) are formidable predators. They grab passing prey with their front legs, pinning it down before stabbing it with stiletto-like mouthparts. These mouthparts act like a hypodermic needle to inject toxins and digestive enzymes into the victim, then allow the bug to suck up a liquid lunch of bug purée. Some assassin bugs are large and quite capable of defending themselves against a molesting finger with a sharp stab from the beak – a bite that I can attest is extremely painful. To help hang on to large or fast-moving prey, some of these bugs coat their legs with the sticky resin that oozes from certain plants. It's such a neat trick that it has evolved independently four times in this group of bugs, so that unrelated 'resin bugs' are found both in the Americas and in south-east Asia.[10]

However, one group of these bugs, *Apiomerus* spp., living in the New World, has come up with another use for pungent, sticky resin. Female bee-killers, as these bugs are sometimes called, due to their effectiveness at using their flypaper legs to pluck bees from the air, also store the resin in special structures under their abdomens. These structures prevent the resin from setting hard, so a mother bee-killer can carry it back to her egg clutch. There she uses combs on her legs to smear the eggs with the resin.[11] Experiments on a species from the south-western deserts of the US show that these resin-covered eggs repel predatory ants.[12] There are times, however, when swarms of ants could come in very handy.

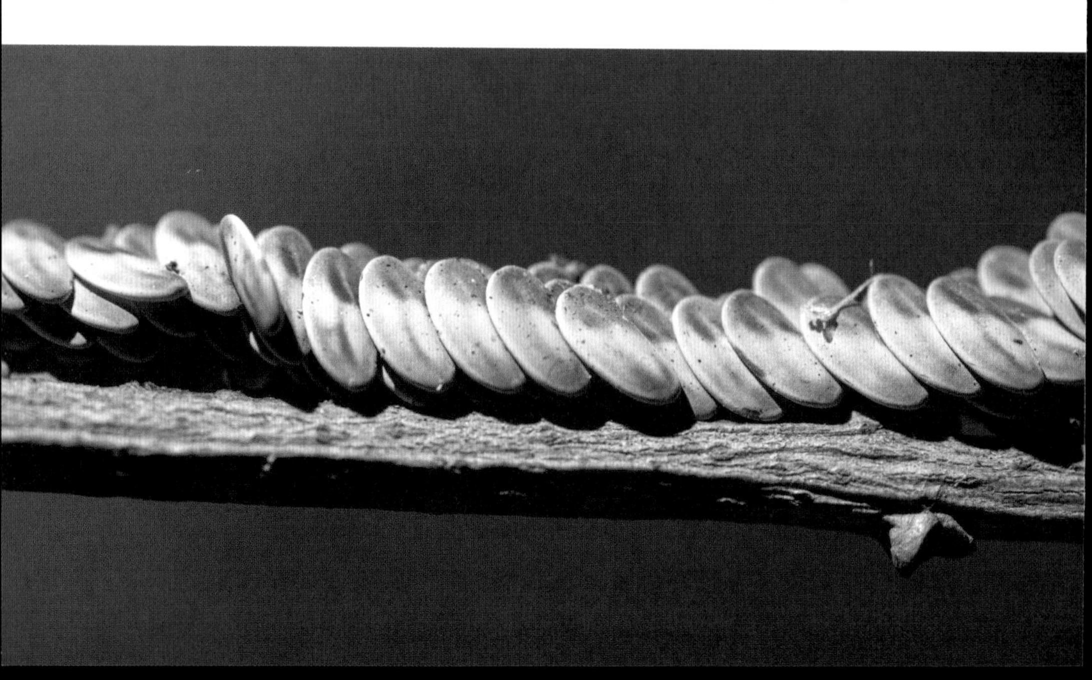

Like many children with a growing fascination for bugs, I used to keep a colony of giant prickly stick insects, *Extatosoma tiaratum*. These 12-centimetre insects are truly spectacular and are very popular pets. They're simple to acquire (since they breed prolifically in captivity) and feed very happily on blackberry leaves. In the wild, they're found in Australia, where they feast on the leaves of *Eucalyptus* trees, and as they munch gum leaves at one end, females produce eggs from the other at the relentless pace of a production line. As each large egg emerges from the tip of the female's abdomen, she cradles it briefly in the flaps of her ovipositor before she flicks it away with what seems to be casual abandon. It falls where it may on the forest floor below – but she has prepared it well.

On one end of the egg is a white knob, called a capitulum. This bears a striking resemblance to structures on the seeds of some plants, called elaiosomes. We met these in Chapter 6, where we discovered that plants pack these structures with tempting treats for ants, which encourages the ants to carry the seeds back to their nest (see page 245). There the elaiosome is eaten and the seed discarded, but left safely below ground and planted in fertile soil. Not only the capitulum but the whole giant prickly stick insect egg looks just like a seed, and clearly tastes like one, too, since ants also pick up stick insect eggs and carry them home. The capitulum is eaten and the egg left underground to hatch, which for stick insects is many months later. Eggs collected by ants are much safer from parasitic wasps during their long incubation than those left on the surface for that time. In this way female stick insects, like cuckoos, found a way of avoiding the hassles of parenthood – by foisting their eggs on other insects.[13]

BABYSITTERS

Ants make such good surrogate parents that a lot of insects use their babysitting services. Many other stick insects lay seed-mimicking eggs that are inadvertently protected by ants, but other insects have evolved even more intimate relationships. In the same forests in eastern Australia where the giant prickly stick insects lurk, imperial blue butterflies, *Jalmenus evagoras*, fly in search of acacia trees, the food plant for their caterpillars. They eat any of around twenty-five species of acacia, but female imperial blues are very choosy over exactly which individual trees they lay eggs on. They prefer trees that have strong colonies of ants living on them – and not just any ants. imperial blue females can identify

The flattened shape of the eggs of a Giant Katydid, *Stilpnochlora couloniana*, allows the female to stack them neatly along a twig.

ant species by sight – a skill that eludes many entomologists – and favour trees occupied by *Iridomyrmex* ants.[14]

The caterpillars secrete fluids from glands covering their bodies that are rich in carbohydrates and amino acids. The ants love these and in return they protect the caterpillars throughout their whole lives. Caterpillars living on acacias with big ant populations are far less likely to be parasitized by wasps, so much so that they can afford to gather in large, conspicuous groups. A caterpillar can even follow the ants' own chemical trails, laid down as the ants run to and from the caterpillars, to find one of these aggregations. The pupae too produce ant food, so these are also watched over by vigilant ants. They are so confident in their security force that the caterpillars pupate in large groups, exposed on the branches of the acacia.

The secretions that bribe the ants into defending the caterpillars cost a good deal of energy to make and use up food reserves, so – to produce enough – the caterpillars need to be feeding on trees with a high protein content. However, a female imperial blue, just like a female silver-washed fritillary, can assess the quality of the foliage of each tree she lands on, and she can make the tricky decision of where best to lay, based on both the quality of the tree and the presence of the right species of ant. She goes to a lot of trouble to make sure her caterpillars start out in the best possible environment.[15]

Many butterflies in the same family (Lycaenidae) use ants as babysitters, but one group has taken this to an extreme. The genus *Maculinea* includes the famous large blue butterfly, *M. arion*, which became extinct in Britain in the late 1970s for reasons that – while not entirely obvious at the time – were clearly linked to a decline in the quality of its grassland homes. It took some painstaking research by Jeremy Thomas and his team at the Institute of Terrestrial Ecology at Wareham in Dorset, to understand how totally dependent large blue caterpillars are on one species of ant and how dependent these ants are on just the right grazing regimes on the grasslands where they live.[16] Once the details of this very specific relationship were worked out, it was possible to recreate the right conditions for the ants and reintroduce the butterfly from similar populations in Europe. Now, once again, the large blue flies over a steadily growing number of grasslands in southern England.

There are perhaps half a dozen species of *Maculinea*, five of which occur in Europe, although their taxonomy, along with that of the closely related *Phengaris*, is still complicated and confused. However, all need particular species of *Myrmica* ant to raise their offspring.[17] Since the

Large blue butterflies, *Maculinea* (*Phengaris*) *arion*, lays their eggs on thyme or, in this case, marjoram, but the caterpillars will only survive if they are later taken into the nest of *Myrmica sabuleti* ants.

right ants can't survive in rank, overgrown and degraded grassland, these butterflies have become conservation flagship species – poster children for promoting the survival of species-rich grassland. Get it right for these exacting butterflies and you'll get it right for a lot of other species as well.

Female *Maculinea* butterflies lay their eggs on just one kind of plant, although each butterfly chooses a different species. The large blue lays on wild thyme (and occasionally marjoram), while Rebel's large blue, *M. rebeli*, which flies in a few Alpine meadows, lays on cross-leaved gentians. The more widespread Alcon blue, *M. alcon*, lays on marsh gentians. The caterpillars spend their first few weeks of life feeding on their food plant, like any normal caterpillar, until they moult into their fourth instar.* Then they drop to the ground and hope that an ant of the right species finds them. These caterpillars all possess dorsal nectar organs that secrete honeydew to attract ants and a battery of other organs (epidermal glands, pore-cupula organs and tentacle organs), which secrete other nutrients along with substances that seem to manipulate the ants' behaviour and prevent attacks.[18]

* Insects, like all arthropods, grow by moulting their exoskeleton. An instar is the stage between moults. After hatching, an insect is in the first instar. After its first moult, it reaches the second instar.

Bribed and pacified, an ant picks up the caterpillar and carries it back to the nest. Here the caterpillars use both chemical and acoustic signals to continue to manipulate their hosts. In some kinds of ant, the queen communicates her exalted status to her workers with sound, and *Maculinea* caterpillars have learnt to speak ant – with varying degrees of fluency.[19] Rebel's large blue caterpillars are such convincing queen mimics that they achieve a high status in the nest. If food runs short, they are fed in preference to the ant larvae and are even rescued ahead of the ants' own brood if the nest is disturbed.[20]

Rebel's large blue caterpillars live on the same diet as ant larvae, fed to them by duped worker ants, but large blue and Alcon blue caterpillars are even worse house guests. They've turned carnivorous. They feed on the ant larvae, while their chemical and acoustic camouflage, along with a very tough skin, keeps them safe from attack. As surrogate parents, these ants are making the ultimate, if unwitting, sacrifice.

While dumping eggs or larvae on ants – which can fiercely defend themselves and their nests – seems like a good strategy, one insect (at least that we know of) has come up with a stranger and as yet not fully explained tactic to avoid having to guard its own eggs. The golden egg

The caterpillars of the Adonis blue butterfly, *Polyommatus bellargus*, feed on horseshoe vetch, but when fully grown, the pupae are buried in chambers connected to an ant nest, where they're constantly tended and protected by the ants.

bug, *Phyllomorpha laciniata*, lives around the Mediterranean, where it's found on just one kind of plant, Algerian tea, *Paronychia argentea*, a distant relative of carnations. Females could just lay their eggs on this plant, ensuring that their young have plentiful food when they hatch, and that's what some of them do. But more often than not, they glue their eggs to the back of another golden egg bug – male or female. This unfortunate individual is then left to carry the eggs until they hatch. Eggs on the move may be less easy for ants or parasitic wasps to find or attack, or perhaps by avoiding extremes of temperature, the carrier bug also protects the dumped eggs from desiccation. While some observations do suggest that mobile eggs have a better chance of hatching than those laid directly on the plant,[21] other studies show that egg-carrying bugs are more prone to attack.[22] So, it's not entirely clear what the advantages are of each of these two strategies and why golden egg bugs don't put all their eggs in one basket.

This kind of behaviour is more easily explained as conventional 'egg-guarding' in giant water bugs, belonging to the family Belostomatidae. Some of these bugs are frighteningly impressive. At up to 12 centimetres in length and armed with the same kind of hypodermic mouthparts as assassin bugs, they fully earn their common name of toe biters. Yet they're also caring parents. Although the bugs live underwater, where they are formidable predators, species of the genus *Lethocerus* lay their eggs in a clump on a plant stem or stick just above the surface. Their mother then abandons them, and it falls on their father to look after them. He crawls out of the water to stand guard, but exposed to the air like this, the eggs would soon dry out. So, periodically, the male must crawl into the pond, then climb back out again to dribble water onto his eggs. He also must defend his clutch against predators. Eggs without a guardian are much more likely to fall victim to ants.[23] But babysitting a clutch of eggs above the surface takes a toll on the male, since it's impossible for him to hunt. Far better if he just carries the eggs with him. In species of the genera *Belostoma* and *Abedus*, the female glues her eggs to the back of the male, and he ferries them around until they hatch.

Just as for the golden egg bug, a mobile clutch may be harder for predators to track down, but in the case of these toe biters, the eggs are the male's own offspring, so it makes evolutionary sense for him to invest time and energy in looking after them. At intervals he performs a series of what look like push-ups with his back legs, a behaviour called *brood-pumping* that serves to oxygenate the eggs. As the embryos grow, they need more oxygen and their father compensates by increasing the intensity of his brood-pumping.[24]

Devoted fathers like these are rare in the insect world, but such paternal care is practised by a few other bugs, notably some assassin bugs (Reduviidae). Species of the genera *Zelus* and *Rhynocoris* lay their eggs in a clump on a plant and the male then guards them until they hatch, attacking any parasitic wasps that try to approach.[25] Not all male *Rhynocoris* bugs guard, however. In some species, such as *R. carmelita*, it's the female that takes on these duties, while in *R. tristis*, living in the same area, it's the male. These two coexisting species have allowed scientists to look at what factors dictate whether it's better for males or females to guard their eggs.

R. tristis lives at much higher densities, which makes it easy for males to find additional females without long, risky searches and also permits males to guard multiple broods without much extra effort. Females, on the other hand, can only lay their own eggs and if they were to guard them, that would prevent them reproducing again until the clutch had hatched. So, in this species, on balance it makes more sense for the female to abandon her clutch to the male, who can easily maximize his reproductive output by mating with several females and guarding multiple clutches. As an additional bonus, if he gets a little peckish with all this extended paternal duty, he can always eat one or two of his numerous offspring.

For *R. carmelita*, living at much lower densities, it's far harder for males to find multiple partners. In terms of energy expended, females generally invest more in the next generation, since they're the ones producing large eggs full of food. Sperm is much cheaper to produce and can be made in vast quantities. Because they're making less of a commitment, there is less incentive for males to invest heavily in raising their offspring. Instead, they play the field and fertilize as many eggs as possible. It's much harder for a male *R. carmelita* to attract enough females to his patch of plant to make it worth his while to guard the eggs, so in this species, the balance tips in favour of each female looking after her own precious clutch.[26] Even in species with male guards, the system depends on circumstances. If a male *R. tristis* dies in the line of duty, the female sometimes returns and replaces him as a guard.[27]

In species with paternal care, the female still needs to know that her partner is going to take good care of her large investment in the eggs. Males already guarding eggs must obviously be dutiful fathers and therefore they attract more females than non-guarding males. In fact, being a 'sexy dad' is such a draw that males sometimes stand guard over eggs that aren't their own just to show females what doting fathers they are. However, because these males are investing so much more effort than is normally the case, they have become the choosier sex. Male *Belostoma* bugs prefer big females who can lay more eggs, so in these species females have evolved to become the larger sex.[28]

HELICOPTER PARENTS

In most of the species we've explored so far, care ends when the young hatch. They're on their own to face the big bad world. However, a few insects, like burying beetles, extend care to the larval stages as well. This can take two forms. One or both parents can simply continue guarding and defending their brood as they did the eggs. Alternatively, they can also provide their offspring with food, either by herding them to suitable feeding places or by bringing food back to their young.

It should come as no surprise that the aptly named parent bug, *Elasmucha grisea*, has taken parenting to this more advanced level, at least in some parts of its range. In the north of Europe most females guard their offspring, but in central Europe only around one in twenty do, for reasons that have yet to be fully explained. This species has also come up with the concept of kindergarten. Parent bugs are a kind of shield bug (Acanthosomatidae), a species that I often find in local birch woods throughout most of the year, although they only mate in spring. The males soon die, leaving the females alone to guard the eggs. Before the eggs hatch, the female smears them with fluids from her mouth that contain the symbiotic bacteria that will help the young bugs digest the plant sap they feed on. When the eggs hatch, and while their mother stands over them, the nymphs eat their eggshells and so ingest the bacteria. Deprived of this vital inoculation they would soon starve and die.

The young nymphs feed themselves but if one wanders too far from its mother, she leans towards it and strokes it with her antennae, a gentle tap on its shoulder to remind it not to stray too far. Lots of creatures sharing their tree would make a quick meal of a young bug. If one of these predators comes too close to the family, the female tilts towards the threat, fanning her wings and jerking her body in warning. If these displays go unheeded, she releases a foul-smelling chemical from scent glands on her abdomen. Not for nothing is this family of bugs also known as 'stink bugs'.

As the nymphs grow, they wander further in search of food, accompanied by their mother – but they grow at different rates. Once they reach

their third instar, they begin to forage ever more widely on their own. So, while some nymphs are still small enough to be herded together, others are beginning to wander off, making it much harder to guard the whole family. In these situations, several families may unite and all the females share the task of looking after this shield bug crèche.[29] There are about twenty species of *Elasmucha*, and several of these (seven so far described) are known to guard their nymphs, some for a lot longer than the parent bug, all the way until their nymphs are fully grown. In some of these species, the nymphs release an alarm pheromone if threatened, which instantly rallies the female to their defence.[30]

The relationship between mother and young is even closer in another family of bugs. Treehoppers (Membracidae) are truly extraordinary-looking creatures, many adorned with bizarre projections from the thorax. In most cases, these make them look like thorns on a plant stem, but some are extravagant beyond belief. I fell in love with this family of bugs during the years of my undergraduate degree. Every summer break my professor, Howard Hinton, would head off to the wilds of Mexico armed with a butterfly net, collecting jars and a revolver (he wasn't afraid of bug hunting in bandit country). On his return at the start of the autumn term his office was lined with intriguing specimens, including an astonishing variety of membracids, many of which looked like nothing on Earth. Later research has shown that their behaviour is just as extraordinary as their looks.

Umbonia crassicornis is one of the less flamboyant species, though still a very convincing thorn mimic. Like many bugs, the female guards her clutch of eggs, then her young nymphs when they hatch. She takes up a position on a stem below her family, where she's well placed to encourage them back up the stem if they stray from the group. If she senses a predator on the stem, she displays her displeasure by fanning her wings and buzzing. If the predator persists, she approaches it head down, thorny back facing forwards, described by entomologist James Costa as looking like a charging rhino.[31] She also aims karate kicks at the threat.

Recently, in Florida, we filmed a thorn bug defending her brood against a paper wasp, *Polistes* sp., a common predator of young bugs. When the wasp ignored a wing-buzz threat, the mother bug sidled up to the wasp, which was much bigger than she was. The next thing we knew, the wasp was on the floor. We had filmed this process with a high-speed camera, which slowed down the action forty times. So, on watching the footage back, we could see what had happened. The bug slowly bent one of her legs away from the wasp, then delivered a kick so swift that

A female thorn bug, *Umbonia crassicornis*, guards her large brood. She defends them with karate kicks aimed at any predator that approaches too closely.

even slowed down it was still a blur. It hit the wasp full in the face and knocked it backwards off the branch. The action was so fast that it must involve a spring mechanism that stores energy in an elastic substance like resilin, in the same way that many jumping insects do (see Chapter 3, page 112). These feisty little bugs, we found out, will even threaten an over-curious human observer.

The thorn bug nymphs can also call for help. If they are attacked, they release an alarm pheromone in the same way as parent bug nymphs do, and to which the female instantly responds. As an alternative, they drum on the plant stem to broadcast vibrational signals, although it takes the whole brood drumming together to alert their mother through this communication channel. If a wasp gets too close, the whole group makes a little jump and in seconds their mother is on the case. The nymphs are so dependent on their protective mother that none will survive if she disappears.[32] However, her family depends on her for more than just protection.

As soon as her nymphs hatch, the female uses her ovipositor to make some slits in the stem near the brood, together with a series of spiral slits a little further down the stem. About half a day after hatching, the young bugs head down the stem to the spiral slits to feed while the female stands guard over them. It's not known how critical these prepared feeding sites are to the survival of the nymphs, but it takes parental care to another level, as parents become both protector and provider.

Again, I can find another good example of this stage of parental care in my own English garden. Turning over old broken pots in a neglected corner I sometimes find an earwig nest – a female surrounded by little white nymphs. In fact, this is what all these broken up pots are doing here, to provide cover for garden insects. Several times over the years, I've collected earwig nests and taken them to our film studio where we could capture the behaviour of maternal earwigs.

The common earwig in my garden is *Forficula auricularia*, sometimes called the European earwig. However, as we'll see shortly, it's not the only species to care for its young. Earwigs do their courting in autumn, when males use their pincers to gently poke and prod their intended. If all goes well, the female might seductively nibble the male's pincers and, if she finds him up to standard, mating follows. The female earwig then finds a cavity in the soil in which she lays a clutch of eggs. She sits on these through the late winter, like a broody hen, periodically licking them to remove any harmful microbes and fungi. Sometimes an earwig doesn't appreciate the exposure we give her on a film set, and if there's

Common earwigs, *Forficula auricularia*, are excellent mothers. They tend their eggs in a hidden nest, licking them continuously to keep them free of mould.

a more secluded spot at the back of the set, she'll pick up the eggs, one at a time, and carry them to a new nest, out of sight of bright lights and prying lenses.

Newly hatched nymphs stay in the nest at first but, like any toddlers, they wander around all over the place, forcing their mother to pick them up in her mouth and carry them back to the rest of the brood. It's just as touching as watching a cat with her kittens. Somehow, their mother finds time to bring food back to the nest but she also feeds her babies directly, mouth to mouth, as some burying beetles do.[33] When they are old enough to forage for themselves, they emerge under cover of darkness. However, it's safer if they remain together in a group, so their mother must still spend considerable time shepherding her little flock. If she's removed, the young wander all over the place and are less likely to survive their nocturnal forays. It's possible, though, that any wandering nymphs might be adopted by other family groups. This has been observed in burrower bugs, *Sehirus cinctus* (Cydnidae), which have a very similar pattern of parental care to earwigs, nesting underground, tending eggs and nymphs and herding them on foraging expeditions. Individual burrower bug nests are often grouped close together, which makes it easy for orphaned nymphs to join a neighbouring unrelated family group.[34]

All the earwigs so far studied make good parents, but one Japanese species, the hump earwig, *Anechura harmandi*, makes the ultimate sacrifice. When the nymphs have outgrown the need for maternal care, they simply eat their mother. This rather shocking behaviour – the technical term is matriphagy – is vital to their well-being. If the nymphs are deprived of their mother's remains, their survival rate is much poorer.[35] It makes a certain amount of evolutionary sense. Female Hump Earwigs only ever lay one clutch of eggs, so when their nymphs have matured, their purpose in life is complete. Other species, like some populations of European earwigs, can lay two clutches if conditions allow, so it makes no sense for a female to sacrifice herself if she may be able to breed again.

Matriphagy is not common in the insect world, even though among arachnids quite a few spiderlings feast on their mothers. However, a moth from the Mediterranean region, *Heterogynis penella*, suffers a similar fate. While the male is winged, females are wingless and legless bags of eggs that never leave their cocoons. The pupal magic that transforms earthbound caterpillars into ethereal winged creatures works in reverse for female *Heterogynis*. What emerges from the pupa is much simpler than the caterpillar, a process called catametaboly or regressive metamorphosis.[36] Like hump earwigs, *Heterogynis* only ever lay one batch of eggs and the first act of the newly hatched caterpillars is to devour their mother, whose body is packed with enough nutritious fats to send them out into the world with a hearty meal.

Another kind of burrower bug makes a similar sacrifice once she's done all she can for her family. *Parastrachia japonensis* is an attractive red and black bug from Japan that feeds solely on the fruits (technically drupes – single-stoned fruits, like cherries) of a shrub called *Schoepfia jasminodora*. Like all bugs, it has stiletto-like mouth parts, but instead of stabbing living victims in the manner of assassin bugs, it stabs these rose-hip-shaped fruits and sucks their juices. Female *Parastrachia* keep their eggs and nymphs safe in an underground nest. The problem is that *Schoepfia* bushes are widely dispersed, too far to herd the family to find fallen fruits. So, a mother *Parastrachia* has no option but to go off and collect food herself. She impales the fruit on her beak, then straddles it. The fruits can weigh twice as much as she does and are bigger than her body, so her feet barely reach the ground. Nevertheless, she manages to struggle back to the nest, perhaps 10 metres away – at her scale it's a very long walk to the shops.[37] As the nymphs get older, they get more demanding, and their mother has to ferry fruits to the nest almost continuously. It's such a hard job that she eventually works herself to death and her young then feast on her remains, their last meal in the nest before they have to leave home and make it on their own.

Parastrachia has another way of provisioning her young, one which is widespread throughout insects (and a few vertebrates, too) – she lays trophic eggs. Trophic eggs will never hatch; they have evolved as food. They are sometimes thin walled and usually packed with nutrients, a conveniently packaged high-energy snack. Originally, trophic egg production may have evolved to prevent cannibalism among siblings, but their use has grown ever more sophisticated. Third instar larvae of a group of beetles that we'll meet in the next chapter, the bess beetles (Passalidae), stridulate to let their mother know when they're hungry, and she responds by laying a trophic egg. Among the clutch laid by *Parastrachia* some eggs fail to develop, and newly hatched nymphs probe these eggs with their beaks, sucking out their contents. Such trophic eggs are important in the nymphs' development. Deprived of this first meal, nymphs do not grow as fast or survive as well as those that have had a nourishing egg breakfast.[38]

FEEDING THE FAMILY

An alternative to being eaten by your children is to find enough food to sustain your brood through its entire larval life without the need for this ultimate self-sacrifice. This is what burying beetles do. A mouse or small bird is plenty to feed the whole family until the kids fly the nest. A parallel case is that of dung beetles, since a pile of dung is a mountain of food for a beetle. However, both dung and dead bodies are often scattered and hard to find, which means that there's a lot of competition for such rich prizes. Being dependent on these kinds of sporadic riches is one reason why some insects build nests – places where scarce food can be safely hidden or more easily defended. Once an insect has a permanent base, the evolution of more elaborate parental care is easier.[39] Dung beetles are textbook examples illustrating a complete progression from simply abandoning eggs in a pile of dung, to ever more complex family life.

I confess to having a particular soft spot for dung beetles. My PhD supervisor, Les Strong, worked extensively on the dung beetles of cowpats and the threats to this essential clean-up crew created by overuse of antibiotics and other treatments for cattle. Years later, I became friends with Sarah Beynon, an Oxford entomologist with an utterly infectious enthusiasm for these beetles. She now runs, among other enterprises, an organic farm in North Wales, where dung beetles thrive on a healthy chemical-free diet. All of these beetles live either in the dung or in the soil beneath it. Unfortunately, in Britain we don't have any of the most famous kinds of dung beetle, those that roll balls of dung. The dogged determination of these beetles to roll a ball of dung several times their own weight, over long distances, has been admired for thousands of years.

Ancient Egyptians linked the dung beetle, or scarab, with Kheper, the scarab-faced god who created the sun anew each morning, then rolled it across the sky. Not surprisingly, dung beetles are commonly depicted in Ancient Egyptian art. One tomb painting I saw in the Valley of the Kings shows a beetle with a ball beautifully depicted in such a way that it could either be dung or the sun. A series of exquisite scarabs carved in lapis lazuli crawl around the necklace that Howard Carter and Lord Carnarvon found around the neck of Tutankhamun. Dung beetles are everywhere in Ancient Egypt, but most impressive is a giant sculpture of one of these beetles in the temple at Karnak. So, when I wanted to film dung beetles doing what they do, for a sequence that also touched on their role in Ancient Egyptian religion, Karnak seemed like the place to go.

We spent a number of fiercely hot June days persuading dung beetles to roll their dung balls past appropriately symbolic backgrounds (what did you do at work today, dear?). The reward was some dramatic footage and some very memorable evenings. We were filming during the 1990 FIFA World Cup, for which that year Egypt had qualified. There was only one TV in the tiny village where we were staying, so the inhabitants ran a cable to the middle of a dusty street to plug in the communal TV and arranged enough chairs for the whole village to watch. In drinking copious quantities of the local beer, it felt good to know we were continuing a 5,000-year-old local tradition. According to legend, Osiris taught the art of brewing beer to the Ancient Egyptians – and they haven't stopped since.

Not all dung beetles roll dung. They have evolved several different strategies to make best use of a fresh pile of dung.[40] Some dig tunnels beneath the dung, from which branch multiple brood chambers. Often male and female beetles work together to drag lumps of dung down to each chamber, where the female shapes it into a brood mass or brood ball. The former is just a pile of dung filling the end of the tunnel, the latter a more carefully worked sphere of dung, which may be coated with a mud render – but both are made to provide food for the larvae. This is the most basic dung beetle life cycle from which others have evolved. The advantage of tunnelling is that there is a ready supply of food close at hand. The disadvantage is that there is also a lot of competition, not just from other dung beetles, but from flies, along with a host of other insects. This may be why in some tunnellers, the male and female cooperate to dig tunnels and bury as much dung as they can as quickly as possible. Usually, the male abandons the nest shortly after building is complete, leaving the female to carry out the remaining parental duties on her own.

Copris lunaris is a large beetle that buries dung either beneath or close by a cowpat or pile of horse dung. It's a good example of a tunneller that has developed more complex care. It is widespread across Europe but scarce in Britain, although I sometimes come across it in the New Forest in Hampshire, where free-ranging New Forest ponies deliver liberal quantities of food for these beetles. A pair starts out by cooperating, the female digging the tunnel while the male passes her parcels of dung that she moulds into brood balls. Then, his job done, he leaves the female to do the hard bit. She lays an egg on the top of each brood ball and keeps a close eye on larval development. If the ball is dislodged, she carefully rights it again, so the egg or larvae remain on top. If the ball is damaged in any way, she repairs it.[41]

Some beetles don't dig tunnels but make brood chambers in the dung itself, separating a ball of dung from the mother lode and creating a cavity around it. These beetles are surrounded by food but face the fiercest competition from other dung eaters. The alternative is to take the dung ball and roll it away from the competition as fast and as far as possible. In these dung rollers, the male excavates, shapes and rolls the dung ball, then releases a pheromone to attract a female. They cooperate to dig a nest burrow once they've rolled their precious dung ball far enough away from the melee, and in some species, such as *Phanaeus daphnis*, the male stands guard over the ball as it rests outside the burrow entrance, an imposing sentry with his long, curved horn. Meanwhile the female breaks pieces off the ball and drags them down the burrow to stock brood cells.

As is common among insects, there are endless and often bizarre variations on these themes, particularly in how to acquire a good supply of dung. Several species of Australian *Onthophagus* beetle specialize in the dung of kangaroos and wallabies. These marsupials are often nomadic and can hop great distances with relative ease, so their dung ends up spread over a large area. In response, *Onthophagus* beetles have evolved large claws that allow them to cling tightly to the rear ends of kangaroos as they bounce along, which gives a whole new meaning to the expression 'down under'. They hang on tight, waiting for their hosts to visit what Australians refer to as the dunny, at which point they are guaranteed the very freshest dung.[42] Even so, marsupial dung is rather dry, so some *Onthophagus* beetles bury the pellets deep enough that they can absorb moisture from the soil.

A similar problem is faced by a pretty little dung beetle, *Scarabaeus denticollis*, with bright orange elytra (wing covers), which makes a seemingly impossible living on the shifting sand of the Namib Desert in south-west Africa. One of the few big mammals to venture into these vast dunes is the gemsbok. These large antelopes are often drawn here by nara melons, succulent fruits produced by nara bushes that grow at the bases of the dunes, wherever their 50-metre roots can reach moisture deep below the sand. As a parting gift, a gemsbok may leave a pile of neatly spherical pellets, conveniently pre-rolled dung balls that *S. denticollis* is quick to remove, using a unique trick.

It has two long spikes on its back legs with which it stabs the dung ball, and which act like axles, allowing the beetle to roll the ball easily while walking on its four front legs. It buries the ball in a tunnel in the sand, but it's not finished yet. These beetles don't eat the dung itself but a fungus that grows on it. The problem is that gemsbok are well-adapted desert antelopes and waste as little water as possible, so their dung pellets are too dry for any fungus to grow. Unlike the Australian *Onthophagus* beetles, it's doubtful that *S. denticollis* could ever reach the moisture far below the sand, so instead it collects bits of vegetation, such as nara flowers, and adds these to its burrow to increase the humidity and thereby allow fungus to grow on the buried dung balls. It's charming to watch a little beetle trundling over the sand taking a flower home to freshen up its nest.

Several other dung beetles concentrate their efforts on neatly packaged antelope dung. In South Africa, *Epirinus flagellatus* prefers the dung of eland, produced in copious quantities by this huge antelope. A quick sniff, and the beetle grasps the dung pellet between its back legs and wheels it off, to bury in some quiet spot where its larvae can grow in safety. However, some of these beetles are going to be sorely disappointed. They're not rolling eland dung at all – but the seeds of a sneaky South African plant, *Ceratocaryum argenteum*.

The seeds of this rush-like plant are as large as a pellet of eland dung, much the same shape and texture, and they even smell like eland dung. The beetles are completely taken in and roll the seeds off to bury them like diligent gardeners.[43] The plant benefits from this faecal mimicry but the

The South African dung beetle, *Epirinus flagellatus*, specializes in eland dung, which comes in handy, ready-to-roll pellets.

beetles don't. The seeds are rock hard and provide no sustenance for the beetle's grubs. While studying the details of this one-sided relationship, scientists found a second kind of dung beetle, *Scarabaeus spretus*, also rolling away the seeds of this plant.[44] The seeds of *Ceratocaryum argenteum* are obviously excellent mimics of eland dung and are irresistible to these beetles.

A world away, the rainforests of Central America are also teeming with diligent dung beetle parents, although here they face different problems. Some species are fond of the dung of howler monkeys – there's clearly no accounting for taste. I've been unfortunate enough to have stood beneath a howler monkey when nature called. That stuff is pungent poo, no question. The dung that reaches the forest floor presents little problem; the beetles simply home in on the smell (hardly a difficult task), then roll away a ball for burial. Often, though, the dung ends up smeared over leaves tens of metres up in the canopy. No problem. The beetles turn up in their droves, carve out a ball each, then simply roll off the leaf, clinging to their precious ball. Dung makes a lousy parachute, but the beetles are small enough to be unharmed by the drop. Thanks to these beetles, smelly monkey dung continues to rain from the canopy long after the howlers have moved on.

Quite a few dung beetles have abandoned dung as a food resource for their larvae. The behaviour patterns evolved to exploit dung – rolling a ball to a secure site, then burying it and remaining with it as the larva grows – can be adapted to a wide range of larval foods. But one North

American species, *Deltochilum gibbosum*, has taken this to a rather bizarre extreme. It collects feathers, folds them up into a ball, then rolls this rather disappointing offering away, often for a considerable distance, before it coats the ball with mud and leaves and digs a pit to hide it in. This unappetizing diet will be all their larvae get to eat.[45]

In Brazil, another dung beetle, *Canthon virens*, has opted for a very niche diet. It has turned predator and specializes in the queens of leafcutter ants, *Atta* spp. On the face of it, this seems like a very foolhardy thing to do. *Canthon* is only about the size of a leafcutter queen's head – a head that possesses some very powerful and sharp mandibles. Undaunted, the beetle struggles onto the ant's back and proceeds to saw away at its neck with its serrated legs until the queen is beheaded. The body is then rolled away and buried. Underground, the beetle makes two or three brood balls from the queen's flesh to feed her larvae.[46]

In the rainforests of Peru, a close relative of the North American feather-roller has also switched to feeding its larvae on flesh. *Deltochilum valgum* is a hunter of millipedes. It too seems ill-equipped for a predatory life, but manages to kill millipedes using its shovel-like head to disarticulate its unfortunate prey. Sensibly, this beetle seems to prefer injured millipedes. It's probable that it uses the defensive secretions of millipedes to track down its prey, and injured millipedes release clouds of the stuff, making them easier to find. The site of the injury also gives the beetle a place to start on the gory process of levering the millipede apart.

The seeds of the South African plant, *Ceratocaryum argenteum*, are such convincing mimics of eland dung that *Epirinus flagellatus* dung beetles are completely fooled and roll the seeds away to bury them.

Once in pieces, the convenient chunks are then rolled away for burial.[47]

Despite being protected by noxious secretions of quinones, millipedes are a popular choice among dung beetles that have abandoned dung. In South Africa, several species of *Sceliages* beetle also use these defensive chemicals as a means of tracking down millipedes.[48] But Sceliages seems even less adept at its alternative lifestyle than the Peruvian millipede hunters. It can't kill a large, healthy millipede. Instead, it needs the help of assassins – millipede assassin bugs. *Ectrichodia crux* is a large and colourful assassin bug, equipped with suction cups on its front legs to help it hang on to the smooth, slippery surfaces of large millipedes – far better equipped to do the job.

Adult bugs measure around 2 centimetres, so are much smaller than giant millipedes, but they leap onto their prey and hang on with their specialized front legs. Stabbing the millipede with their stiletto mouthparts, they inject toxins and digestive enzymes. Needless to say, the millipede objects to this and writhes and bucks. The bug holds on, like a rodeo cowboy, until the toxins take effect, and then it can slurp up the partially digested remains. Smaller nymphs work together to overpower these big millipedes; there are often so many of them that the millipede is entirely hidden under a mass of dozens of bright red assassin bugs. After they've eaten their fill, there's still plenty left and now *Sceliages* can move in.

The beetle either scoops out the softened flesh to roll into a ball, or dismembers the millipede and rolls away chunks. These will be buried and eaten or, if intended as a brood ball, the female scoops out the flesh and forms it into a ball, lays her egg on it, then covers the ball with a layer of mud. It's possible that *Sceliages* is simply drawn to the carcass of a millipede, no matter how it was killed, but there appears to be more to it than this. It seems that *Sceliages* and *Echtrichodia* have formed some kind of relationship, although exactly how this works is still not clear.

Dung beetles go to a lot of effort to provide a safe nest site and plenty of food for their offspring. However, such diligence comes at a cost. Parental care raises the likelihood of offspring reaching maturity, but the time and effort this takes limits the number of offspring that can be reared. Among dung beetles, the costs and benefits of care are illustrated by an example of extreme parental devotion. *Kheper nigroaeneus* is a large dung beetle found in South Africa. Like many dung rollers, the male abandons his partner after the nest is complete. The female then lays just a single egg on the brood ball and remains with her only child, all the way through larval growth and pupation, until she sees it emerge as an adult.[49] This strategy improves the chances of her young surviving; up to 80 per cent make it to adulthood, comparable with the best care shown by birds and mammals, but each female may produce only three young in her entire life.

Dung beetles and burying beetles are parallel cases. Both depend on resources that, although large, may be dispersed and hard to find, which makes them worth hiding or defending. The ecology of these beetles has helped promote the development of parental care, and in the case of the more varied and numerous species of dung beetle, has provided a good framework for looking at how different stages of care may have evolved and at what the costs and benefits might be. In an evolutionary sense these animals are continually weighing up these advantages and disadvantages to make decisions on how best to maximize their reproductive output over their lifetime – their ultimate purpose in life.

The burying beetles in Michelle Scott's lab provide a window on how hard it is to be a burying beetle parent. Like many dung beetles, male burying beetles usually decide to leave at some point early in the process, though a variety of circumstances dictates the exact timing. It may even be disadvantageous for the male to hang around too long. He must also feed himself off the carcass that he and his partner have buried, but if this is only small, he'll deprive the young of food, and his presence may reduce larval survival. On the other hand, if the female dies, the male will stay on and is quite capable of being a successful single parent.

Because males, energetically speaking, invest less in the next generation, if the carcass is large enough it makes sense for the male to mate with several females. So, he continues to call for females even after the first has joined him – though if the original female spots him doing this, she knocks him off his perch. For her, it's far better if he's entirely focused on helping raise her offspring. The ecology of burying beetle family life does have one other intriguing consequence.

Burying beetles face stiff competition from flies, and in extreme cases this may force them to breed communally. Several pairs work together to bury the carcass quickly, then each rears its offspring on it. In this case their behaviour extends well beyond just parental care. Now all sorts of decisions must be made about cooperation or conflict. If one female is larger, she may destroy some of the eggs of the smaller female. Even the insects' physiology changes. Egg production increases in the dominant female, while it slows down in the subordinate.[50] Parental care has become something more complex – the first glimmerings of social living.

On America's West Coast, there's another opportunity for burying beetles to develop a more social way of life. Every year, many of the rivers emptying into the Pacific fill with salmon of several species, all heading upstream to spawn, having spent several years feeding and growing in the ocean. It's a once-in-a-lifetime journey for these fish, since most will die after spawning in the upper reaches of the rivers. Their carcasses provide a banquet for a host of animals, from bears and wolves down to flies. Even the trees in the surrounding forests feast on the nutrients released as the carcasses rot, and one of the creatures that helps fertilize the forests is a species of dung beetle, *Nicrophorus investigator.* The beetles bury fragments of decomposing fish, or even whole corpses, on which they raise their young, though this is such an enormous task that many pairs must work together to hide their fish supper before it ends up inside a bear. These groups raise their young communally and in one survey, nearly half of the salmon carcasses were home to around 100 larvae, and some held up to 750.

In the next chapter we'll meet a group of beetles (Passalidae) in which social behaviour like this has fuelled the development of true colonies, and in later chapters we'll explore the highly complex world of the two great insect societies, those of the ants, bees and wasps (Hymenoptera) and the termites (Blattodea). It's within this latter group that we can most easily see how small families can become extremely big families. To that end, I've omitted from this chapter one group of insects that shows an illuminating range of parental strategies, from minimal care all the way to carrying their young with them and even feeding them from structures on their abdomen, an insect equivalent of suckling. These are the cockroaches, and modern molecular phylogenetics has shown that cockroaches and termites are so closely related that they all belong together in the same order. In other words, termites are nothing less than highly social roaches.[51] So, we'll meet these maternal roaches in a later chapter.

However, before visiting these super-societies, there is a surprising variety of other kinds of social behaviour to explore in the insect world. James Costa coined the phrase *The Other Insect Societies* for this intriguing collection of semisocial and subsocial insects, and there proved to be more than enough material for him to publish a voluminous and definitive yet very readable book on the subject.[52] That was back in 2006; since then, all kinds of other remarkable stories have emerged that continue to reshape our perceptions of the social side of insect life.

9

Living Together

Insects as social creatures

Their [locusts'] numbers… are so vast, that they quite darken the sun; while the people below are anxiously following them with the eye, to see if they are about to make a descent, and so cover their lands.

Pliny the Elder, AD 77[1]

The mangroves had… a small kind of caterpillar, green and beset with many hairs: they sat on the leaves many together… like soldiers drawn up… if these wrathfull militia were touchd but ever so gently they did not fail to make the person offending them sensible of their anger, every hair in them stinging much as nettles do.

Joseph Banks, 1770[2]

The little town of Seventeen Seventy lies on the Queensland coast halfway between Rockhampton and Bundaberg. It is named for the year in which Captain James Cook anchored here, his second venture ashore in Australia, a week's sailing from his first port of call in Botany Bay, near what is now the bustling harbour city of Sydney. Travelling with Cook on board His Majesty's Bark *Endeavour*, the naturalist and botanist Joseph Banks soon disembarked to explore the mangroves along the shores of Bustard Bay. It was here that he encountered his 'wrathfull militia' (see preceeding page), probably a group of caterpillars of a moth called *Doratifera stenora*, .[3] It belongs to a family of moths (Limacodidae) with some of the strangest-looking caterpillars on the planet, most of which have stinging hairs.

I've come across the stinging rose caterpillar, *Parasa indetermina*, from the same family, in the eastern forests of North America, and it certainly lives up to its name. It's brightly coloured, with seven pairs of fleshy horns each crowned with a tuft of hairs, all of which are filled with a stinging fluid; it takes just the slightest touch to snap these hairs and receive a nettle-like sting. *D. stenora* caterpillars have even more horns and hairs and, to make matters worse, they live in large social groups, lining up in close-packed formation to feed on the leaves of the spotted mangrove, *Rhizophora stylosa*. For a naturalist like Joseph Banks, it must have been an intriguing, if unpleasant, encounter.

(*previous page*) The caterpillars of the lackey moth, *Malacosomia neustria*, stay together as a tight-knit family group throughout their larval life.

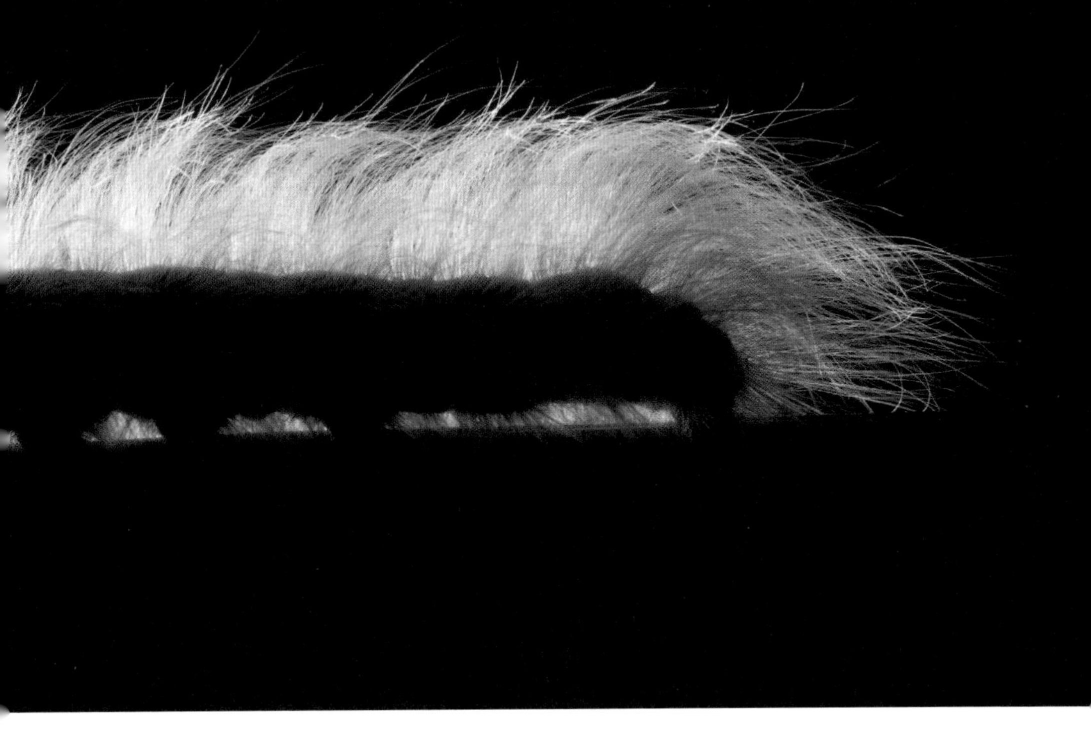

Mention social insects and few would think of caterpillars, yet so far we have discovered more than 300 kinds of caterpillar that lead complex social lives. Moreover, there are many other surprisingly social insects: social sawflies (Hymenoptera), social aphids (Hemiptera), social thrips (Thysanoptera), social grasshoppers (Orthoptera) and social beetles (Coleoptera). However, in the popular imagination, as well as in the scientific literature, it is two other highly social groups that have drawn most attention.

The ants, bees and wasps (Hymenoptera) and the termites (Blattodea) form the largest, most complex and most conspicuous societies of all, and not surprisingly dominate studies on social insects. They represent a major turning point in evolution, a unique stage in the development of life on Earth, which is why they've attracted such interest. We too will explore the lives of these insects in the following chapters, but before then there's much to ponder among all the other social insects. Their societies are often formed in different ways from those of the hymenopterans and termites, so they help us put these evolutionary innovations into some kind of context. Equally, they all have a fascination of their own. How these societies differ from those of the high societies of ants, bees, wasps and termites is best summarized by plunging into a few biological definitions – the sooner done, the sooner finished.

Ants and the social bees and wasps, together with termites, are

Many caterpillars, such as this *Philotherma jacchus*, are protected from parasites and predators by a thick coat of hairs. These defences are even more effective when caterpillars band together.

referred to as *eusocial* insects, meaning 'truly social'.* Other forms of sociality have, in the biologist's insatiable desire to classify, been given other names. *Subsocial* insects are those we met in the previous chapter, whose social lives revolve around the bond between parents and young. *Communal* insects are those in which members of the same generation live together, like Banks's stinging caterpillars. *Quasisocial* insects are in essence communal insects that also show cooperative care for their brood, while *semisocial* insects show all the features of their quasisocial kin but also have some form of division of labour among the colony members.

The 'truly social' insects have all these attributes and in addition two adult generations overlap, allowing the adult offspring to help their parents in defence and brood rearing. Names matter – not in the sense that you should memorize these terms to impress friends at dinner parties, but in that all of them – quasi, sub, semi – suggest a somewhat imperfect form of society, inferior to the teeming colonies of eusocial insects. Of course, such a simplistic view is hardly valid. Each species has adopted a strategy for living dictated by its own evolutionary and ecological circumstances. It's a strategy that works perfectly well for the current situation in which each individual species finds itself. In fact, some scientists now argue that these definitions are not all that helpful as they obscure many of the subtleties of insect social life. So, let's follow the lead of Ed Wilson, the father of social insect studies, who argued for a much broader definition of sociality so as not to exclude some very intriguing and very enlightening stories, and broaden our view of the social interactions of insects – and let's start by exploring the world of hopper herds and communal caterpillars.

HOPPER HERDS

Willian Henry Hudson was born in Argentina in 1841 and spent his early years fascinated by the local plants and animals. In 1892, he published *The Naturalist in La Plata*, in which he recorded his observations of social grasshoppers. He wrote:

> There is in La Plata a large handsome grasshopper (*Zoniopoda tarsata*), the habits of which in its imago and larval stages are in strange contrast, like those in certain Lepidoptera, in which the caterpillars form societies and act

* Eusocial insects form the most complex societies in which several generations of insects live and work together. Often, individual workers are specialized for different tasks, both in their behaviour and in their morphology.

Despite their large size, the adults of the grasshopper *Tropidacris collaris* are hard to spot in vegetation with their subdued green and brown colouration.

> in concert. The young are intensely black, like grasshoppers cut out of jet or ebony, and gregarious in habit, living in bands of forty or fifty to three or four hundred.[4]

This grasshopper belongs to the New World family Romaleidae, the lubber grasshoppers, the nymphs (hoppers) of some of which are gregarious. The gaudy grasshoppers (Pyrgomorphidae), a family of largely Old World distribution, also has many species that pass their youth in large hopper herds. Such social aggregations give the young grasshoppers a variety of advantages over their solitary kin.

The western horse lubber grasshopper, *Taeniopoda eques*, is an impressively large and handsome black grasshopper, covered in a filigree of yellow lines, which lives in the desert of the American south-west. I've filmed them on several occasions in the Chihuahuan Desert, which is no easy place for a grasshopper to make a living. Baking in the heat of a summer day, it gets seriously chilly at night. The black colouration of the adults helps them absorb heat in the early morning, and their large bulk helps them retain it, but the smaller hoppers face a real problem with such fluctuating temperatures. At the end of the day, they band together and leave their food plants to head to communal roost sites a metre or two up in the vegetation.[5] These sites are chosen so they intercept the early morning sun, but huddling close together might allow the hoppers to warm up more quickly than if they spent the night alone. This, as we'll soon see, is certainly the case for many communal caterpillars.

It's equally likely that clumps of young hoppers serve a defensive function. Horse lubber nymphs are jet black, with a bright yellow line along the back and yellow stripes around the head and thorax. Yellow and black are classic warning colours in nature, advertising the fact that their bearers are packed with nasty-tasting toxins. Crowding together amplifies the signal and leaves a predator in no doubt that the hoppers are best left alone. Many kinds of pyrgomorph and romaleid sport such aposematic colouration, backed up by potent chemical defences. Romaleids spray toxins from their front spiracles, while pyrgomorphs release their chemical weapons from glands on their abdomens. In either case the effect is the same – to persuade predators to leave them alone. Banding together makes bright colours all the more conspicuous and easily remembered so the lesson, once learnt, won't soon be forgotten.

Related to the horse lubber, *Chromacris colorata* is an even more colourful grasshopper from Central and South America. Adults are painted in intricate patterns of blue and yellow, while the hoppers have black bodies with bright red heads. They too strengthen their warning signals by moving in large, coordinated groups that march in such close synchrony that they look like one much larger animal.[6]

From the same part of the world, *Tropidacris* grasshoppers are among the largest in the world. Like those of many romaleids, the nymphs feed in crowded and conspicuous groups on exposed leaves, as if daring the world to take them on. It might seem from all these examples that displaying bright colours warning of potent chemical defences and living in large, easily seen groups are twin parts of a defensive strategy, but this may not always be the case.

A recent study of *Chromacris psittacus* suggests a different reason for hoppers adopting a communal lifestyle. In this species, the nymphs feed on a few toxic plants in the nightshade family (Solanaceae). They use the chemicals in their diets to create their own defensive arsenal, and like many of their kin have bright warning colouration. The adults, however, become cryptic and solitary, although they do have brightly coloured hindwings that serve to startle any predator as the grasshopper makes its escape. Strangely, the adults feed on exactly the same plants as the hoppers and have the same chemical defences, so why do the young band together in brightly coloured gangs, while their parents rely on not being seen? Is it just to enhance their colourful warning or could there be other reasons?

There is a cost to eating toxic plants. *Chromacris* hoppers are surprisingly slow growing, probably because they have difficulty handling

The nymphs of the grasshopper *Tropidacris collaris* are packed with foul-tasting chemicals and advertise this fact with bright colouration and by hanging around in conspicuous gangs.

the plant's defences and digesting their food. However, there's also a cost to the plants. They invest a lot of energy in making the chemicals required to dissuade herbivorous insects from feasting on their leaves – energy that could otherwise go into growth and seed production. To avoid wasting these expensive chemicals, the plants manufacture toxins largely in response to leaf damage, deploying their defences exactly where they're needed. Plants can detect which leaves are under attack from munching mandibles and respond accordingly.

So, perhaps attacking a leaf en masse allows all the hoppers to feed quickly before that leaf becomes unpalatable, after which they simply move on to pastures new and play the same trick again. If the nymphs were scattered over the whole plant, all the leaves would soon become unpalatable. Forced into herds by their food plants, the hoppers then evolved their warning colouration to reduce predation on these otherwise exposed groups.[7] In other words, group living may have been a strategy to cope with plant toxins, at least in this species, with a defensive role tacked on later.

The communal larvae of the willow leaf beetle, *Plagiodera versicolor*, face a different problem with their food, at least when they're newly hatched. Their tiny mandibles have a hard time cutting through the cuticle of the willow leaves on which they feed. The effort needed is visible as each larva rocks back and forth, trying to make some inroad into the leaf. Eventually, one succeeds in breaking through, at which point all the other larvae cease rocking and pile into the breach, which is soon expanded to accommodate all. This communal effort makes a difference. Larvae from larger groups grow more quickly than those from smaller groups, the reward for all working together to attack the leaf.[8]

In the Old World, pyrgomorph grasshoppers are often even more gaudily coloured than the romaleids. In Africa, adults and hoppers of elegant grasshoppers, *Zonocerus elegans*, and variegated grasshoppers, *Z. variegatus*, look as though they're wearing harlequin costumes. Once again, the brightly coloured nymphs march in tightly coordinated groups, which can often reach impressive sizes. These substantial numbers are a direct result of the female variegated grasshoppers' penchant for laying their eggs close together. Each egg pod, buried in the ground, can contain from a few dozen to up to several hundred eggs, and there may be thousands of egg pods packed into each square metre – creating a ready-made army when the hoppers dig their way to the surface.[9]

However, these swarms of *Zonocerus* grasshoppers fade into insignificance in comparison with the marching hordes of hoppers of the twenty or so species of grasshopper generally referred to as locusts. In these species, both hoppers and adults are gregarious, although 'gregarious' seems something of an understatement. On occasions, these insects congregate in swarms on such a scale as to defy imagination. Hoppers marching over the ground become a relentless tide of insatiable insects, advancing up to a kilometre and a half each day and devouring everything in their path. They grow rapidly and soon turn into winged adults that take to the air. Now they can travel much further on the prevailing winds to drop in on unsuspecting farmers and destroy their livelihood in a matter of hours.

BIBLICAL PLAGUES

The marching bands of locust hoppers might look like greatly swollen versions of the gregarious hoppers we've already met, banding together for mutual benefit, although social interactions within these locust bands are not, in reality, quite as harmonious. Hoppers are not above a spot of cannibalism, frequently biting at the abdomens of those marching in front. A hopper only begins to march when it sees nymphs approaching from behind, and hoppers prevented from seeing behind by painting over the rear part of their eyes don't bother marching at all, which suggests that it's fear of being eaten rather than the need to eat that drives the hopper bands in their relentless march across the land.[10]

Locust swarms are among the biggest aggregations of animals on the planet. In 2020, a plague of desert locusts, *Schistocerca gregaria*, threatened the livelihoods of whole populations across a broad swathe of Africa and Asia, people already struggling to cope with the COVID-19 pandemic. In Kenya alone, the swarm covered 2,400 square kilometres, probably the largest ever to hit the country.[11] Half a million locusts can eat enough food in a single day to have fed 2,500 people, and these swarms contain hundreds of millions or billions of insects. Across Africa, parts of Asia and Australia locust swarms strike with little warning.

Nor is the New World safe. In 1988 a vast swarm of desert locusts descended on the Caribbean, the first ever recorded transatlantic migration. The insects had ridden the easterly trade winds to beat the previous long-distance record set for their species in 1954, when a swarm reached the British Isles. Reconstructions of the weather patterns revealed that the 1988 crossing could have taken no less than ninety-three hours.[12] Locusts fly only during daylight hours, so the swarm must have been forced to settle on the ocean as dusk fell. The first locust to land will drown. As more and more pile in, a raft is slowly created from the accumulating dead bodies and later arrivals can then ride safely on the backs of all those drowned insects, ready to take to the air again the following morning. They would have to repeat this costly and self-sacrificing process several nights running, so only a really large swarm can hope to make it across the Atlantic. This swarm was still huge when it arrived in the Caribbean, but it was later estimated that its numbers must have been at least five times greater when it left Africa.[13]

This may have been the first recorded transatlantic flight by a locust swarm, but such events have certainly happened before. The desert locust is the only member of its genus to live in the Old World; all other *Schistocerca* grasshoppers live in the Americas. Recent genetic studies indicate that all New World *Schistocerca* derive from a single common ancestor, more than likely from an ancient swarm that was large enough to have survived the journey across the Atlantic.[14] In this case, extreme social living has allowed desert locusts to colonize a new world and diversify into around forty new species that, luckily for the inhabitants of the western hemisphere, are less inclined to swarm than their Old World ancestors. Even so, North America has seen locust swarms far larger than any in the Old World.

On 15 June 1875, Albert Child was standing on the porch of an office on the edge of the small frontier town of Cedar Creek in Cass County, Nebraska. As he gazed out over the endless prairie, he saw the horizon darken as clouds built up. But these clouds weren't bringing much-needed rain to parched frontier farms. Instead, they brought death and destruction. As the cloud approached it became clear that it was moving under its own power and Albert couldn't see any end to it in either direction.

Then, suddenly, it was on him, a trillion beating wings and biting jaws. Another swarm of Rocky Mountain locusts, *Melanoplus spretus*, had descended from the mountains to plague hard-pressed frontier farmers on the plains. However, Albert was no farmer – he was a scientist, so he set about trying to measure the scale of the swarm. What he found was truly astounding.

Using the telegraph to send messages up and down the line, he found the swarm front to be unbroken for 110 miles. With his telescope he estimated the swarm to be over half a mile deep and he watched it pass for five full days. He worked out that the locusts were travelling at around 15 miles an hour and came up with the astounding fact that the swarm was 1,800 miles long! It covered 198,000 square miles, over half a

Family groups of caterpillars of brown-tail moths, *Euproctis chrysorrhoea*, often band together in such numbers that they become a plague, defoliating trees and creating a public health hazard with their irritating hairs.

million square kilometres. If transported to the East Coast, it would have covered the states of Connecticut, Delaware, Pennsylvania, Maryland, Maine, Massachusetts, New Jersey, New York, New Hampshire, Rhode Island and Vermont.[15]

Albert Child had recorded the largest ever swarm – the biggest aggregation of animals ever seen on Earth – estimated to be thirteen and a half trillion insects strong. In contrast, the largest swarms of desert locusts in Africa cover no more than a few thousand square kilometres.

Child was a meteorologist, busy measuring weather and wind conditions at the time the swarm hit. So, scientists could later work out that a particular combination of drought and the strength and position of a conveyor belt of air, called the Great Plains Low Level Jet, gave rise to history's largest ever swarm of animals – a 'perfect swarm'.

While Albert's swarm of Rocky Mountain Locusts was the largest on record, countless others descended on hard-pressed farmers through the mid- to late nineteenth century. In fact, just a year earlier, in 1874, vast swarms overran all the plains states from Canada to northern Mexico. Some were so large that they blotted out the sun, described by one farmer in Kansas as 'like a great white cloud, like a snowstorm, blocking out the sun like vapor'. Within hours all the crops had been devoured, and the locusts moved on to eating the wool off the backs of sheep. As the locusts reached the end of their adult lives, their dead and dying bodies lay inches deep on the ground, so slick that trains couldn't gain traction on rails. They left nothing but desolation in their wake, eating everything 'down to window blinds and the green paint'. There are even horrific descriptions of sleeping soldiers being covered from head to foot in seconds by 6-centimetre-long insects desperate for anything to eat. Occasionally, swarms would come to grief in places like the Great Salt Lake of Utah, leaving local people to describe vast mounds of putrefying insects lining the shoreline and creating an unbearable stench for many miles downwind.[16]

The Rocky Mountain locust was a force to be reckoned with, adding to the many challenges faced by settlers and farmers living and working on the western frontier. The locusts seemed to be an unstoppable plague that might even halt the westward expansion of the United States. Was Manifest Destiny to be challenged by a mere insect?

The scale of these outbreaks is astounding enough, but what makes this story utterly astonishing is the fact that by the start of the twentieth century, the Rocky Mountain locust had apparently vanished from the face of the Earth – from record-breaking abundance to extinction in just a few decades. The last swarm was recorded in 1892, and the last ever live specimen was collected in 1902.

How could something so abundant be wiped off the face of the planet in such a short time? The answer lies in a curious bit of biology common to many swarming grasshoppers, and which marks another difference between these 'locust' species and the gregarious grasshoppers we began with. However, this wasn't discovered until over two decades after the demise of the Rocky Mountain locust.

Boris Uvarov was born in 1886 in what is now Kazakhstan and worked as an entomologist in Russia until a chance meeting with a British army medical entomologist resulted in an invitation to move to London in 1920. There he was invited to join the rather grandly named Imperial Institute of Entomology, which had a pressing interest in locust swarms. At the start of the twentieth century, the unpredictable appearance of devastating locust swarms across Africa and Asia, apparently from nowhere, was just as threatening to European ambitions of empire and expansion in this part of the world as the Rocky Mountain locust was to American dreams.

Working on the migratory locust, *Locusta migratoria*, in the northern Caucasus region, Uvarov made a startling discovery. He realized that what were believed to be two species of grasshopper in this area, the solitary *L. danica* and the swarming *L. migratoria*, were in fact one and the same species. This grasshopper existed in two distinct forms, different in both appearance and behaviour. One was a solitary grasshopper, shunning the company of others of its kind and living unobtrusively, munching on wild grasses and other plants. But this benign creature was capable of a Jekyll and Hyde-style transformation. Uvarov discovered that when their population reached a critical density, the grasshoppers transformed into the dreaded migratory locust, a process today called gregarization.[17]

We now know that other swarming grasshoppers undergo the same phase shift when conditions are right. Several wetter than usual years preceding the 2020 outbreaks of desert locusts allowed numbers of the harmless solitary phase to build up unnoticed. until they were triggered to transform into the swarms that wreaked havoc across Africa and Asia.

There are at least two triggers that cause this transformation. One is a gregarization pheromone, a chemical signal that, once produced in big enough quantities by the ever-increasing numbers of hoppers, causes them to seek each other's company and aggregate into huge marching hopper bands.[18] The other is tactile stimulation. As numbers build, hoppers inevitably bump into each other more often and, when this becomes frequent enough, it stimulates the phase shift. Specifically, continued stimulation of the outer face of the hind femora spurs the transformation, a discovery made by the painstaking experiment of touching eleven different body regions of a desert locust every five seconds for four hours! No other body region elicited the locust's Jekyll and Hyde act.[19] This tactile trigger seems to be the most important one in transforming innocuous grasshoppers into biblical plagues.

The Rocky Mountain locust led a similar double life. When not standing in the way of Manifest Destiny across much of western North America, it retreated to a handful of Rocky Mountain valleys where the solitary phase lived a quiet life in the fertile grassy plains bordering mountain rivers. However, these same valleys became prime real estate in the late nineteenth century, when they were converted to farms to feed a burgeoning population drawn to the area by the wealth of minerals newly discovered in the Rockies.[20] Ploughing destroyed the buried egg pods of Rocky Mountain locusts at a time when they were at their most vulnerable – when they existed in small numbers in a small number of areas.

The idea that Rocky Mountain locusts have an obscure solitary phase led to an intriguing debate among entomologists as to whether the locust really was extinct or whether it still existed in its solitary phase, perhaps described as a different species, just as the solitary and gregarious phases of migratory locusts were, until Boris Uvarov uncovered the truth. For some time, *Melanoplus sanguinipes* was fingered as the alter ego of the Rocky Mountain locust, but modern genetic studies suggest that this is not the case.[21]

Although *Melanoplus spretus* is long gone, well-preserved specimens of the Rocky Mountain locust can still be found, entombed in the ice of several glaciers descending from the high peaks of the Rockies. In the distant past, swarms of locusts must have occasionally foundered on the glaciers, where some of the bodies became encased in ice. They then began a slow journey downslope until their remains were revealed at the melting snout of the glacier. Entomologists have been able to extract DNA from these deep-frozen remains, which has helped to identify the Rocky Mountain locusts as a distinct species, separate from any of the extant grasshoppers in the mountain valleys. It seems that the scourge of the American frontier really has vanished, despite the swarms of trillions that tormented the Great Plains 150 years ago.

Although social life opened up a new world of opportunities for desert locusts, and allowed them to diversify into many species, it failed to protect the Rocky Mountain locust from oblivion, even in this most extreme of forms. However, for most other social grasshoppers, group living affords benefits that include defence, easier feeding and perhaps temperature control. Those same benefits have driven other kinds of insect to adopt social lives. In fact, the caterpillars of many moths and butterflies, and the larvae of sawflies as well as those of leaf beetles have, in many cases, evolved far more sophisticated societies than those of grasshoppers.

CATERPILLAR SOCIETIES

As I write this, I have just finished my daily round of caterpillar feeding. The exact species residing in my studio varies with the time of year or on what I need for filming or photography, although, whichever kinds I have, they all need fresh food every day. In addition, good hygiene is a vitally important part of caterpillar husbandry – which means disposing of surprisingly large amounts of caterpillar poo. As we saw in Chapter 2, caterpillars are basically eating machines, but it's not until you try to keep them well fed and dispose of all their waste material that you realize just how voracious these little larvae really are.

At the moment, the task is made a little harder because the caterpillars of one species I'm culturing, the North American bullseye moth, *Automeris io*, are covered in spines and bristles that inflict nasty stings. This species is also highly social, the young larvae hanging out in large gangs and covering the branches and twigs that I have to replace each morning. So, any careless move results in a very painful experience, presumably not unlike that described by Joseph Banks at the start of this chapter.

However, both Banks and I can be grateful that we haven't bumped into the caterpillars of *Lonomia obliqua*, a species of moth from parts of Brazil, Paraguay, Uruguay and Argentina. Like my bullseyes, these caterpillars are covered in rosettes of sharp, hollow spines filled with venom, but in this case the venom can be deadly to humans. It causes uncontrolled internal bleeding and if that blood leaks into the brain, it will prove lethal.[22] The caterpillars are very gregarious and rest in large clusters on the trunks of trees. Unlike many such toxic creatures, they don't advertise the fact, and instead are coloured in sombre shades of brown and black, which makes them hard to spot against the bark of a tree. Any accidental close encounter is therefore likely to be with a great many caterpillars, increasing the amount of venom injected and

the likelihood of serious consequences. In fact, these creatures have now made it into *Guinness World Records* as the world's most dangerous caterpillars.

However, my bullseye moth caterpillars have been more than worth the pain just to observe their social behaviour. As tiny caterpillars they rest in large groups along branches until hunger drives them to find a fresh leaf. They then set off in long lines, sometimes in single file, sometimes several abreast. Like the elephants on parade in Disney's 1967 *The Jungle Book*, the head of each caterpillar touches the rear of the one in front. More than that, the caterpillar behind continually wags its head from side to side, brushing the backside of the one in front. These head movements stimulate tiny tubercles on the lead caterpillar's rear end, and it needs this constant encouragement to continue walking. If contact is lost, the caterpillar stops and waits until the one behind catches up and resumes contact. This caterpillar conga line snakes along twigs and onto leaves, where the caterpillars fan out and begin feeding, still packed tightly side by side.

The youngest caterpillars are dull shades of brown and green but as they grow, they moult into brighter and brighter shades of luminous lime green – fair warning for any creature foolish enough to see them as food. Like some of the social hoppers, large caterpillar groups may also facilitate feeding.

The pipevine swallowtail, *Battus philenor*, is a spectacular black butterfly with iridescent blue hindwings, found widely across North America. Its caterpillars feed on the vines that give the species its name – pipevines of the genus *Aristolochia*. Across most of their range, the caterpillars keep themselves to themselves, but those in California are noticeably more social. The most obvious difference between this population and those across the rest of North America is that the species of pipevine that California swallowtails feed on, *A. californica*, is covered in a dense coat of hairs, evolved to protect its leaves from hungry caterpillars by making them less easy to hang on to.

Caterpillars in large groups are better able to overcome these defences, perhaps by producing larger and thicker silken mats than a single caterpillar could on its own, which helps them retain a foothold on the fuzzy leaves – and the more the merrier. Caterpillars in large groups grow 25 per cent larger than those in smaller groups,[23] although in this species, group size is fairly modest – perhaps a few dozen individuals are all it takes to crack a pipevine's defences. Some communal caterpillars, though, live in much larger crowds.

Caterpillars of the Io moth, *Automeris io*, travel to fresh leaves in long lines in which each caterpillar stays in contact with the one in front by brushing its head against tubercles on the leading caterpillar's rear end.

The pine white butterfly, *Neophasia menapia*, lives across western North America, its caterpillars, unsurprisingly, feeding on pine needles. Historically, its populations reached colossal sizes. In the late nineteenth century, travellers venturing into the American West came across clouds of adults flying around the tops of pines and firs, and swarms of caterpillars on the lower branches. Even today, pine white butterflies sometimes congregate in such large numbers that they partially defoliate pine trees across large tracts of forest. However, this isn't just a plague; the caterpillars interact with each other in distinct social groups.

The entomologist William Henry Edwards obtained some eggs of *Neophasia menapia* from one of these western expeditions and reared the caterpillars in captivity so he could observe their behaviour. In his three-volume *Butterflies of North America*, the third part of which was published in 1897, he wrote: 'They fed in clusters, as many as could lie close together encircling the leaf, their heads making a ring of black beads.' He also observed that the caterpillars remained gregarious throughout their larval life,[24] although more recent studies indicate that they may go their separate ways as they grow larger. While Edwards was making his observations in America, across the Atlantic, another pine defoliator with a much more complex social life drew the attention of the French entomologist Jean-Henri Fabre, whom we last met in Chapter 7.

Fabre made ingenious and meticulous studies of a wide variety of insects, from wasps and cicadas to caterpillars. In 1916 he published

The Life of Caterpillars,[25] in which he devoted several chapters to his observations on the caterpillars of the pine processionary moth, *Thaumetopoea pityocampa*, famous for trekking over the forest floor in long lines, nose to tail.

> The Pine Caterpillar is even more sheep-like, not from foolishness, but from necessity: where the first goes all the others go, in a regular string, with not an empty space between them.
>
> They proceed in single file, in a continuous row, each touching with its head the rear of the one in front of it. The complex twists and turns described in his vagaries by the caterpillar leading the van are scrupulously described by all the others. No group of Greek theoroi winding its way to the Eleusinian festivals was ever more orderly. Hence the name of Processionary given to the gnawer of the pine.

Columns of processionaries can reach extraordinary lengths. Around the same time that Fabre was making his observations, the Rev. Canon T. G. Edwards measured one column extending for 26 feet.[26] By carefully guiding the lead caterpillar onto the rim of a basin, the columns could be made to march in a never-ending circle, a somewhat heartless experiment but that nonetheless begged an important question. How could the caterpillars follow their leaders so faithfully?

They lay down silk as they march, and Fabre thought that it was this silken trail that guided them. Continuing his ancient Greek analogies, he

A caterpillar of the vapourer moth, *Orgyia antiqua*, is covered in a battery of different kinds of irritating hairs. Although not strictly social, they sometimes congregate in groups of several hundred – illustrating how more elaborate social behaviour could evolve.

somewhat poetically compared the caterpillar's behaviour to the myth of Theseus finding his way out of the Minotaur's maze.

> In the Cretan labyrinth, Theseus would have been lost but for the clue of thread with which Ariadne supplied him. The spreading maze of the pine needles is, especially at night, as inextricable a labyrinth as that constructed by Minos. The Processionary finds his way through it, without the possibility of a mistake, by the aid of his bit of silk.[27]

In fact, the truth is more interesting still, and marks pine processionary Moths, along with several other species, as more sophisticated social creatures. The caterpillars lay down a chemical trail, a pheromone, in just the same way that foraging ants do, and it's this that guides each caterpillar.[28] Communication is a cornerstone of more complex societies, and social caterpillars have independently evolved a similar method to ants for passing vital information between group members.

Social caterpillars of another moth family, the Lasiocampidae, have been closely scrutinized on both sides of the Atlantic and have revealed surprisingly sophisticated communication systems. On my side of the pond, in the UK, I often rear lackey moths, *Malacasoma neustria*, just to watch the behaviour of their caterpillars. The adult moths might seem nothing to write home about – nondescript brown moths with brown stripes – but their behaviour is as fascinating as their caterpillars'. Allow them to mate in a flight cage, then house the females in a breeding cage with pencil-sized twigs (mine always seem to prefer sprigs of hawthorn if given a choice). They will reward you by laying a band of eggs right around the stem, each egg neatly tucked up against its neighbours, perhaps a hundred or more of them in a necklace around the twig, waiting until early next spring when they hatch into caterpillars that are both brightly coloured and utterly absorbing to spend time with.

The caterpillars soon spin a silken mat or tent on which they rest when not feeding. When hunger strikes, they scatter among the branches in search of fresh leaves. Those that feast well lay a chemical trail back to their bivouac that other caterpillars then follow, reinforcing the trail if there is still plenty left to eat.[29] In this way, the caterpillars tell each other where the best meals are to be had, just as ants and honeybees do. When they've exhausted the leaves around them, they decamp and build a new tent among fresh leaves.

Across the Atlantic in North America, eastern tent moths, *M. americanum*, look almost identical to Lackey Moths, both as adults and caterpillars, but have an even more elaborate social life – one of the most sophisticated yet discovered among the many kinds of social caterpillar.

They build more substantial permanent tents from which they make long forays into the surrounding vegetation. When young, the caterpillars feed only on nutritious young leaves at the tips of branches – if all they have to eat are older leaves, they'll die. It's vital for eastern tent moth caterpillars to be able to track down a supply of fresh, succulent leaves, and they've evolved an efficient way of searching for them.

When looking for food, a caterpillar leaves a light, spotty exploratory trail, so it knows where it has been. Once young leaves have been discovered, it walks slowly back to the tent, dragging its abdomen along the twigs and branches, and leaving a powerful recruitment trail that leads right on to the surface of the tent. When other caterpillars bump into this trail, they follow with enthusiasm and soon there is a line of caterpillars weaving its way through the branches. As well-fed caterpillars head back to the tent, the trail becomes a busy two-lane highway, until the leaves are exhausted. The trail, no longer bolstered by satiated caterpillars, evaporates, and the search starts over again.[30]

Both lackey and eastern tent caterpillars hatch early in the spring when young, tender leaves are just bursting from buds. There's plenty of food but the weather is unpredictable, often dropping below freezing. Below about 15°C the caterpillars can't digest leaves and would soon starve to death despite the spring plenty. This is where tents come in handy. Those of the eastern tent caterpillar are made from layers of silk dividing the interior space into separate compartments. Near the outer surface of the tent, the compartments act like little greenhouses into which chilled caterpillars can retreat to warm up and digest their most recent meal. Even basking in spring sunshine on the outside of the tent is better done as a group. Large groups of caterpillars become much warmer than single ones, sometimes raising their temperatures as high as 35°C above that of their surroundings.[31] If they become too warm, they can retreat to the deeper, cooler layers of the tent to chill out.

This social arrangement is a big win for each and every individual in the group and even in chilly spring weather a 2-millimetre hatchling can grow to 4 centimetres, and more than a thousand times its original mass, in just over a month. The thermal advantages of group living are so critical that many other early spring caterpillars play the same trick.

The New England buck moth, *Hemileuca lucina*, is a rather handsome black and white silk moth, lovingly called the 'sky panda' by Sam Joffe, founder of the Caterpillar Lab in New Hampshire. Its caterpillars are jet black and bask in large groups in the early spring sunshine. Sunbathing as tightly packed as holidaymakers on a popular beach, each caterpillar can be up to 5°C warmer than a caterpillar basking alone. In these large groups heat is transferred laterally between close-pressed bodies; together the whole mass acts like a much bigger single creature with a smaller surface area in relation to its volume than a single caterpillar,* which slows down heat loss in the early spring air.[32]

Building communal silk shelters is a good way to make an early start in the season, before the competition heats up. It even allows some caterpillars to get going shortly after the year has turned. The Glanville fritillary, *Melitaea cinxia*, is named after Lady Eleanor Glanville, a seventeenth-century butterfly collector who discovered the species. She is well known in lepidopterists' circles for being posthumously declared insane by her family, who were contesting her will on the very reasonable grounds that no one in their right minds would go chasing around the country collecting butterflies.

Caterpillars of Glanville fritillaries hatch in the summer, then join forces to spin a nest in which to shelter over winter. Lady Glanville collected this butterfly in Lincolnshire, in the east Midlands of England, during a time when it was more widespread across Britain, but when I wanted to photograph this species, I had to head to England's far south, to the south coast of the Isle of Wight. It has now disappeared from mainland Britain apart from a few short-lived colonies that occasionally pop up on the adjacent coast of Hampshire. There was also a colony (probably established from illegally released captive-bred butterflies) along the Bristol Channel, much closer to my home, but I haven't seen them there for a few years. The caterpillars venture forth from hibernation as early as mid-February when the British climate, even on the south coast of the Isle of Wight, is rarely kind.

On one of my visits to their favourite haunts on landslip sites where their food plant, ribwort plantain, *Plantago lanceolata*, grows in profusion, I took a thermistor probe as well as my usual assortment of camera gear. This device consisted of a thin wire connected to a sensor, which gave a readout of the temperature at the tip of the probe. The probe was small enough to insert inside the silken tents of the tiny caterpillars where, even under cloudy skies, the interior was several degrees warmer than the air temperature. However, when the sun shone, the nest interior quickly climbed to 15°C above the outside temperature as the shelter acted like a miniature greenhouse.

* Surface area of an object increases as the square of its length, but volume increases as the cube of its length, so as the linear dimensions of an object get bigger, its volume increases much more rapidly than its surface area.

Caterpillars of the Glanville fritillary, *Melitaea cinxia*, begin life in a silken nest made by the whole family group. In later life, when larger, they become more solitary.

The nest is also warmed by the bodies of the caterpillars, which all cuddle up on top of their shelter in a black clump to absorb the sun's heat. Even when the air temperature is cool, these huddled masses can regulate their temperature very precisely in their preferred range of 33–34°C. Once a caterpillar reaches its optimum working temperature, it nips off to grab a few mouthfuls of plantain, then dashes back to its family before it cools down too much.[33] Yet, despite their communal efforts to control their microclimate, Glanville fritillaries are on the northern edge of their range in Britain. The southern coast of the Isle of Wight records more hours of sunshine than anywhere else in Britain. In 2020, Shanklin, on the south-east coast of the island, basked in 262 hours, 19 minutes and 2 seconds of glorious sunny weather. Glanville fritillary caterpillars start feeding and growing in late winter, when every second of sunshine counts, and this may explain why this stunning little butterfly has retreated to Britain's sunniest spot. I don't know whether, when Lady Glanville was out and about, our notoriously cloudy islands had more sunshine further north.

The black-veined white, *Aporia crataegi*, has entirely vanished from the UK; another victim of the British climate, it was probably killed off by a period of cool and wet weather in the 1920s. Like the Glanville fritillary, it hibernates as small caterpillars, all hunkered down together in a cosy silken nest woven into the branches of their food plants, hawthorn and blackthorn. Temperature is just as important to the caterpillars of black-veined whites as it is to those of Glanville fritillaries, so too little sunshine at critical times of the year may have pushed them over the edge.

By the reckoning of the Glanville family, Winston Churchill must have also been insane, since he was another keen butterfly enthusiast, encouraging many kinds to live in his gardens at Chartwell in Kent. Black-veined whites were his favourite species and in the 1940s, when I would have thought he had more pressing things on his mind, he reared hundreds of these butterflies in the hope of reintroducing them. Unfortunately, he failed, though his efforts weren't helped by his overzealous gardener, who destroyed most of the nests as he trimmed up straggly hawthorn bushes. Now, however, a recent study by Butterfly Conservation, the UK's leading conservation charity for butterflies and moths, suggests that, with the global climate warming, we might see the return of black-veined whites. Perhaps Glanville fritillaries, too, will reclaim their lost ground.

With a large enough labour force, communal nests can reach enormous sizes. In Central America, coconut caterpillars, *Brassolis isthmia*, use their silk to weave whole palm leaves together. These shelters may be a metre long and contain up to 2,000 caterpillars. They leave a hole at one end of their construction, the only way in or out, so the nest provides protection against predators as well as the elements. Since the adult butterflies lay only about 400–500 eggs, it's clear that these silk and leaf mansions are the work of several broods of caterpillars that have joined forces.

In many species, female butterflies make it easier for their broods to join up into larger colonies by laying their eggs communally. Even the small tortoiseshell, *Aglais urticae*, a common butterfly across the UK (though less common than it was), often lays its eggs like this, behaviour called social oviposition by entomologists. This can result in piles of a thousand or more eggs.[34]

Perhaps the most impressive shelter is made by the madrone butterfly, *Eucheira socialis*, from the highlands of Mexico. These structures, called *bolsas*, are usually the work of several unrelated families of caterpillars.[35] At first, the caterpillars chew on the leaves of madrone trees, then fold weakened leaves together with silk to make a loose shelter, their so-called primary bolsa. As they grow older, they make a bigger, tougher, secondary bolsa that surrounds their earlier nest in multiple layers of durable silk. By the end of the season, bolsas can be strong and dense enough to hold water. There is an entrance facing down at the bottom of the nest, which protects the caterpillars from rain and makes it easy for them to remove droppings and dead caterpillars.

The one thing the bolsa doesn't do is warm up the caterpillars, even though these tough little creatures feed throughout the winter, which gets pretty cold in the Mexican highlands. The caterpillars of madrone butterflies positively shun heat and their bolsa is thick enough to provide some cool shade from the sun. During the day, they shelter in the coolest parts of their nest, then, as the sun sets, the hardy madrone caterpillars set out to feed. They remain active even when the temperature drops below zero, although if it gets really cold, they become immobilized and have to wait until morning to warm up, just enough to crawl back to the cool depths of their bolsa.[36]

By early May, my lackey moths are growing ever bigger and ever more voracious. To cut down on time spent feeding and cleaning, I've released some groups onto a large hazel tree in the garden where I can sit and watch them during coffee breaks. Time spent with lackey moths is never wasted. They're now strikingly colourful, with blue, orange, black and white stripes along their hairy bodies. Their heads are powder blue with two black spots that look like cartoon eyes, which makes them decidedly cute. They rest draped over the larger branches, packed as tightly together as they can get. If I accidentally move too close, all the caterpillars rear up and rhythmically flick their heads from side to side. This group defence is disconcerting enough to a large vertebrate, but is probably most effective in deterring parasitic flies and wasps. Combined with the 'safety in numbers' factor – the fact that, in a group, each individual has a lower probability of being the unlucky victim – this is another good reason for these caterpillars to seek each other's company.

True altruism is rare in a natural world of selfish genes. Strictly speaking, it's only in those insect societies that have sterile workers, who have no opportunity to breed themselves and therefore work for the benefit of related colony individuals, that altruistic behaviour has evolved. These are the subjects of the next two chapters. Beyond these insects, apparently altruistic acts are really selfish behaviour in disguise. For example, caterpillars that share information about the tastiest leaves seem to be depriving themselves of food by telling everybody else about it. However, this behaviour does have a direct benefit for the 'altruistic' individual. For lackey moths and tent moths, it's all about synchrony. The advantages to group living are maximized when all the caterpillars are of a similar size. Equal access for all to plentiful food helps maintain this synchrony, and each individual gains from this. Lackey moths all grow at the same rate and even moult at the same time, leaving a forest of cast skins, like an army of ghost caterpillars, arrayed along a branch.

Caterpillars of the genus *Datana*, for example Drexel's datana moth, *Datana drexelii* have an even more gymnastic defensive display than my lackey moths. When threatened, these North American caterpillars, decked out in bright longitudinal stripes, contort their bodies into a 'U' shape. Even for an entomologist expecting this reaction, it is still alarming, so I presume it's a real shock for an unsuspecting caterpillar eater. The most effective group defence I've witnessed was in the suburbs of Sydney – a display from creatures that most would assume were yet more moth caterpillars, but which belong to an entirely unrelated order.

Sawflies are hymenopterans, the order that includes the ants, bees and wasps. Unlike these better-known members of the group, with highly social adults, the adults of sawflies are solitary, although their plant-feeding larvae often engage in elaborate social lives. These larvae look very much like moth or butterfly caterpillars, but can be distinguished by counting their legs. Both groups have the standard insect complement of three pairs of true legs at the front of the body, but because they have

The family that displays together stays together. The communal caterpillars of the lesser willow sawfly, *Nematus pavidus*, react to threats by all arching their bodies in close synchrony.

such long, thin bodies, they need some additional support behind. Both kinds have evolved fleshy protuberances, called *prolegs*, edged with rows of hooks to attach to leaves and branches. Butterfly and moth caterpillars never have more than four pairs of prolegs, whereas sawfly caterpillars have five or more.

Like many butterflies and moths, sawfly caterpillars are brightly coloured and sometimes exist in such numbers that they become pests. Every year in my garden, my collection of ten or so species of Solomon's seal, *Polygonatum* spp., is stripped of its leaves if I don't notice that numbers of the Solomon's seal sawfly, *Phymatocera aterrima*, are building up. I don't use insecticides to control insects, preferring to let a diverse insect ecosystem develop and find its own balance. To this end, I'm happy to sacrifice a few leaves, so I remove most of the sawfly larvae by hand but leave a few to add to garden diversity. Some kinds of sawfly larvae, however, wouldn't take such an assault lying down.

The larvae of spitfire sawflies or steel-blue sawflies, *Perga affinis*, known locally as spitfire bugs, live in social groups on *Eucalyptus* trees throughout eastern Australia. They all travel together to find fresh leaves, in such a tightly packed cluster that they seem to hinder each other's progress. Their tails are draped around the bodies of their neighbours as they all edge forwards in a perpetual cuddle. The group moves slowly but if filmed in time-lapse to speed up the whole process, their liking for each other's

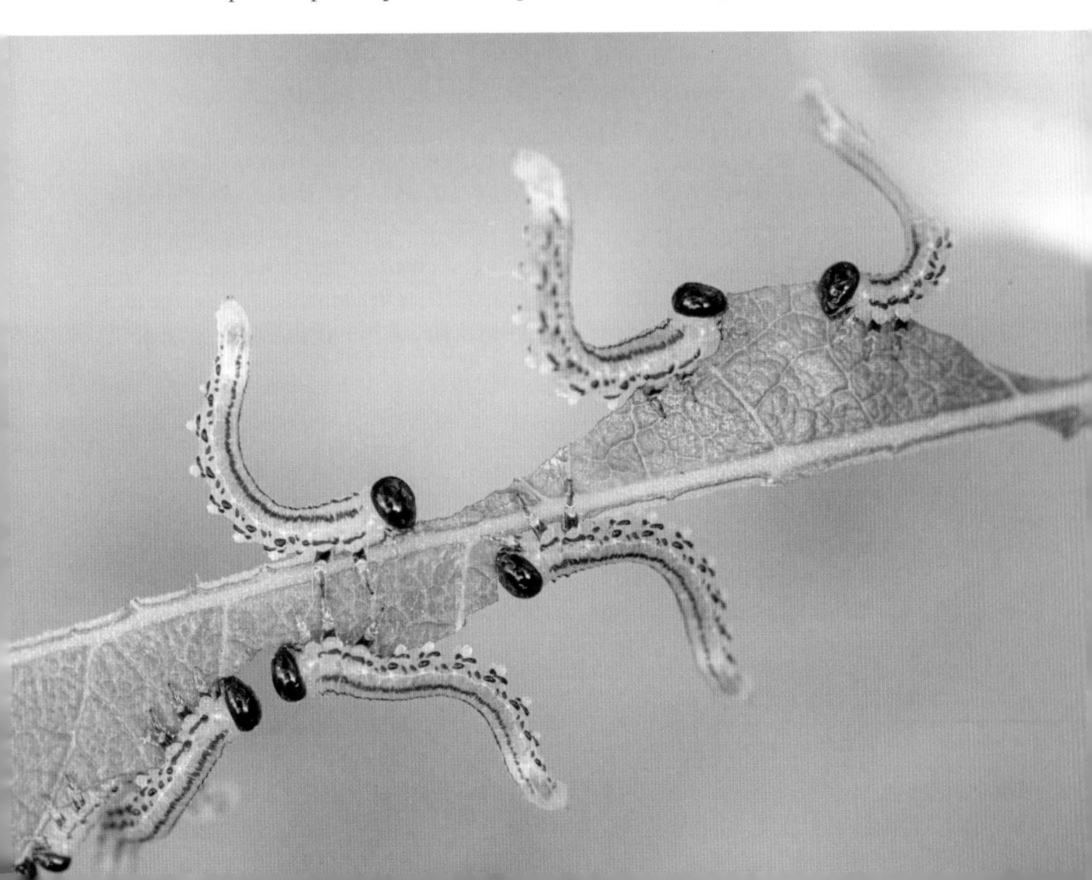

company becomes more apparent. The whole group moves as a single organism, like an alien slug surging along a branch before pausing, ready to surge again. Occasionally, I saw an individual lag behind and become separated from the group. A lost larva immediately sends out a distress message using the sawfly equivalent of Morse code. It taps the tip of its abdomen on the branch, transmitting vibrations as a call for help. The whole group responds as all the other larvae tap their abdomens in near synchrony, a powerful homing signal to guide a wayward larva back so it can once again snuggle up to its family.

The sawfly larvae stick so tightly together because this makes their defensive strategy all the more dissuasive. The larvae feed on *Eucalyptus* leaves that contain some very pungent oils. Eucalypts manufacture these chemicals as a deterrent to leaf feeders, although a surprising number of insects seem to have found ways to circumvent them and now relish eucalypt leaves. The spitfires not only sidestep the eucalypts' chemical defences; they also turn them to their own advantage. They have brush-like structures on their mandibles that absorb eucalypt oil, which is then channelled to special storage organs branching off their guts. This system may have originally evolved as a way round the plant's defences, siphoning off the oil and allowing the larvae to digest the more palatable parts of the leaf. Now, however, the larvae use these oils in their own defence. When threatened, they all rear up in a menacing display and regurgitate the oil. They each hold a droplet in the mouth, a palisade of pungent oil, ready to be smeared on anything foolish enough to venture too close. If the threat is heeded, they swallow the oily droplet again and store it ready for use another time.

Just like many of the social caterpillars, several broods of sawfly larvae often amalgamate to create much larger bands and an even more imposing threat. In the 1920s, entomologists working in South America came across an unidentified species of sawfly that merged into such large, tightly packed groups that they reached a foot in length and 4 inches in breadth, moving slowly over the ground like a 'huge slug' or 'gigantic planarian' (flatworm).[37]

These larvae may well have been looking for somewhere to pupate, which they do with the same fervour for close company as they exhibit during their larval life. Working together, they dig into the earth to form a chamber in which to build their cocoons, which are packed together in two layers, in such a way that the whole mass looks like a honeycomb. Pine processionary caterpillars also pupate together like this and some of the longest processions, as described by the Rev. Edwards, are those wandering in search of suitable ground in which to pupate. By working together, the caterpillars can break through surprisingly hard ground to find a safe place to transform into adult moths.

Living together gives insects an edge in lots of different ways, although defence seems to be one of the main driving factors, true for the caterpillars of both moths and sawflies, as well as the larvae of a few other insects. The larvae of many leaf beetles (Chrysomelidae) and tortoise beetles (Cassidae) make good use of the tactic of facing any threat with a unified front. Social leaf beetle larvae, like those of sawflies, have an arsenal of noxious chemicals at their disposal and often adopt a defensive formation when resting, to make themselves more intimidating. They form a circle, all resting with their heads facing to the outside, where they can meet a threat coming at them from any direction.

The larvae of *Coelomera* leaf beetles also adopt a defensive circle, but in this case they rest with their rear ends facing out. They have a hard shield over the tip of the abdomen and in addition secrete a noxious fluid from the anus – so it makes sense to face the hostile world with the end that is best defended. The world that these leaf beetles inhabit is particularly hostile. There are around thirty-five species of these beetles in Central and South America, all of which feed on *Cecropia* trees. We saw in Chapter 6 that to defend their leaves from attack, *Cecropia* trees have enlisted the help of particularly fierce ants of the genus *Azteca*. A *Cecropia* tree provides the ants with shelter and food, and in return the ants attack anything trying to eat its leaves. So, every night, after a day of feeding, the *Coelomera* larvae have to circle the wagons to fend off raids by marauding ants.[38]

Tortoise beetle larvae have a more unusual way of warding off attacks. Many have a long appendage projecting from the rear of the abdomen. This works a little like a toasting fork – with a somewhat unappetizing treat impaled at the tip. A tortoise beetle larva carefully attaches its dung to this structure, more properly known as a *furca*. The furca is hinged at the base and is highly mobile, so it can be thrust in any direction into the face of an approaching predator. Like some leaf beetles, many tortoise beetle larvae are social and rest in a circle with a palisade of foetid furcae pointing outwards towards the enemy, like the mass of spear points in an ancient Greek phalanx.

In all of these cases, social behaviour has individual benefits, but in any insect community founded on this selfish basis, there is a limit to how elaborate social behaviour can become. Larval societies and hopper herds can only proceed so far along the spectrum of sociality. The communities of ants, bees, wasps and termites achieve their extraordinary complexity because they are truly altruistic, societies in which some individuals have given up the right to breed and in which these individuals lay down their lives for queen and colony. No society like this has ever evolved from the kinds of social interaction we've been exploring, in which aggregations of often unrelated individuals cooperate purely for selfish gain.[39] Even so, some of these societies foreshadow fundamental features of the eusocial insects.

Some caterpillars, like those of madrone butterflies, build complex nests, a behaviour in which they resemble eusocial insects. However, they mirror these more complex societies in an even more intriguing way. Some caterpillars take on different roles within the colony, a simplified form of the sophisticated division of labour that characterizes the societies of ants and termites, in which different castes often have grossly differing anatomies, specialized for different jobs. Among madrone caterpillars, the males take on the lion's share of nest building, which leaves them less time for feeding, and which, therefore, should put them at a disadvantage. However, colonies experimentally manipulated to contain more males produce heavier pupae, both of males and females, than those with an equal sex ratio or with more females. So this crude form of division of labour somehow still benefits all, and in nature, colonies usually have ratios skewed towards more males.[40]

There's also some evidence that there are behavioural differences between individual western tent caterpillars, *Malacosoma californicum*, in which some are more active explorers. It's hard to know whether these really are permanent differences or whether they simply reflect how well fed some caterpillars are at the time of observation. However, it does seem that pine processionaries, like madrone caterpillars, have a gender-based division of labour. Female caterpillars are far more likely than males to lead their marching columns.[41]

Spitfire bugs also have born leaders. Only about one in five larvae routinely take up the most dangerous positions, either leading the column or bringing up the rear. The rear guards are just as important as the leaders. As the colony heads off to find a fresh gum tree leaf, it's their job to cut through the stem of the leaf they have been feeding on to remove the telltale signs of damage.

Exposed at the front and rear, these larvae may be more prone to attack by parasitic flies – so why take the risk? An intriguing suggestion is that these leaders may be fitter than their companions, with stronger immune systems that can more effectively neutralize the parasites' eggs.

Or, alternatively, they may simply be more active explorers, as suggested for western tent caterpillars.[42] In either case, colonies need their leaders. All the individuals in a colony grow bigger when the group consists of a mix of leaders and followers, so, in ways not yet fully understood, all benefit individually from this division of labour.[43]

Leaders are important to the group because they possess local knowledge. Larvae will only lead their group if they're familiar with the area – if they know where they're going. If they do know the area, they'll even lead groups of larvae they have never met before. Yet such leaders won't guide groups into the unknown. If they are moved to an area they don't know, they refuse to take the lead. In the absence of a leader, though, followers, even ones that know the local patch, never take up a leadership role.[44] It seems it takes a certain character to be a leader – even for an insect.

For social caterpillars, living together has lots of benefits but it has one drawback. It makes them a tempting target for that most ingenious of predators, one that's smart enough to find its way around any defences – us. Mopane worms are the caterpillars of a kind of emperor moth, *Gonimbrasia belina*, which lives in southern Africa. I've seen mopane trees draped in uncountable numbers of these caterpillars, enough to defoliate most of the trees in an area, but each is also a bite-sized packet of high-quality protein. Traditionally, the caterpillars were harvested in many parts of Africa, providing a valuable dietary supplement. Now, they are harvested commercially in a multimillion-dollar industry in both South Africa and Botswana. The caterpillars are usually collected from the wild, which inevitably raises fears of overharvesting, although some local communities are taking steps to create a more sustainable approach and protect what, for them, has been an invaluable seasonal harvest for many centuries.

Once plucked from a mopane tree, each caterpillar is pinched at the rear end to break the gut, then squeezed like a tube of toothpaste to remove a partially digested mush of mopane leaves. They are then cooked and left to dry in the sun, spread on large sheets. Now the sheer abundance of mopane worms becomes apparent, as great heaps of drying caterpillars are scattered all around a village. They can be stored dry and eaten as a snack, which needs strong teeth since they are rock hard. Or they can be rehydrated and cooked in a variety of more appealing ways. Moreover, if all the preparation is just too time-consuming, the local shops stock canned mopane worms in tomato or chilli sauce.

Elsewhere in the world, the social caterpillars of many other moths

and butterflies are important parts of the human diet. In Mexico, caterpillars of madrone butterflies are so eagerly sought that the butterfly has become extinct in some areas. To maintain a fresh supply, people often collect the tough bolsas of this species and hang them around the eaves of their houses. In this part of the world, sixty-seven kinds of moth and butterfly caterpillar, particularly the easy-to-collect social species, are eaten by humans.[45]

The larval societies of grasshoppers, butterflies and moths, sawflies and beetles have reached varying levels of sophistication, including some elements that anticipate the complex communities of eusocial insects. However, unlike eusocial colonies, it's rare for these larval societies to continue into adulthood. A few types of beetle though have developed societies that are much more like those of ants or termites, in which adults also remain together – not that many of us will come across any of these obscure beetles by chance.

SECRET SOCIETIES

Although some bess beetles (Passalidae) reach the imposing size of 8 centimetres, they all spend their lives hidden in galleries tunnelled through rotten wood, making them little known beyond those few entomologists studying them. They are also largely tropical. However, one species, the horned passalus, *Odontotaenius disjunctus*, is widespread in North America, where there are plenty of entomologists to scrutinize its secretive life.

Its eggs are laid in a nest of shredded wood and once the larvae hatch, their parents feed them baby food in the form of wood they've chewed into a more easily eaten and digested pulp. The adults remain with their offspring throughout their larval life, just like the burying beetles and some of the dung beetles that we met in the previous chapter. When the larvae are ready to pupate, the adults help them to construct their pupal chambers, the larva moulding the inside of the chamber while the adults shape the outside. Then, when this next generation emerges as adults, some may stay and help rear their younger siblings, in just the same way as the workers of bees, wasps, ants and termites do.[46] However, unlike many eusocial workers, these helper beetles are fertile and will eventually leave to set up their own galleries and rear their own families. It takes these beetles a long time to reach sexual maturity – up to eight months in some species – so, until then, they may as well remain in their parents' galleries and help out.

However, the most remarkable feature of these beetles – and one still not fully understood – is their complex acoustic communication system. Both larvae and adults make sounds by stridulation, although in different ways; the larvae scrape their legs over pegs on their abdomens, while adults rub their abdomens over the insides of their folded wings. Somehow, using these simple techniques, they manage to create a whole symphony of different noises, a complex language of distinct sounds used in many different situations. In the well-studied North American species *O. disjunctus*, seven different sounds are produced and used in various

An individual fully grown caterpillar of the lackey moth, *Malacosoma neustria*, is a thing of beauty, made all the more impressive as even at this stage they rest in tight-packed groups.

combinations in thirteen different contexts. These sounds are clearly important in holding these tiny societies together, although we're still a long way from speaking fluent bess beetle. Furthermore, it turns out that it isn't just one language. Adults respond to any disturbance to their galleries with calls that are assumed to be alarm signals. Several species often share a rotten log, but the alarm calls of each species are distinct. Each species communicates with its own kind in its own dialect.[47]

Flat bark beetles (Silvanidae) number only a few hundred species and are named for the majority that live under the bark of dead trees, quietly feeding on fungi or detritus. However, a couple of species have adopted an intriguing social life. They were discovered in Guyana living on a plant called *tachigali*. These are well-known ant plants, providing homes for ants in enlarged, hollow leaf stalks. In return the ants protect the plants. Not all leaf stalks are taken up by the ants and if a flat bark beetle finds an unoccupied stalk, it moves in. If it is joined by a member of the opposite sex, the pair start to raise a family.

The beetles begin by chewing the pithy interior of the stalk, enlarging their home, before laying eggs. They also allow scale insects to move in and plug themselves into the inner walls of the leaf stalk where they suck the plant's sap. Like ants, these beetles have discovered the trick of milking scale insects for their honeydew, the sweet excretions produced as they process large volumes of sugar-rich plant sap. Whether they 'farm' these insects as many ants do is still not clear, but the beetles do avoid covering them in their droppings by carefully depositing their frass (excrement) in long lines between rows of scale insects. It seems the least they could do. The adult beetles stay with their larvae and when their offspring emerge as adults, sons and daughters often remain in the nest. In fact, they have even been seen mating with their own brothers or sisters within the nest, and raising their own larvae, so turning an inbred family into a little society.[48]

There are few, if any, complex social interactions among these beetles, so calling it a society is a bit of a stretch, although the parents do stay with their young, which lands them squarely in the subsocial category of insects that we defined at the start of this chapter. Quite a few other kinds of beetle teeter on the edge of sociality in this manner.

Bark and ambrosia beetles (Curculionidae) dig tunnels below bark or through wood, which they infect with a fungus as a food source for their larvae. The female then lays her eggs at intervals along this tunnel. On hatching, the larvae burrow away from the central tunnel at right angles, cleverly avoiding each other's feeding tunnels, and in the process creating an artistic pattern that's revealed if the bark is peeled from old trees. Most adult beetles never see their larvae, but a few create longer-lived colonies that may even cross the boundary into eusociality.

The horizontal borer, *Platypus incompertus*, is found only in Australia, where it burrows into *Eucalyptus* trees. The galleries it makes last for decades and within them an extended family slowly builds up, generation after generation, descended from a single founding female. Many adults live together in these old nests, but surprisingly only one female is fertilized and produces eggs. All the other adults are unmated females, daughters that have failed to leave the nest and find a mate, the males having all left to pass on their genetic legacy in new colonies.

These spinsters help maintain the galleries in which they were born, working so hard that eventually they lose the tarsal segments from their legs – literally worn down by housework. Now, they can't leave the nest, even if they wanted to. They're obliged to be workers for the rest of their lives. Hidden inside these gum trees are strong parallels with the great insect societies of the Hymenoptera and Blattodea. Overlapping generations, division of labour between a single reproductive 'queen' and unmated workers, and cooperative brood care; these obscure little beetles seem to pass the test for admission into the hallowed echelons of eusocial insects.[49] Nor are they the only overlooked insects to achieve this honour.

You don't often hear the term 'samurai aphid' – but samurai aphids really do exist, tiny selfless soldiers willing to make the ultimate sacrifice in defence of their families. They were discovered in Japan by Shigeyuki Aoki in the early 1970s, and since then many other kinds of aphid soldier have been described from other species. These aphid societies present a special case among social insects because of the way aphids reproduce.

As any gardener knows, aphids can build up to plague proportions almost overnight but, rather than spraying them with insecticides, take a closer look. Over the summer, most are wingless creatures, each plugged into your precious plants with its sharp, tubular mouthparts. At the back end you may see an even tinier aphid struggling to escape. Aphids give birth to live young that look like miniature versions of their mother, all of them females and all of them born pregnant, so they soon give birth in turn to their own live young. These young aphids don't just look like their mothers – they're absolutely identical in every respect. They are clones, produced by virgin mothers, and soon produce their own clones to quickly smother your broad beans in what amounts to a single individual, broken down into lots of separate insects.

Eventually, as summer turns to autumn, aphids start producing both males and females, which migrate to a different species of plant, known as their primary host, where they mate in the normal way, and lay eggs. Their offspring will eventually migrate back to a secondary host, where they will once again build up their clone army.

The self-sacrificing behaviour necessary to maintain a colony of eusocial insects is predicted to evolve if members of that colony share a lot of their genes. The sterile workers and soldiers can't breed themselves, but if they share many genes with the breeding queen, they can still help those genes multiply by helping their queen and their colony. Aphid clones share every last one of their genes, so it's not surprising that eusocial aphids have evolved. In fact, the puzzle is why there are so few social aphids – limited to only about 1 per cent of known species.[50] We'll see in the next chapter that ecological circumstances may be a more important factor than genetic relatedness in driving the evolution of social behaviour among bees and wasps, so perhaps the same is true for aphids.

Aphid soldiers often assume different forms on their primary and secondary hosts. Among the original samurai aphids, *Colophina clematis*, discovered in Japan, soldiers on the secondary host are nymphs that are frozen in time. They never moult from their first instar and will therefore never breed. On the primary host, soldiers are second instar nymphs that may later moult and go on to breed.[51] There is much variation on this theme but all aphid defenders are child soldiers that, if they aren't permanently locked into early instars, usually delay their moults to allow them to perform their defensive duties for longer – an insect version of the draft.[52]

Aphid soldiers take many forms. Some have enlarged front legs and mouthparts with which they attack surprisingly large predators like hoverfly larvae. A troop of soldiers descends on a fly larva. They each grasp it with powerful front legs, then use stiletto mouthparts to stab it. There are even horned aphids, sporting a pair of sharp projections from the front of their heads. They too have strong front legs and are often more heavily armoured than the reproductives in their colony. They attack in a similar manner to other soldiers, first taking a firm grasp of their victim, then head-butting it, trying to pierce it with their sharp horns. One particularly plucky species of horned aphid, *Ceratoglyphina styracicola*, has even been seen to rally its troops against vertebrate predators.[53]

As they feed, many aphids cause their host plant to form a gall, which serves both as a protective fort for the little colony and an unlimited supply of food. Having a home to defend is a good reason to evolve a soldier caste and – as we'll see in the next chapter – fortress defence is considered to be one of the factors behind the evolution of the mega-colonies of advanced eusocial insects. Likewise, all social aphids are gall makers and fortress defenders, although not all gall-making aphids are social, so there are clearly other factors at work. Nevertheless, it does look like living in galls is a prerequisite for social life among aphids.

In the 1990s, two decades after the discovery of social gall aphids, the American evolutionary biologist Bernard Crespi was studying the galls of different insects in Australia. These galls were made by thrips (Thysanoptera), tiny insects, usually no more than a millimetre in length, with long, feathery wings. They are entirely unrelated to aphids but share many similarities in their biology, as Crespi was about to discover. He busied himself with puncturing tiny holes in the galls of one species and carefully introducing members of a related species. Thrips poured out of the breach in their gall to grapple with the invaders. Like soldier aphids, these defenders had strong front legs, and they were also wingless. They were a distinct caste of soldiers.[54]

Later work showed that seven thrip species, all in the genus *Kladothrips*, which make galls on Australian acacia trees, produce soldiers, and satisfy the conditions to be considered eusocial, albeit at a primitive level. The galls are induced by a foundress thrip that then lays eggs which develop into specialized soldiers. Unlike those of some of the aphids, these soldiers do eventually breed, mating with their siblings to produce a generation that is adapted not for defence but for dispersal – to find a new acacia tree and begin the cycle again.[55]

There's one final group of spectacularly obscure insects that lives an unexpectedly complex life. These insects are as little known to most people, many biologists included, as thrips, but are even more intriguing. They are the webspinners (Embioptera). The whole order contains an estimated 1,500 species, varying in size from a third of a centimetre to 3 centimetres long. Only around 400 species have been scientifically described – and all are very similar. Indeed, one taxonomic work listed their main distinguishing characteristics as 'lacking distinguishing characteristics'.[56] However, all look as though they're wearing boxing gloves since their front tarsi are swollen and bulbous.

These are silk glands that produce the extensive webs which give webspinners their name and in which they spend their lives. More familiar silk-spinners, like silk moths, have only a single pair of silk glands. Webspinners have 150 glands in each leg, connected to fine silk ejectors that allow them to spin silk in copious quantities, like tiny versions of Spiderman. They live in tunnels running through their webs, which has forced them to develop some very distinctive characteristics. Some kinds have wings, which would be terrible encumbrances in confined tunnels, so webspinner wings can be deflated when not in use. Pre-flight checks include pumping up the wings with blood to make them stiff enough for flight. However, like aircraft stored on an aircraft carrier, the wings can also be folded away. Once deflated, the wings can even be folded backwards, so a webspinner can run through its tunnels in either direction with equal ease.

Webspinners lay their eggs in the tunnels, usually in a compact nest over which the female stands guard until her young hatch. She may even feed her larvae, and some females live long enough to see their grandchildren. The silk nests often contain several breeding adults, although so few people have looked closely at these insects that we don't even know whether these are related females or not, so it's hard to know exactly how their social systems work.

One species, *Rhagadochir virgo*, from Africa, is – like many aphids – capable of virgin birth, so the colonies that build up are clonal. Individuals of this species live crowded together and even move as coordinated groups, but whether their close genetic relationship predisposes them to develop even more complex social lives, no one yet knows.[57]

Webspinners are one of the least studied groups of insects, though what is known of their lifestyles suggests that they might hold clues to the factors driving the evolution of social behaviour among other insects. The exact opposite is true of the group of insects we are about to meet in the next chapter. Bees and wasps have evolved eusociality on several separate occasions, and their lifestyles encompass every stage, from solitary species, through those with colonies in single figures, to teeming communities of thousands. Bees are also vital pollinators of commercial crops and produce valuable honey. Not surprisingly, they've been the subjects of detailed studies for many centuries. So much has been written about these creatures that it's a hard task to compress their story into a single chapter. What follows, therefore, is inevitably a canter through a wide range of the remarkable behaviour of bees and wasps, both solitary and social, and the insights this gives us as to how the great selfless societies of the insect world arose.

Caterpillar societies, such as those of lackey moths, *Malacosoma neustria*, provide many benefits, from protection against parasites and predators to more effective temperature control.

10

For Queen and Colony

Bees and wasps – the journey from solitary to social

I… will confine myself to one special difficulty,
which at first appeared to me insuperable, and actually fatal to my whole theory.
I allude to the neuters or sterile females in insect communities:
for these neuters often differ widely in instinct and in structure from
both the male and fertile females, and yet from being sterile they
cannot propagate their kind.

Charles Darwin, 1859[1]

The life of bees is like a magic well.
The more you draw from it, the more it fills with water.

Karl von Frisch, 1937[2]

The sun blazed down from a hazy blue sky. The thermometer was already pushing towards 35°C and the day had barely begun. It looked like another pleasantly warm day lay ahead in Arizona's Sonoran Desert. This place might be defined as a desert due to its sparse rainfall – somewhere between just seven and a half to forty centimetres a year – but if the word *desert* conjures up an endless vista of empty dunes and blowing sand, think again. I prefer to see the Sonoran Desert as a cactus forest, the canopy made of tall-growing saguaro and cardon cacti, the shrub layer of prickly pears and chollas, and the ground layer of pincushion and barrel cacti.

This spiky landscape is softened by a host of shrubs and trees belonging to the pea family (Fabaceae), such as various kinds of mesquite, whose roots penetrate deep below the surface to seek out scarce water, and palo verde, one of the most beautiful trees in the desert when it flushes green with its tiny leaves. Among these more conventional-looking trees, ocotillos occasionally reach 10 metres in height but spend most of the year looking like a bunch of dead sticks shoved into the ground. With the first rains, they sprout long lines of tiny leaves and beautiful crimson flowers. These are just the most conspicuous of some 2,500 species of plant living here. Far from being a barren desert, the Sonoran teems with life – both plant and animal.

This extraordinary place is home to 130 kinds of mammals, more than 400 different birds, 100 reptiles – which thrive in the dry heat – and an astonishing 750 different types of bee.[3] I was here to film just one of those bees, *Diadasia rinconis*, usually called cactus bees because their life cycle is closely tied to the flowering of the cacti. More of these intriguing bees later, but for the moment I was temporarily distracted in my task by a much more impressive hymenopteran busy feeding on flowers that lined a damp gully. It was a wasp, well over 4 centimetres in length, with a blue-black body and contrasting bright crimson wings – a tarantula hawk, *Pepsis grossa*, and a stunning beast in more than one sense. Adult tarantula hawks feed on nectar and pollen, but their larvae feast on the flesh of the desert's equally impressive tarantulas; a female tarantula hawk's job is to catch, then stun this formidable prey, so her larvae can feed on it.

I've watched tarantula hawks hunting tarantulas on several occasions, and each encounter has only increased my deep respect for this wasp. Once, I followed a tarantula hawk as it ran in fast zigzags over the ground in the heat of the morning for some considerable time until it came to a hole a couple of inches across, with an entrance lined with silk. The wasp

(*previous page*) Honeybees live in colonies of thousands although the great majority of bees are solitary creatures.

The Sonoran Desert in Arizona is bee heaven; more than 700 different kinds live among the flamboyant flowering cacti.

and I both knew what this meant – there was probably a tarantula in residence. The wasp confirmed this by tapping the silk with its antennae and, tasting the scent of tarantula, promptly disappeared down the hole.

For a few seconds all was quiet, then a tarantula nearly twice the size of the wasp burst out of the hole, hotly pursued by the fearless wasp. On the ground these long-legged wasps are lightning fast and this one circled the spider as nimbly as any matador, darting in to touch the spider, then retreating as the spider reared up and bared its centimetre-long fangs. Before long, the intense heat began to exhaust the spider, which would normally remain hidden in its deep burrow until the cool of night. One last lunge from the wasp and she grabbed the spider, clamping her jaws around the base of a leg. The pair tumbled over the ground, which left the spider's belly exposed and vulnerable. Now the wasp arched its body, extended her frighteningly large sting, some three-quarters of a centimetre long, and stabbed the wasp between its legs – and that was that.

The spider wasn't dead, merely paralysed. Unable to resist, it was dragged by the wasp into a burrow she'd previously excavated about 10 metres away where, hidden from my prying eyes, the wasp laid a single egg, then sealed up the burrow. This one large spider would provide enough food for her larva to grow to full size, and being paralysed rather

Cactus bees, *Diadasia rinconis*, feed almost exclusively on cactus flowers, particularly the huge flowers of prickly pears, *Opuntia* spp.

than dead, the unfortunate tarantula would remain fresh almost until the last bite.

These wasps are so big and so ferocious in a fight that I always keep a respectful distance from any tarantula hawks I meet. So far, I've avoided being stung. On the other hand, Justin Schmidt, from the Carl Hayden Bee Research Center, just down the road in Tucson, has gone out of his way to find out what those poor tarantulas experience.

Ant, wasp and bee venom is remarkable stuff, with many potential biomedical applications. As part of a survey of venom types and actions, Schmidt decided that one way to categorize them would be to record their effects on a human subject.[4] Since the only human subject crazy enough to participate in this experiment was Schmidt himself, the Schmidt Pain Index was born. His starting point was a comparison of the haemolytic* properties of venom from a small range of species with the degree of pain it inflicted. He then expanded his survey and subjected himself to stings from a representative array of ants, bees and wasps from all over the world. He placed them on a scale from one to four, along with a subjective, if evocative, description of each painful experience.

Based on his work, I feel I've made the right decision in not antagonizing any tarantula hawks. Schmidt reckons he has now been stung by most of the world's stinging hymenopterans and the tarantula hawk ranks a four on the scale – one of only three species to be thus honoured. The other two, which any sensible person should avoid at all costs, are bullet ants, *Paraponera clavata*, and the somewhat scarily named warrior wasps, *Synoeca septentrionalis*, both of which we will meet in the following pages. Schmidt describes the sting of a tarantula hawk as 'blinding, fierce, shockingly electric', although, if it's any comfort, the pain only lasts for a few minutes.[5]

The power of the pain inflicted by *Pepsis grossa* is nevertheless enough to persuade most predators, with the exception of Schmidt, to leave the wasp well alone; there are only a handful of records of anything at all being foolish enough to hunt a tarantula hawk. It seems they live a carefree life, with no predators of their own to worry them as they go about stocking their burrows with paralysed tarantulas. Their venom serves two purposes: the first to paralyse prey for their offspring; the second to inflict excruciating pain on anything that threatens them. Wasp venom originally evolved to paralyse the wasps' invertebrate prey

* Haemolysis is the breakdown of red blood cells and the release of their contents into the bloodstream.

since this is how a great many solitary wasps provide for their young. Some wasps, however, have been able to add other components to their venom that inflict terrible pain on us vertebrates, without compromising the venom's ability to paralyse invertebrate prey.

Social wasps don't paralyse their prey. They simply butcher it on the spot, so they've been able to brew up some very nasty concoctions in their venom glands without worrying about any compromises – chemical weapons that they use to great effect in defending their nests. Venom, originally evolved to subdue prey, now serves solely as a weapon. This ability to defend their fortress nests with potent venom has been a factor in helping wasps evolve their complex societies.

For people only too familiar with the buzzing metropolis of a yellowjacket's* or hornet's nest, it may come as a surprise that most wasps are solitary. Only three out of thirty-seven wasp families contain social species.[6] The order to which they belong, Hymenoptera, also contains the ants, all of which are social, and the bees that, like the wasps, live varied lifestyles from solitary to highly social. In the deep past, ants evolved from a solitary wasp-like ancestor, but no solitary ant survives today. So bees and wasps provide us with the best opportunities to explore how complex societies may have arisen.

For the sake of taxonomic completeness in exploring the lives of bees and wasps, I should also mention the sawflies, whose communal larvae we encountered in the previous chapter, and the horntails or woodwasps, whose larvae bore through wood. They also belong to this same order, but are very different creatures from the more familiar ants, bees and wasps. For this reason, the order Hymenoptera is divided into two broad groups: the sawflies and woodwasps (Symphyta), with plant-eating larvae; and the ants, bees and wasps (Apocrita or 'wasp-waisted'), the majority of which are parasitoids or predators, although some wasps and quite a few ants have switched to a vegetarian diet, as have almost all bees.

Feeding on plants, symphytans have little need for the sophisticated cocktail of proteins that makes up apocritan venom, but at least one species does manufacture a chemical to manipulate its host. As it inserts its eggs into a pine tree, the sirex woodwasp, *Sirex noctilio*, inoculates it with a symbiotic fungus that grows to form the only food its larvae will eat. At the same time, it also injects mucus produced in a pair of glands

* In America members of the genera *Vespula* and *Dolichovespula* are known as yellowjackets. In Britain these insects are just called wasps, although 'wasps' does, somewhat confusingly, also refer to all the numerous solitary wasps that are not closely related to the yellowjackets.

Bees are vegetarian wasps that feed their larvae on pollen and nectar. Worker bees, such as this buff-tailed bumblebee, *Bombus terrestris*, soon learn which flowers provide the best source of food.

in its abdomen. The mucus contains chemicals that cause the pine leaves to accumulate starch that would normally be circulated to sustain the whole tree and, eventually, the needles drop prematurely. The mucus weakens the tree and aids the fungus in overcoming the pine's defences, to the benefit of sirex wasp larvae. Often, the combined effects of mucus and fungus kill the tree, providing plenty of food for both fungus and woodwasp larvae, but making this species a serious pest in places where it has been introduced.[7] These toxins evolved to manipulate the woodwasp's host, in this case a plant, but it foreshadows the complex and subtle concoctions produced by its wasp-waisted kin.

KOINOBIONTS – THE INSPIRATION FOR ALIEN

Among the Apocrita, there are two broad types of parasitoids. The first are those that insert their eggs into a host that is then left to carry on its life as normal, at least until it is eaten from the inside. In case it comes up in a pub quiz, these are called *koinobionts* and, depending on the species, the prey may be briefly paralysed to allow easier egg laying. The second kind, *idiobionts*, like the tarantula hawk we met at the start of the chapter, inflict permanent paralysis on their prey, though it remains alive and fresh while the wasp's larva feeds on it. Both lifestyles are gruesome enough to have caught the attention of Charles Darwin, who used their behaviour in a riposte to criticisms of his *The Origin of Species* as being

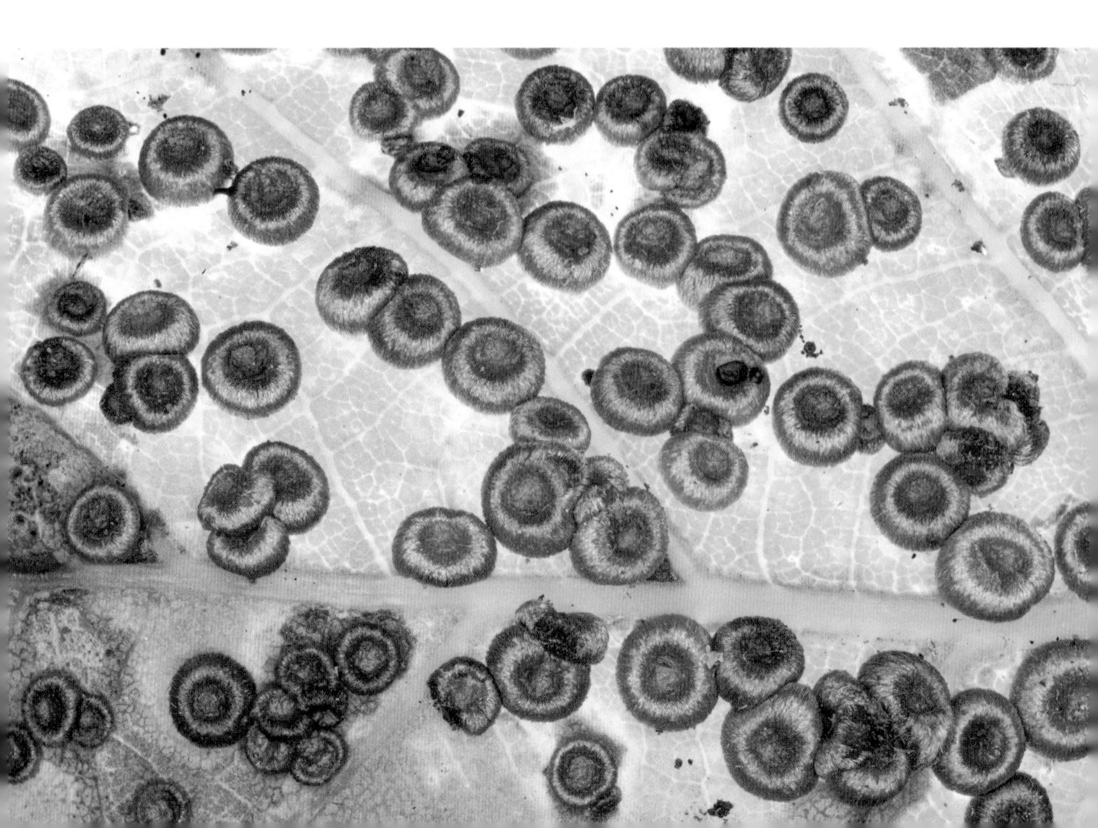

anti-religious. In a letter to the American botanist Asa Gray, Darwin pointed out that a close look at the natural world reveals little evidence of a benign and benevolent Creator, and especially not if you happen to be the victim of a parasitoid.

> But I own that I cannot see, as plainly as others do, & as I shd wish to do, evidence of design & beneficence on all sides of us. There seems to me too much misery in the world. I cannot persuade myself that a beneficent & omnipotent God would have designedly created the Ichneumonidæ* with the express intention of their feeding within the living bodies of caterpillars, or that a cat should play with mice.[8]

Yet, despite their macabre behaviour, these wasps provide entomologists with the opportunity to study everything from biological pest control to the evolution of sociality. Some parasitoids, as we discovered in Chapter 1, are so small that their larvae can grow to full size inside the egg of another insect, and they are so effective at tracking down eggs that several kinds are now bred in bulk and used as biological control agents.

To observe one of these microscopic insects in action, I recently reared some common clothes moths, *Tineola bisselliella*, a species that most people go to great lengths to get rid of. Mine, however, were fed a diet of fine, aged wool until they emerged as moths, when I moved them to a box lined with black card. This allowed me to see their tiny white eggs, which would normally be invisible on a woollen garment. Using a single hair mounted on a cocktail stick, I was able to arrange these eggs in a tiny cell, made from the coverslips of microscope slides, and place the whole thing under my binocular microscope. To this, I added a small cardboard sheet containing commercially bred *Trichogramma evanescens* wasps.

When the wasps hatched, they were barely visible to the naked eye, smaller than a speck of dust, yet they had all the sensory abilities needed to find my egg collection. I watched as one female climbed onto the pile of eggs and walked slowly over them, presumably sizing up their potential as homes for her larvae. Eventually, she curled her abdomen around and pushed her sharp ovipositor into the tiny egg to lay her own even tinier egg. It's hard to imagine the scale at which this whole drama unfolds – it would be completely invisible if it weren't taking place beneath a microscope.

---

* Ichneumonidae is a family of parasitoids wasps with roughly 25,000 species. They are sometimes called Darwin wasps – I'm not sure if he'd be flattered.

While many wasps are parasitoids of a wide variety of insects and arachnids, the gall wasps (Cynipidae) specialize in plants. They induce plants to grow galls which house and feed their larvae. Each species creates a distinctive gall – these silk button galls belong to *Neuroterus numismalis*.

Other wasps insert their eggs into the larvae, pupae and adults of almost every kind of insect, with different species of wasp specializing on different hosts. We saw in the first chapter that this means that there must be an unimaginable number of these parasitoids out there, which may make the Hymenoptera the most diverse order of all insects. It also means that the threat from parasitoids is so pervasive that it's been a major factor in the evolution of parental care. Those species that guard their eggs suffer less from wasp attacks. Although insect eggs can't do much about parasitoids on their own, the other life stages of insects can and do fight back. The communal head-lashing behaviour of lackey moth caterpillars that I described in the previous chapter (see page 352) is enough to put off some parasitoids.

More drastically, some aphids commit suicide if they become infected. As we saw in Chapter 9, aphids can reproduce without mating, cloning themselves so they end up living among a crowd of genetically identical siblings. Aphids that are playing host to a wasp larva take more risks than their siblings, exposing themselves to predation, behaviour that eliminates both the aphid and the wasp larva. From the genes' point of view, the sacrifice is worth it since it benefits the remaining clones, which carry an identical set of genes to the suicide victim.[9] The parasitoids themselves don't take these defensive tricks lying down, and an arms race has developed between parasitoids and their hosts as each tries to outdo the other. Some years ago, in Kentucky, I discovered one of the most remarkable results of this conflict.

Lexington, Kentucky, lies in the middle of horse country. The drive there took us through rolling green fields surrounded by immaculate white fences, where horses grazed that looked as though they were worth more than I would earn in a lifetime of making wildlife films. Our destination was the University of Kentucky and the lab of Kenneth Yeargan, which would become our home for a few days while we tried to film another micro-drama.

One of Ken's projects focused on the parasitoids of green cloverworms, the caterpillars of the black snout moth, *Hypena scabra*, which can sometimes become pests of legumes, particularly soya beans. The young caterpillars are tiny and very hard to spot on soya bean leaves, at least for us. However, a species of wasp, *Diolcogaster facetosa*, which parasitizes these caterpillars, has no trouble at all, although it does have help. Plants that are under attack broadcast a distress call by releasing chemicals from their damaged leaves. The wasp tracks these chemicals back to their source with unerring accuracy, to find a tiny, defenceless green

Many wasps are small enough to parasitize the eggs of other insects. *Trichogramma evanescens* is so effective at finding the eggs of common clothes moths that it's now bred commercially as a biological control agent.

cloverworm. If the caterpillar senses the wasp's arrival, it reacts with a trick of its own. It attaches a thread of silk to the leaf, then jumps off, spinning more silk as it drops to make a lifeline that it can use to climb back to the leaf when the danger has passed. Meanwhile, back on the leaf, the wasp was searching frantically backwards and forwards, like a cat that's lost a mouse it's just caught.

As we watched this story unfold in a corner of Ken's lab, cameraman Kevin Flay was doing a lot of cursing. These creatures were so tiny and moved so fast that they were impossible to follow with a macro lens – but more fun was to follow. The wasp found the base of the silken thread and tracked it to the edge of the leaf, where it did something that left both of us open-mouthed in amazement. It hooked some long hairs on its back legs around the thread and abseiled down it with all the skill of an experienced mountaineer.[10] Once it reached the caterpillar, the wasp quickly impaled it with its sharp ovipositor and laid its egg inside, before flying off to respond to more plant distress signals. The doomed caterpillar climbed back up the thread to carry on feeding, but its day wasn't over yet.

Ken revealed another wasp in his collection, an even tinier beast – sorry Kevin – an ichneumon wasp called *Mesochorus discitergus*. Released onto the leaf, it soon found the cloverworm. By now I was beginning to feel sorry for the poor caterpillar as it once again anchored its lifeline and leapt off the leaf. Just like *Diolcogaster*, *Mesochorus* searched along the leaf

edge until she found the caterpillar's thread, but instead of sliding down it this wasp grasped the thread in its front legs and hauled it in hand over hand – or at least tarsus over tarsus – like a sailor hoisting sails. The wasp hauled quicker than the caterpillar could spin more silk, and the unfortunate creature once again found itself at the mercy of a tiny wasp.

However, *Mesochorus* isn't a parasite of green cloverworms – it's a parasite of *Diolcogaster* larvae. To define *Mesochorus* precisely, it's a hyperparasitoid that feeds inside a *Diolcogaster* larva feeding inside a green cloverworm.[11] If the world of parasites wasn't complicated enough, there are parasitoids of hyperparasitoids, too. Sometimes one hyperparasitoid will eat another, even a member of its own species. Next time you see a caterpillar, you might think of it more like a Russian doll, packed with ever smaller larvae, each inside the other.[12]

Caterpillars, while not entirely defenceless, are easy prey for parasitic wasps, but every insect, no matter how dangerous, is fair game for one wasp or another, although those wasps that tackle more threatening hosts need a few other clever ploys to ensure that they come out on top. Some wasps from the family Diapriidae adopt the high-risk strategy of parasitizing the larvae of army ants, *Eciton* spp. These ants have no permanent nest but make temporary camps between hunting expeditions

When the larvae of parasitoid wasps are fully grown, they burst out from their host's body to spin a silken cocoon and pupate.

to protect their larvae, constructed from thousands of interlinked ant bodies. If a diapriid wasp wants to reach the larvae it will have to run the gauntlet of thousands of biting, snapping jaws. To survive in the hostile environment of an army ant bivouac the wasps try to blend in, but their large, membranous wasp wings are a dead giveaway. Female diapriids have adopted the extreme tactic of allowing the ants to chew off their wings so they might more readily pass for ants – if no one looks too closely.[13]

Before we leave the world of micro-wasps to consider creatures you can actually see, I should mention two more reproductive quirks of these little creatures. Some parasitoids indulge in polyembryony. Once laid, an egg begins to divide into multiple cells, as happens in all animals, although in this case the division results in cells of vastly different sizes. When it reaches a certain stage of growth, it produces multiple clones of itself, sometimes dozens. Polyembryony itself isn't too unusual; it's how identical twins are produced in humans. What *is* unusual about these wasps is that their offspring are not identical. Some develop into fat larvae that go on to emerge from their host, turn into adults and reproduce. Others develop into sterile soldiers that attack any parasitoids they find swimming in the host's body that aren't from their own family. It's a form of social behaviour like that of the clonal aphids we met in the previous chapter, although this time all taking place inside a single caterpillar.

The parasitoid's second reproductive quirk is shared by all ants, bees and wasps. They can choose the sex of their offspring – unfertilized eggs becoming males and fertilized ones creating females. Some of the parasitoids that produce soldier larvae lay two eggs in each host, one male and one female. Multiple larvae grow from each egg by polyembryony, but only the female egg produces clone soldiers that, in this case, also attack and kill male larvae from their own family. Rather than defending the family against other parasitoids, they are more concerned with actively adjusting the sex ratio of adults that emerge.[14] This bizarre fratricide comes about because the females all arose from a single egg, so they are all perfect copies of each other. They are more closely related to their sisters than they are to their brothers, so it makes sense, at least genetically, that they help each other at the expense of their brothers. A single male can fertilize many females, so, in this case, a female-biased population makes evolutionary sense. About 95 per cent of the adult wasps that emerge are female, so the lucky 5 per cent that are males really do have their work cut out.

HUNTING WASPS

The bigger, more spectacular wasps that hunt and paralyse their victims have an equally broad range of prey. In the eastern US, I've watched cicada killers, *Sphecius speciosus*, hauling cicadas much heavier than themselves back to burrows in sandy soil. Some straddle their paralysed prey and drag it, often for many tens of metres, occasionally making short hops into the air before crashing back to earth. Others, perhaps with slightly smaller prey, fly directly to their burrow entrance with their prey suspended beneath them, looking like heavy-lift helicopters.

Spider-hunting wasps (Pompilidae) specialize in catching all kinds of spiders, while the European beewolf, *Philanthus triangularis*, catches honeybees, either at flowers or in flight. In the last few decades, this yellow- and black-striped wasp has become more common on the sandy heaths of lowland Britain, much to the horror of beekeepers but to the delight of us wasp watchers. On those same southern British heathlands, there are two species of sand wasp, the more common red-banded sand wasp, *Ammophila sabulosa*, and the smaller and much rarer hairy sand wasp, *A. pubescens*. Both of them collect caterpillars for their larvae but in different ways.

A red-banded sand wasp ferries caterpillars to her nest until she's happy that she's accumulated enough to satisfy the appetite of the single larva that will occupy this burrow. Then she lays her egg, seals the burrow, and kicks sand around the entrance to hide her handiwork and avoid drawing attention to the well-stocked larder of fresh caterpillars hidden below. A hairy sand wasp, on the other hand, places a single caterpillar in her hole, then lays her egg. She stops the burrow entrance with a temporary seal, then waits until her larva has finished eating the first caterpillar. At this point, she unstops the hole and stocks it with a second caterpillar. She'll keep up this *progressive provisioning* until her larva is fully grown and ready to pupate. In contrast, the red-banded sand wasp's technique is called *mass provisioning*; the distinction, as we'll see shortly, is important.

The processing power of a wasp's poppyseed-sized brain should give software engineers pause for thought. Hairy sand wasps may have several nests on the go at the same time and can keep track of which ones need restocking. Some species even qualify as tool users. Those that close their nest permanently after stocking tamp down the soil as they backfill the hole. Some pound the soil with the tips of their abdomens, others use the fronts of their heads, vibrating their bodies as they press like a piledriver. More impressive still, *A. urnaria* and *A. aberti* grasp a

small stone in their jaws and use that to pound the earth. They even pick up several stones, one after the other, until they are satisfied that they've found the best tool for the job. The great golden digger wasp, *Sphex ichneumoneus*, and several other species have also been observed using tools like this.[15]

The behavioural complexity of solitary wasps is astounding but the most remarkable hunter that I've seen in action is the jewel wasp, *Ampulex compressa*, a beautiful creature sometimes called the emerald cockroach wasp, which tells you all you need to know. The wasp is like a little gemstone, iridescent blue and green, and moves over the ground with fast, jerky movements as it searches for its prey – large cockroaches. When she finds a suitable host for her larva she does what all the hunting wasps I've described so far do – she wrestles with her prey until she can insert her sting and inject paralysing venom. Then this wasp does something far more sophisticated than just drag her prey to her burrow.

She runs around to the front of the roach, which is now just standing there like a zombie, then closes her jaws over the base of one antenna and walks backwards, sliding her jaws along as she goes, as if she was

Cicada killers, *Sphecius speciosus*, specialize in catching and paralysing cicadas, often bigger than they are, which they drag back to a prepared burrow before laying an egg on the unfortunate victim.

measuring off a particular length. About halfway along she stops and bites hard to sever the antenna at that point. Sometimes her whole body quivers with the effort but in short order she's trimmed one antenna, and now she repeats the process with the second.

After all this effort she's ready for a drink so she returns to the base of an antenna and clamps her jaws around it again. This time, as she walks backwards, she pinches the antenna and, like squeezing a tube of toothpaste, she extracts a drop of haemolymph from the cut end, which she promptly drinks. Other wasps sometimes drink a few drops of sustaining haemolymph if their host's body is punctured in the struggle, or if a leg should end up being removed, but the jewel wasp does this very deliberately.

Suitably refreshed for her next task, she picks up an antenna at its cut end and uses it like reins to lead the roach to her nest. The wasp's sting was first placed into the nerve centre at the base of the roach's front legs to slow it down, then a second sting injected venom with the precision of a neuroscientist into the roach's brain that robs it of its willpower. Unable to resist, the roach walks down into the nest and once the wasp has laid her egg, she carefully seals up the tunnel entrance with small stones and dirt, and leaves the roach to its fate.

Most of these wasps prepare their nest holes in suitable ground before setting off to paralyse prey. Some wasps dig like little dogs, excavating loose sandy soil with their front legs and kicking it out in a broad spray behind. Sand wasps have a different technique. A female disappears down her tunnel, only to emerge a few seconds later with a handful of soil clasped in her front legs and wedged beneath her head. She disposes of this some distance from the nest, avoiding the telltale spoil heaps that accumulate around some wasp burrows. Beewolves go to even more trouble to create a safe and secure nest by lining their burrows with a secretion from their antennal glands. This contains a culture of a symbiotic bacteria, *Streptomyces philanthi*, which serves to prevent less benign bacteria and fungi from gaining a foothold in the nest. The larvae inoculate themselves with this culture as soon as they hatch, and even use it to protect the silk of their cocoons.[16]

By digging their tunnels first, all these wasps can choose the best site for a nest: for example, in a place where it won't flood or on a south-facing bank where it will remain warmer. However, the wasp needs to be able to find the nest again after hunting perhaps hundreds of metres away. She does this by performing an orientation flight when she leaves the nest. The wasp takes off and turns to face the nest entrance. She then flies in an arc, keeping her nest hole centred in her vision. Gradually, the wasp increases the distance from the hole and simultaneously also increases the length of each arc and her flight speed. In so doing, she is keeping her relative velocity constant. What can a wasp glean from such intricate flight patterns?

The parallax effects viewed at different distances out from her nest will tell her the relative positions of landmarks, whether they are far or near. She is also able to remember snapshots of the position of these landmarks in relation to her nest for future reference. When she starts looking for her nest again, she first uses the more distant landmarks to navigate to broadly the right area, then switches to using closer landmarks until she has matched her view with the snapshots she took when she left.[17]

An easy, if slightly heartless, experiment shows the importance of these landmarks. Place a conspicuous object close by a nest while the wasp is inside, then move it after she's performed her orientation flight. When she returns, she'll search in vain for her hole where she had memorized it in relation to the object. When I did this for a film I was making, I soon felt sorry for the hard-working wasp mother and replaced the object in its original place as soon as the camera stopped rolling – at which point a much-relieved wasp quickly found her hole again and dragged another caterpillar into its depths.

The navigational accomplishments that evolved in solitary wasps are a vital prerequisite for the evolution of social behaviour. All social wasps build nests that are sometimes large and elaborate and are often constructed over many generations. After having located rich sources of food that may lie a long distance from the nest, the workers of these species need to be able to find their way back home, and then be able to relocate the food source again when they next leave the nest, orientation skills that they inherited from solitary ancestors.

In searching for possible paths to sociality, scientists have seen other clues in the behaviour of solitary wasps. This falls into distinct categories. To begin with, there are wasps that find their prey first, then incapacitate it and dig a nest hole wherever they happen to be. Such wasps don't have to expend energy dragging large prey to a nest hole that could be a long distance away, but they have to put up with nesting wherever they happened to find their prey, which might not be prime wasp real estate. On top of that, they often lose their precious prey to other wasps that are not above a spot of thievery, while they're preoccupied with digging their burrow.

Then there are wasps that dig their nest first and stock it with a single large prey, like the tarantula hawks we met earlier. Moving large prey is tough work, but at least the nest has been excavated in a place of the wasp's own choosing – an optimal spot for her larva. Other wasps, like red-banded sand wasps, tackle smaller, often less dangerous prey. These are much easier to ferry over long distances, although the wasp will need a lot of trips to stock each nest. Finally, we've already seen that the hairy sand wasp feeds its larva fresh caterpillars one at a time throughout its larval life. In addition, this wasp often has several such nests on the go at any one time, each at a different stage, so it requires some serious brain power to remember where they all are and how hungry each larva is.

Arranged as I've done above, this sequence of wasp behaviours used to be seen as a series of stepping-stones of increasing behavioural complexity, each stage adding traits that are vital for the final transition to a full eusocial life. Each requires more cognitive power, helping to shape the formidable brains of social insects. Switching from a single large victim to many small prey items is exactly what social wasps have done, and progressively provisioning the nest creates extensive contact between mother and young, absent in the other categories but ever-present in eusocial insects. Unfortunately, these neat and tidy 'just-so' stories rarely hold up, and this one has failed to survive the advent of modern genetics.

When scientists recently mapped these behaviours onto a family tree of wasps, it became clear that there was no such neat evolutionary progression. It seems that in one family of wasps (Sphecidae) at least, mass provisioning with multiple small prey items was the ancestral state; more importantly, there were many reversions from one style to another along different branches of the family tree, although switching from progressive provisioning to another style was rare. Progressive provisioning, therefore, seems to be a more profound step and perhaps really does stand closer to the brink of more complex social behaviour. But how and why did some bees and wasps cross that threshold? It's time to get back to those cactus bees we left at the start.

THE VEGETARIANS – BEES AND POLLEN WASPS

First, what exactly are bees? Most people make a clear distinction between bees (cute and furry – love them) and wasps (mean and stripy – hate them). So be prepared for a bit of a shock. Bees are just vegetarian wasps. It's been known for some time that in the distant past, the larvae of a group of wasp-like insects adopted a diet of pollen and nectar and became bees. However, a recent study using genetics to plot a new family tree of wasps and bees showed that bees are actually part of an extant wasp family, the Crabronidae. Unfortunately, these same studies show that this family is a bit of a hotch-potch, containing separate lineages of unrelated wasps. In other words, these molecular studies tell us that crabronids do not all derive from a single common ancestor, making it a family that taxonomists call paraphyletic.[18] Or, to put it another way, it's a family in need of serious revision. Nevertheless, it doesn't alter the fact that, just as birds are dinosaurs, so bees are wasps.

The new family tree indicates that bees became bees sometime between 110 and 125 million years ago, during the Cretaceous period. This was a time when flowering plants were diversifying, making it a good time to switch to a larval diet of pollen and nectar, and this change of diet spurred a rapid diversification of bees. Today, there are some 20,000 different kinds of bees buzzing around flowers across the world.[19]

Bees aren't the only wasps to have gone vegetarian. Pollen wasps (Masarinae) have also switched from hunting prey for their larvae to gathering pollen and nectar. Globally, there are around 300 species of pollen wasps, living in North and South America but reaching their greatest diversity in South Africa. To reach nectar hidden in flowers, pollen wasps have evolved long tongues, just like many bees. Tiny *Quartinia* pollen wasps from South Africa are extreme cases, with tongues longer than the rest of their body.[20]

Then there's *Krombeinictus nordenae*, a wasp known from only a few specimens found in the late 1990s in Sri Lanka by Karl Krombein and Beth Norden. Although both are immortalized in the name given to the wasp, it's unlikely to gain them fame. Hardly anyone, most entomologists included, have any idea that this creature exists, yet its complex behaviour parallels many more 'advanced' wasps and bees. *Krombeinictus* nests in hollow stems and raises just one larva at a time, feeding it progressively on a mix of pollen and nectar. When its larva is ready to pupate, its mother moves it to the base of the hollow stem to create space to allow another larva to be reared.[21] This virtually unknown wasp exhibits well-developed parental care and possesses all the cognitive skills that allow it to harvest pollen and nectar efficiently from a variety of flowers.

By taking advantage of flowering plants, bees have diversified into a great many species.[22] So why hasn't the switch to partnering with flowering plants had similar benefits for the pollen wasps, which number just a few hundred species? One difference between bees and pollen wasps is in how they transport pollen. Bees have evolved a variety of ways of carrying pollen on their bodies – including pollen baskets made of long hairs on their legs, or the more straightforward method of coating their hairy bodies with the stuff. Pollen wasps ingest pollen in order to carry it, and this might limit the amount they can transport compared to bees. But it may just be that bees evolved earlier and have therefore had more time to diversify.[23]

Back in the Sonoran Desert, I'm watching cactus bees collecting pollen from the huge flowers of prickly pears, so intense a yellow in the desert sun that they are almost dazzling. Each flower is packed with a forest of tall, pollen-bearing anthers, and a cactus bee dives right in and swims through the anthers, showering herself with pollen in the process. After a few laps around the inside of the flower, she scrapes most of the pollen onto a specialized structure on each of her hind legs, the scopa, a dense patch of hairs that holds the pollen firmly in place. Then, with legs bulging with pollen, the bee sets off to fly back to her nest.

The bee's burrow is in a patch of sandy soil 50 metres away, and she's not alone. The whole area is alive with cactus bees. Just like some of the wasps we've seen, these bees choose the optimum spot to dig, where the soil is warm and loose but has just the right amount of moisture. Too little and her larva will desiccate; too much and the pollen store will go mouldy. Such places are few and far between, so the bees congregate in huge numbers at the most favourable sites. These are solitary bees, although it's hard to believe that as I watch hundreds of them coming and going. Yet each female is only concerned with her own nest. She might live among a crowd but that's just because it's the most desirable neighbourhood.

Just how important soil conditions are is demonstrated even more dramatically by another solitary bee. Drive through the Walla Walla Valley in Washington State and you'll pass through large tracts of land covered in alfalfa crops. Interspersed among all this verdant growth are dry, dusty patches of bare soil that seem like a waste of space. However, if you'd been paying attention to the road signs, you'd have a clue as to why these apparently neglected patches are a vital part of farming here.

As you approach each of these areas, you'll see signs reading 'Slow: Alkali Bee Area'. The bare patches are *bee beds*, scraped clear of vegetation and maintained in just the right condition by local farmers for alkali bees, *Nomia melanderi*. 'If you build it, they will come', a mysterious voice whispered to Kevin Costner in *Field of Dreams*. In this 1989 film, Costner played a baseball fan, inspired by the voice to build a ballpark in the

A male ivy bee, *Colletes hederae*, shelters in his burrow. Although solitary, many thousands of these bees dig their burrows close together where the soil is just right – in this case, a warm, south-facing bank of dry sandy soil.

middle of nowhere. Had he been an alfalfa farmer those words would ring with an even greater truth. Get the soil right and the bees will come.

Alkali bees need silty soil with just the right moisture levels, as well as being free of vegetation. They also prefer soils that are occasionally sealed by a crust of salt as minerals dissolved in water in the soil are left behind as the water evaporates. When farmers create these conditions the bees swarm in, creating a landscape of miniature volcanoes as they excavate their nest tunnels. The farmers of the Walla Walla Valley don't construct this bee-friendly landscape out of the goodness of their hearts. These little bees are the best pollinators of their alfalfa crops – far better than honeybees. The bees from a half-hectare bed can pollinate 40 hectares of crops and help to generate 50 tons of seeds. This symbiotic relationship between farmers and alkali bees has persisted for at least fifty years, over which time the bees' population has multiplied ninefold to an estimated 17 million bees – quite a labour force. On a single one and a half hectare site there were estimated to be more than 5 million bees, so many that it's hard to think of them as solitary.[24]

The cactus bee colony seems sparse by comparison, but it's no less fascinating. Females are busy digging out nests, scraping soil into piles with their front legs, then shuffling backwards to use their back legs to

level the heap. Those that have already dug out their tunnels are now starting work on the infrastructure, a mud turret that extends upwards, then sideways from the tunnel entrance. The purpose of these home extensions is not entirely clear, although they may have several uses. They prevent flooding from the occasional monsoonal storms and, in between, when it's dry and windy, they prevent dust and sand from blowing into the nest. Some species of bees impregnate the turret with glandular secretions, which perhaps give each nest a unique fragrance, so making it easier for each bee to find her nest in the bustling neighbourhood.[25]

The turrets may also serve a defensive function. A lot of insects, including other bee species, take advantage of all the hard work carried out by cactus bees and similar hard-working species. They sneak in and lay their own eggs on the pollen mass. Their larvae eat the pollen ball and kill the bee larva, either directly or by starving it to death. These cuckoos in the nest are called cleptoparasites and around a fifth of all bee species make their living like this – sponging off their hard-working cousins.[26] Wasps are equally guilty; notable social parasites include the so-called velvet ants (Mutillidae), a family of around 7,000 species whose females are wingless and look very much like ants as they race around at high speed in their search for nests to parasitize.

Many velvet ants are brightly coloured, a warning that they can inflict a very painful sting, but one of the most extraordinary-looking ones I've ever come across was causing trouble among the cactus bees we were trying to film. The wingless female of the thistledown velvet ant, *Dasymutilla gloriosa*, is covered in long white hairs and moves over the ground at such high speed that it looks like a bit of thistledown blown on the wind. We watched as the cuckoo approached a hole and carefully examined the entrance. Satisfied that it would provide a good surrogate home, she disappeared inside, only to reappear in a few seconds having done the dirty deed. Holes capped by turrets may make the entrance less obvious to these ground-hunting cleptoparasites, but it certainly helps guard against attacks from the air.[27]

For these poor cactus bees also face airborne assaults by the precision bombers of the insect world – bee-flies (Bombyliidae). A bee-fly uses all her aerial skill to approach a bee's nest, then hovers rock-steady a few inches out as she lines up on the open hole. With a deft flick of her abdomen, she launches an egg towards the hole. She may hit a hole in one as the egg bounces down the hole, or it may come to rest close to the hole. In either case, the larva soon hatches and quickly makes its way down to the bee's larder. The turret makes it much harder for a bee-fly egg to find its target or for a tiny, newly hatched larva to find a way in.

There's little else these bees can do to deter parasites. They have no option but to leave their nests unguarded while they collect pollen and nectar. Some of the sand wasps use a small pebble to block their tunnels when they leave to go hunting to try to deter nest thieves, but nesting in large aggregations might also help. There's safety in numbers, or in the predator dilution effect, as scientists prefer to say. That is, the chances of a parasite picking on your particular nest are smaller in a crowd of thousands. However, there may be more direct benefits, too.

WORKERS UNITE – THE BENEFITS OF A COOPERATIVE

I was out in Arizona to film cactus bees for a BBC documentary on bees and wasps (*For Queen and Colony*, 1992), but it was a colony of bees much closer to home that allowed us to capture on film a clear advantage of living with lots of neighbours. A hundred metres down the road from the BBC Natural History Unit's Bristol headquarters, I found an old wall held together by mortar that had softened with age. It was no match for the tough mandibles of the local mining bees, *Andrena* sp., which had excavated a few large holes where the wall met the pavement, and were now digging deeper tunnels to house their larvae. Many of these nest tunnels were excavated off the larger entrance hole so, although these are still solitary bees, each solely concerned with her own nest, they had a communal entrance.

On the wall outside the hole, I spotted some very different-looking bees, more wasp-like with bare, brown- and yellow-striped bodies. They were cuckoo bees, *Nomada* sp. I'd seen these bees before, while filming tawny mining bees, *A. fulva*, in a nearby garden lawn. Tawny mining Bees are one of the most distinctive solitary bees, covered in a fuzz of hair that glows day-glo orange in the early spring sunshine. They dig tunnels in soil, often in lawns, creating a little volcano around the nest entrance. Large numbers sometimes descend on favoured lawns, and gardeners who pride themselves on neat and sterile bowling green lawns sometimes try to get rid of them, although in my opinion they make a charming addition to any garden.

There may be thousands of tawny mining bees on a large lawn, but they've all built their own little volcanoes spaced at discrete distances from their neighbours. Watch patiently and you're almost certain to see a cuckoo bee lurking around the entrance to a burrow in a decidedly suspicious manner. She waits until a mining bee leaves, then nips into the hole, safe in the knowledge that she'll be undisturbed for a few minutes while the mining bee is away collecting pollen. The cuckoo bees hanging around the crumbling wall aren't having it so easy.

Several of them are skulking close to the busy mining bees and as a bee leaves, the cuckoo bee moves in – but just as she is about to enter the large entrance hole, another bee rushes out. The cuckoo bee backs away until this bee leaves, then makes another attempt to gain entry. This time a mining bee returning laden with pollen brushes past her and again the cuckoo bee retreats. By now she's looking very edgy. I watched for another half an hour, sprawled out on the pavement and drawing stares and comments from the passers-by stepping around me, but still the cuckoo bee hadn't managed to sneak into any mining bee holes. Tenement living with a communal entrance meant there was always a mining bee in the way. Even though the mining bees were all going about their own business, they were acting as inadvertent guards for their neighbours.

Among these mining bees, this sociality is purely coincidental. I've seen these same bees nesting in individual holes, just like the cactus bees. So, although the communal advantages of the set-up in that old Bristolian wall were simply a consequence of the place the mining bees had chosen to nest, it was still a graphic illustration of the value of guards in protecting the hard-won resources that accumulate in the nest. This simple level of sociality is so effective that some bees make it more of a deliberate strategy.

Back once more in the bee heaven of the Sonoran Desert, I'm in pursuit of carpenter bees (*Xylocopa* spp.). These blue-black, monster-sized bees chew holes in the old woody flower stems of yuccas and agaves, and although they're beefy bees with powerful, wood-chiselling mandibles, these holes take a lot more time and effort to bore out than just digging in soft sand. There are many kinds of carpenter bee, and some prefer to hew nests in timber, including beams and rafters. However, in all cases, suitable sites for their carpentry may be few and far between. Once excavated, their nests are stocked with a paste of pollen and nectar in the same manner as those of countless other solitary bees – and then carpenter bee life gets complicated.

Carpenter bees live for a long time – some for well over a year. In some species, one or more adult daughter bee might opt to remain in their mother's nest rather than face the difficulties of carving their own. These stay-at-homes are unlikely to get a chance to breed in their natal nest, at least not at first, since their mother remains dominant and continues to forage for pollen and build new brood cells. Nests might

be hard work to make but they are robust, so if the daughter bides her time until her mother finally dies, she will inherit the nest. However, in at least one species, *X. pubescens*, young female bees don't always wait patiently for their inheritance. They hang around waiting for the perfect moment to pick a fight with their mother and, if they emerge the victor, take control of the nest. It seems they have a fifty-fifty chance of winning the tussle, which is probably better odds than trying to find a new site and chiselling out their own nest elsewhere.[28]

However, before this struggle for dominance, the daughter rarely leaves the nest. There's little point, apart from for a quick feed, since she isn't provisioning brood cells and doesn't need to collect pollen. Instead, she positions herself in the nest entrance, where she acts as a guard while her mother is out and about.

Sometimes these nests are occupied by females of the same generation. In such cases, a female has taken over an existing nest instead of going to the effort of building her own. Rather than leave, the usurped female often stays in the nest, perhaps because she's unlikely to find anywhere else suitable for a new nest and, in any case, the effort to dig a second nest would be exhausting. The new female becomes dominant and destroys much of the original female's brood. However, she may allow a few of the original female's offspring to survive as an added incentive for the

In sunny weather, thousands of male ivy bees, *Colletes hederae*, leave their tightly packed burrows to fly in their thousands in search of newly emerged females. It's hard to think of these bees as 'solitary'.

original female to stay. After all, if she can be persuaded to hang around, she can provide a useful service as a guard.[29]

Carpenter bees can behave like other solitary bees and breed in isolation, or they can form these simple, often just two-bee societies. Any one species can adopt any of these strategies, and just which one they opt for depends on local environmental conditions, such as the availability of nest sites and food resources or the prevalence of pollen-thieving parasites. This flexibility has earned carpenter bees the title of facultatively social bees, although they're not the only ones.

In contrast to the impressive carpenter bees, sweat bees (Halictidae) are tiny and pass unnoticed by most until they become an annoyance. They get their name for their liking for salts in sweat. In the tropics, where sweat production is copious, they swarm around your eyes and mouth in great clouds, making it hard to concentrate on anything else.

Some sweat bees form small colonies that resemble those of the familiar honeybees and bumblebees more than those of the carpenter bees, with small numbers of workers that assist in rearing their sisters. Yet, even within a single species, some may be solitary while others are social. As is the case with carpenter bees, this flexibility allows these

bees to respond to local circumstances. One species studied in the UK, *Halictus rubicundus*, is social in the south but solitary in the north.

However, these little bees have a great deal of flexibility in how they live their lives. Move colonies from the north to the south and, in some cases at least, solitary bees become social. Again, environmental factors are the driving force. It takes time for a founding queen to raise a brood of workers, then, with their help, raise a second generation of bees that eventually leave the nest to breed themselves. In the shorter northern summer, the season simply isn't enough long enough to fit in a more social lifestyle.[30]

Nonetheless, some sweat bees have adopted a full-time social life, producing several generations of workers. Although their tiny colonies are usually short-lived, the colonies of one species, *Lasioglossum marginatum*, may survive for four or five years.[31] This kind of sociality has evolved independently at least twice among the sweat bees.[32] But more remarkably, it has 'unevolved' many more times. Social *Lasioglossum* bees have reverted from a social to a solitary existence on at least twelve occasions in their evolutionary history, and other kinds of sweat bee have done the same.[33] The conclusion to be drawn is that eusociality is hard to establish but easily lost.[34]

Even so, several other groups of bees have developed some form of sociality, either communal breeding, as practised by some of the carpenter bees, or eusociality, leading in one branch of the bee family tree to the bustling colonies of honeybees. Among bees as a whole, sociality has arisen independently perhaps four or five times, but it has been lost again many more times.[35]

SOCIAL WASPS

This story has parallels among the wasps. There are two broad groups of hunting wasps, apioids (which include the bees) and vespoids (which include many solitary species such as potter wasps, as well as the highly social paper wasps, together with hornets and yellowjackets). Let's start with the apioids. Apart from a variety of bees, most apioids, such as the sand wasps and cicada killers that we met earlier, are solitary hunters. However, quite a few of them share communal nests and – like some of the carpenter bees – are able to mount guards to defend against parasites.

The nest burrows of an Australian digger wasp, *Cerceris australis*, might contain a hundred nest cells, the work of several females, and two or

A male carpenter bee, *Xylocopa caffra*, from tropical Africa. Females of this species chew out nests in agave or aloe stems.

three generations of these wasps overlap in these extended subterranean nests.[36] It's not uncommon to find two to four females of the related Australian species, *C. antipodes*, sharing a nest. Usually only one female is provisioning cells, while the others stay in the nest as guards to fend off marauding velvet ants, which plague wasps just as much as bees. It seems that, as one female finishes a period of provisioning, another takes up that role, so all the females get a chance to breed. In this case, these cooperating females are probably related, maybe sisters that all emerged from the same nest that they are now using themselves.[37]

Unfortunately, nature never stands still. As nest security improved, so parasites found ways to breach it. The digger wasp *C. rubida* is the only species in Europe currently known to have communal nests like the Australian species. Unfortunately, one of its main parasites, a fly, *Pterella grisea*, is adept at avoiding guard wasps. She waits at a safe distance from the nest until she spots a female wasp heading home, then flies an intercept course to fall in behind the wasp and track its flight. If the wasp spots that she has a tail, she pulls some fancy aerobatic manoeuvres in an attempt to shake off her pursuer, but the fly is pretty nifty too, and the wasp rarely succeeds. As the wasp disappears into her hole, the fly executes a close fly-past and drops a tiny wriggling larva, which makes its way into the burrow. The larva of *Pterella grisea* is so small that it manages to pass under the wasp's radar, whereas the guards would surely evict any adult fly that tried to gain entry.[38]

Although most of the apioid wasps are solitary, that doesn't mean they don't put a lot of effort into their parental duties by preparing nests and ferrying food to their larvae as they grow. Often, hunting wasps have only short lifespans and, combined with time-consuming parental care, this means that they can't raise many offspring. One species, *Sceliphron assimile*, averages under ten young in its whole life. Short lives in particular make it difficult for these wasps to take a step towards more complex societies.[39]

However, among this branch of the wasp family tree one group has become truly social. *Microstigmus* wasps are tiny creatures that live in the New World tropics, where they build inconspicuous nests, like little bags, usually hung from the undersides of leaves. Each species chooses its own particular kind of plant, though all are picked for their hairy leaves. The wasps scrape off the plant hairs to make their nests. *Microstigmus* wasps are unusual in that the adults can spin silk from abdominal glands, which they use to bind the plant fibres together. Their colonies are tiny, perhaps just half a dozen females, although usually only one of them lays

Cactus flowers produce copious pollen, and their nectar has high concentrations of sugars, so they're very attractive to many kinds of bees.

eggs. The others, which in many cases are probably daughters of the egg-laying 'queen', head out to hunt tiny prey such as thrips or springtails.[40]

The vespoid wasps, too, are mostly solitary hunters. This group includes the potter wasps, which build little flask-shaped nests from mud. In the past, this branch of wasp-kind also encompassed, among others, spider-hunting wasps (Pompilidae) and scoliid wasps (Scoliidae), which parasitize the larvae of scarab beetles. It also included the ants, all of which are, of course, highly social. However, modern genetics points to this being another hotch-potch paraphyletic group. Nevertheless, the group still contains the vespid wasps (Vespidae), among which complex sociality has arisen on two separate occasions.[41]

The hover wasps (Stenogastrinae) are curious creatures. They live in south-east Asia and build strange paper nests in such bizarre shapes, with fluted entrances or spiky projections, that they would not be out of place in a museum of modern art. Their social life is just as intriguing. Usually, the nest is started by a single female, and her daughters stay on to become workers in a similar manner to yellowjackets and hornets. However, whereas the job of worker is a more or less permanent position for a yellowjacket, among hover wasps it's a more temporary arrangement.

Yellowjackets scrape up soft wood to make their carton nests.

On hatching, some females may leave the nest immediately to start their own nest. Others stay to help rear more sisters, but after a time, these wasps also leave to breed on their own. As we saw among the sweat bees, the strategy that each female follows is dictated by environmental conditions that, at times, favours cooperation over going it alone.

Social living arose a second time in the vespid wasps, this time resulting in the much more complex societies of paper wasps, yellowjackets and hornets, although there is still much variation in how these societies work. The success of these wasps in building bigger societies rests on a few factors, not least their ability to produce better-quality paper or carton, allowing them to build bigger and more robust nests. The mandibles of hover wasps are not so suited to extracting the long plant fibres that yellowjackets use.

Yellowjackets return time and again to places that provide high-quality fibres, and sometimes fence posts or the sides of an old shed are covered in scrapes and scratches where the wasps have been chewing out nest material. The wood is turned into a ball of pulp and carried back to the nest, where the wasp adds it to the edge of the nest as skilfully as any bricklayer laying mortar. Fibres from different sources are often of different colours, so the gradual growth of the nest is recorded as layers in shades of buff and grey.

Using stronger building materials also allows these wasps to better protect their homes. When the founding queen starts her nest, she first builds a thin stalk, the petiole, to hang her first batch of cells from. As the nest grows, this narrow petiole remains the only connection between the nest and the rest of the world. Ants are a big problem for many of these social wasps and, just as a few Spartan soldiers could hold off the entire Persian army in a narrow pass,* so the thin petiole is easy to defend. Added to that, some wasps smear the stalk with chemicals that repel ants.

Ants are mainly a threat to the helpless wasp larvae because these wasps generally don't store food in their nest, so there's nothing to steal. The young are fed mouth to mouth on a diet of liquid protein from prey thoroughly mashed up by the workers. Because wasps have that tiny wasp waist, they can't swallow big bits of food, so subsist on nectar and the juices from pulped prey, some of which they regurgitate to feed the larvae. Surprisingly, however, a few wasps do store honey – just like bees.

* At the Battle of Thermopylae in 480 BC, 7,000 Greek soldiers held off a Persian army, reputed to number a million, for three days by occupying a narrow coastal pass. Given the size of some ant colonies, wasps face similar odds.

Seventeen species of honey wasps, *Brachygastra* spp., live throughout Central and South America and a few, such as the Mexican honey wasp, *B. mellifica*, extend into the southern United States. Colonies of these wasps can be 20,000 strong, a big enough workforce to allow them to collect and store copious quantities of honey in empty cells in their carton nests. This makes their nests an even more attractive target for thieves, but the honey wasps are well able to defend their precious nest. Unlike most wasps, but like honeybees that also must guard huge honey stores, these wasps have barbed stings that remain lodged in a would-be nest robber, pumping venom and continuing to cause acute pain.[42] Despite this, wasp honey is regarded as a delicacy by local people in parts of Mexico.

Warrior wasps, *Synoeca* spp., live in similar areas to honey wasps and, although they don't have large honey stores to protect and their colonies are a lot smaller, they defend their nest with far more aggression. The nest itself is a peculiar structure that looks a lot like an armadillo climbing a tree. That's because the brood cells are hidden beneath an outer layer of brittle carton that forms a hemisphere attached to a tree trunk or large boulder – but beware of admiring their handiwork too closely. First, you'll hear an ominous rattling from inside the nest as the workers all drum their abdomens in time with each other on the inside of the nest wall. If you fail to heed that, bands of workers rush out and drum on the outside of the nest – but not for long. They soon launch an attack.

In the interests of completeness, Justin Schmidt has provoked warrior wasps to attack and once he'd recovered, he jotted down his description: 'Torture. You are chained in the flow of an active volcano. Why did I start this list?' He gave it a four on his one to four scale.

Honey and warrior wasps share one feature of their life cycle. They are both swarm-founding wasps. Most social wasp colonies begin when a lone queen sets off to start her own colony. She is faced with the daunting task of starting the nest, then feeding her brood entirely on her own. It's not until her first daughters emerge that she has a workforce to take on guarding, building and nursing duties. A lone queen faces many dangers, and most colonies fail before they even start. Some wasps though, again like honeybees, send their queens into the big, bad world with a retinue of workers.

Among some swarm-founding wasps, for example *Provespa* spp., a single queen accompanies the worker swarm, but in all the others, multiple queens head out with their workers to found new colonies.[43] Swarm-founding wasps have passed a point of no return; a queen is

A yellowjacket's nest is made of a carton of chewed wood. Wood pulp from different sources shows up as differently coloured bands in the outer protective layer of the nest.

never on her own, so there is never a solitary phase in their life cycle. This prevents these wasps, as well as swarm-founding bees, from ever reverting to a solitary existence, such as we saw happening so frequently among the sweat bees. They depend on each other entirely – on the ties that bind several generations of insects into a single integrated whole.

SOCIAL BEES

There are many parallels with all these varied levels of wasp sociality in the branch of the wasp family tree that contains the bees. Solitary queen bumblebees, *Bombus* spp., like yellowjackets and hornets, must face the hardships of starting a colony on their own. In part, this solitary start to social life is imposed on these bees by where they live. Bumblebees are more frequently found in cooler temperate climates – indeed, one species, the Arctic bumblebee, *B. polaris*, is found right around the Arctic, where it survives freezing temperatures.* Yellowjackets and hornets are also common in temperate regions. Such a seasonal climate makes it hard

* A second species, *B. hyperboreus*, also lives in the Arctic, It's a cuckoo-style parasite of the Arctic bumblebee, usurping its nest and forcing Arctic bumblebee workers to rear her own brood. So even crossing the Arctic Circle provides no escape from social parasites.

Inside a yellowjacket nest, carton cells contain fully-grown larvae in the older top layer and newly-laid eggs in the more recently built bottom tier.

to maintain a permanent colony, because food resources become much scarcer or vanish altogether over winter. Bumblebees store both pollen and honey in their nests, but too little to see a whole colony through the cold winter. In all these cases, only a mated queen overwinters, by hibernating in a hollow log or old rodent burrow. When spring comes, she faces the world alone until she can rear her first workforce.

Honeybees avoid this problem by building up such vast stores of honey that there's plenty to feed even their much larger colonies over winter. By maintaining such numbers, they can afford to send out large swarms during the summer to found new nests. Their large honey stores also allow honeybee colonies to last for many years. However, sometimes, if conditions allow, even colonies founded by solitary queens persist for longer than one season – occasionally with spectacular results.

Deep in the pine forests of central Florida, an ominous humming gives away the location of a yellowjacket supernest. In the warm climate of the Sunshine State, these wasps can find enough to keep their colonies alive and growing over winter. They produce queens that still must start new colonies on their own but, once founded, the colony just keeps on growing – and growing. The one we found was wrapped around the trunk of a tree and extended to nearly 2 metres in height and 1 metre in diameter at its base. It must have contained tens of thousands of workers. Due to climate change, such supernests are being more commonly reported around all the Gulf states and they often reach enormous sizes. One discovered recently filled the entire interior of an old, abandoned car. Another almost filled a derelict shed – but all of these colonies are living right on the edge of existence. An unusual and persistent cold snap in the winter of 2020–2021 killed off most of the nests in Alabama and Louisiana, although further south in Florida some at least survived. If warming continues, though, these southern states can look forward to some seriously impressive wasp nests.

European yellowjackets were accidentally taken to New Zealand in the first half of the twentieth century. A queen German wasp, *Vespula germanica*, arrived hidden in aircraft parts in the 1940s and, in the absence of any competitors, soon created a serious problem. The common wasp, *V. vulgaris*, may have arrived even earlier, but didn't start becoming a problem until the 1970s. New Zealand has no native social wasps, so their arrival at the hand of humans was bound to be an ecological disaster.

In the mild New Zealand climate, wasp nests can persist for many years, and the biggest of them can house tens of thousands of wasps. The largest ever found, containing 4 million cells, was 4 metres high. A single hectare of land can contain dozens of these nests. The New Zealand bush soon began to fill with foreign wasps. Today, the southern beech forests at the top end of South Island have the highest densities of social wasps anywhere in the world.

These forests are full of sap-feeding scale insects, and they often drip with honeydew. Many of New Zealand's unique birds have come to depend on this sugary solution as part of their diet, but millions of wasps soon began to mop up the honeydew. This endless supply of food fuelled the growth of these supercolonies and spelt trouble for the native birds. Several species, including the kaka, a once common New Zealand parrot, have declined in numbers because they now have to compete with wasps for honeydew.[44]

New Zealand had no social bees either – at least until 1839, when two honeybee hives arrived in the care of an immigrant, Mary Bumby. The New Zealand bush proved just as hospitable to honeybees as it would a century later to European wasps. Honeybees make honey, a crop worth over $300,000 a year to New Zealanders, so these foreigners were welcomed with open arms; hives can now be found all across both North and South Island. While honeybees don't seem to impact native birds in the same way as wasps, they are outcompeting the native solitary bees for nectar and pollen. Huge and expensive efforts are underway to tackle the wasp problem and alleviate pressure on charismatic native birds, but

no one is suggesting getting rid of honeybees. New Zealand's native bees will just have to work out how to survive on their own.

Thanks to their valuable honey, western honeybees, *Apis mellifera*, have been transported all over the world. Originally, this species lived in Europe and Africa, but it's just one of eleven different kinds of honeybees, the rest of which live across Asia. The most spectacular are undoubtedly giant honeybees, *A. laboriosa*, which live at high altitudes in the Himalayas and build massive combs that hang high on sheer cliff faces. Thousands of years of beekeeping – for both their honey and their pollination services – have made honeybees perhaps the most familiar of all bees.

Although honeybees are now cultured worldwide, other bee species are also kept for their honey. Stingless bees are tropical species and are arguably a lot more successful than honeybees, at least in terms of their diversity. There are fifty times as many stingless bees as honeybees. Their nests, sometimes underground, sometimes in hollow trees, are marvels of insect architecture. They collect resin oozing from trees and mix this with wax to create their raw materials for building. Viewed from above, some species build elegant spirals of cells, but not content with this artistic achievement, they add a complex network of struts that arch over the brood cells to make their nests look like a set from some science-fiction film. Many also build a long tube extending out from the nest entrance, the end of which is permanently manned by a circle of guards.

Although stingless bees don't store as much honey as honeybees, there's still enough of it in the nest to make it a valuable resource worth defending. If their nest is threatened, squadrons of bees issue from the entrance, equipped with blobs of resin that they daub onto predators to gum them up and slow them down.[45] Some stingless bees even store small balls of resin near the nest entrance, handy to grab as the defenders sally out.

Their worst enemies are often other stingless bees, sometimes even members of their own species. I've seen battles erupt between two colonies of the Australian stingless bee, *Tetragonula carbonaria*, and it's total warfare. Thousands of bees from one colony tried to gain entry to a neighbouring nest, from which thousands of defenders poured out. The air above the nest entrance was black with bees, latching onto each other in flight, then tumbling to the ground. The ground below was littered with small knots of bees biting at each other with their powerful mandibles, and with the dismembered carcasses of earlier casualties. Thousands of workers die in these pitched battles because the stakes are

An underground nest of buff-tailed bumblebees, *Bombus terrestris*.
The large, egg-shaped cells contain larvae.

so high. In this case, the invading bees are after more than just the honey supplies. If the invaders win, they'll bring one of their young queens and take over the whole nest.[46]

Some stingless bees have become full-time thieves. The twenty-one kinds of robber bees, *Lestrimelitta* spp., along with one species of *Cleptotrigona*, no longer visit flowers at all and depend for their own supplies of honey on stealing it from someone else. Robber bees are bigger than other stingless bees, with more powerful mandibles, and can quickly overcome defenders. They're such a threat that several victimized species have evolved a specialized soldier caste.

Among ants and termites, physical differences between different castes of workers, including the existence of well-armed soldiers, are commonplace, but are unknown in wasp colonies and very rare among bees. However, among South American stingless bees, soldiers have arisen on several separate occasions, driven, it seems, by the evolution of robber bees.[47] Soldier bees are bigger and tougher than foragers from the same nest, giving them a better chance to defend the colony's honey stores against robbers.

The honey itself is stockpiled in specially constructed honeypots, sometimes in a separate area of the nest. These are made from the same mix of wax and resin as the other parts of the nest, and the resin imparts a distinctive flavour to the honey, which – for some honey connoisseurs – raises it above honeybee honey for taste. It also has antibiotic properties. Stingless bee honey is widely collected by indigenous people; Australian Aboriginals sometimes attach a small white feather to a foraging stingless bee, to make it easy to follow back to its nest. In the Amazon, I've followed (or at least tried to follow) Kayapo Indians who chase foragers at breakneck speed through the forest to track down a tasty cache of honey.

It's far easier to keep stingless bees in hives. Stingless bee culture is traditionally practised in many places in Central America. Colonies are housed in hollow logs, which are stacked up close to the house. The end of the log can be unplugged to access the honey. Sadly, these traditions are declining in most areas as fewer young people bother to learn the arcane skills of stingless bee management. On the other hand, stingless beehives are becoming a much more common sight in Australia, where meliponiculture (as opposed to apiculture) is growing in popularity. Their honey is harvested, but the bees are also invaluable as pollinators. In southern Queensland, macadamia orchards are planted out in neat patterns with wide rides running through them. Lining these rides at intervals are dozens of hives of stingless bees. The hanging flowers of the macadamia trees are buzzing with the bees themselves, all hard at work pollinating the plants and supporting the valuable macadamia nut harvest.

The evolution of social behaviour, involving selfless acts on behalf of certain individuals, is a 'special difficulty' in biology, which is how Charles Darwin referred to it in *The Origin of Species*, in the quote that opens this chapter. Natural selection is essentially a selfish process – for Darwin, a struggle for existence among selfish individuals or, as more recently envisioned, a struggle among selfish genes to replicate themselves. Any gene that promotes selfless behaviour should doom itself to failure by leaving fewer copies of itself in future generations than one promoting selfish behaviour. Yet the teeming colonies of ants, bees, wasps and termites, where altruistic behaviour is common, were well known to Darwin.

Although the majority of wasps and bees are solitary creatures, social behaviour in the Hymenoptera has evolved independently several times. As we've seen, bees have adopted a social lifestyle on five separate occasions.[48] Ants, all of which are social, also belong to this order, so it seems that something about the Hymenoptera predisposes them to become social. Outside of this order only the termites and a few other obscure groups of insects have developed eusocial colonies. Seeking an answer to this is one reason why the behaviour and genetics of bees and wasps have been so closely scrutinized.

One idea that gained widespread currency relates to the way in which bees and wasps (and ants) reproduce. Firstly, in many species, females only mate once, so all their eggs are fertilized by one male, which limits the potential genetic variability of the offspring compared to cases where several genetically different males provide sperm. Secondly, as we saw earlier, female bees and wasps can choose the sex of their young. Eggs fertilized in the normal way become females, with the usual double set of chromosomes. Eggs that remain unfertilized and consequently have only one set of chromosomes become males. Animals with the normal double set of chromosomes are called *diploid*; those with just one set are referred to as *haploid*. So, this peculiar way of breeding is known as *haplodiploidy*, and it creates some interesting genetic outcomes. The logic is a little complicated, so I'll avoid inflicting a headache this late in the chapter, and simply jump to what this means for the next generation. The end result of haplodiploidy is that daughters are much more closely related to each other than they are to their own mother or would be to their own daughters. On average, they share three-quarters of their genes with

their sisters instead of the usual 50 per cent shared between siblings of animals in which both males and females develop from fertilized eggs.

A gene's only mission in life is to leave as many copies of itself as possible in the next generation. The bodies of animals and plants are merely the vehicles through which genes achieve this. A gene can be successful by cloning itself directly, in other words by the body it inhabits producing lots of offspring, but it can also leave more copies of itself indirectly, by helping other bodies that contain identical genes. In 1964, the evolutionary theorist William Hamilton formulated an idea, backed by mathematical models, that showed how individuals could evolve altruistic behaviour, apparently in contradiction to Darwinian natural

A honeybee swarm. Honeybee queens never have to face setting up a colony on their own. When they leave the nest, they're accompanied by thousands of workers who scour the area for a suitable place to build a new nest.

selection, as long as they were closely enough related to each other. This became known as kin selection and the peculiar sex determination mechanism of the hymenopterans, with the high levels of relatedness among sisters, was seen as the reason behind their frequent flirtations with sociality.

The idea that kin selection is the mechanism behind the birth of complex social systems gained huge popularity through the second half of the twentieth century. Unfortunately, as more examples of sociality were studied, it became clear that kin selection couldn't account for all the new discoveries. Some of the eusocial colonies that we looked at in the last chapter are not haplodiploid. Termites, too, reproduce in the conventional diploid way. Furthermore, some female bees, wasps or ants mate several times – so increasing genetic variability in the colony. Also, the genetic similarity of sisters among haplodiploid insects is offset by their low relatedness to their brothers, with which they share only a quarter of their genes. Finally, as noted above, by far the largest number of hymenopterans are solitary, yet all are haplodiploid. For all these reasons, kin selection is no longer regarded as the central driving force behind the evolution of sociality.

Today, emphasis is placed on ecological factors, such as those that drive some sweat bees or carpenter bees to adopt a social life while others remain solitary, and even for the same species to switch back and forth between solitary and social living. Many of their adaptations to solitary life also predispose bees and wasps to becoming social. For example, having a nest stocked with hard-won food resources makes it worth defending. As we saw in the previous chapter, all the social aphids and social thrips are gall makers that likewise have a home to defend. Fortress defence such as this seems to be an important step on the road to sociality – since a fortress is more easily defended by an army than by a lone sentry.[49]

There are still many aspects of the evolution of social life that are hotly debated. Recently, E. O. Wilson, regarded by many as the father of insect sociobiology, has suggested a new theory based around selection at group level. Group selection has been around, on the fringes of science, for many years as a manifestation of the 'for the good of the species' idea, which I still frequently hear expressed by non-biologists. In this form, it flies in the face of tried and tested natural selection – it is tantamount to some kind of natural magic because there is no known mechanism through which it could operate. New formulations of the idea have narrowed it down and defined it more rigorously, backed by mathematical modelling.

Nevertheless, it was greeted with massive scepticism by a great many biologists when the ideas were first published. Watch this space...

As a mechanism for the origins of social behaviour, some form of group selection is still a long way from widespread acceptance. On the other hand, once eusocial colonies have formed, then the whole game changes. Only the queens lay eggs – the sterile workers can be seen as extensions of her body, their behaviour more akin to the way that your liver cells cooperate with your kidneys to keep your whole body working. Now a form of natural selection *can* work at the group or colony level, since a colony behaves like a single organism – a superorganism. This is why the evolution of eusociality has been called a major transition in evolution. It opens up all sorts of new possibilities that have been exploited by the highly social bees and wasps and, in the most dramatic and varied ways, by two other groups of social insects – the ants and the termites.

Moss carder bees, *Bombus muscorum*, love the flowers of clover, rich sources of pollen and nectar for their larvae.

II

Superorganism

Ants and termites – from colony to supercolony

As a thinker and planner, the ant is the equal of any savage race of men; as a self-educated specialist in several arts, she is the superior of any savage race of men; and in one or two high mental qualities she is above the reach of any man, savage or civilized!

Mark Twain, 1906[1]

You must consider a termitary as a single animal, whose organs have not yet been fused together as in a human being. Some of the termites form the mouth and digestive system; others take the place of weapons of defence like claws or horns; others form the generative organs... The insects themselves should always be thought of as the blood-stream and organs of a single animal.

Eugène Marais, *The Soul of the White Ant*, 1925[2]

INSECT ARCHITECTS – THE CITIES OF TERMITES

Only an engineer would consider a JCB a precision implement, but that was the tool of choice for Rupert Soar, then of Loughborough University's School of Mechanical and Manufacturing Engineering, in the UK, and his colleague, biologist Scott Turner, from the State University of New York (SUNY). They were trying to get to grips with Namibia's skyscraper-building termites, while at the same time, in the early 2000s, I was making a TV series on bio-inspired engineering (biomimicry).[3] The massive termite mounds looked like a promising story, so that was how we ended up in the African bush with Rupert and Scott – and a very large digger.

Most people are familiar with termites that chew through wood and cause serious damage to buildings in warmer climates – but many kinds don't live like this. Instead, like those we went to film, they build impressive clay edifices often in such abundance that they look like an insect version of a city of high-rise buildings. These termite mounds are rock solid and, although they rise many metres into the air, the main living quarters lie beneath the equally hard-baked Namibian earth; so, all in all, it seemed like the digger was a good idea. The bucket sliced down through one half of the tower, then continued the excavation below ground level, proceeding as carefully as one can with a JCB. Eventually, the whole nest was bisected to reveal the below-ground living chambers and above them a tower interlaced with a complex branching network of tunnels, surrounding a much larger central chimney.

The problem that Rupert and Scott were trying to solve was how these mounds functioned. Scott had previously inserted a whole variety of instruments into the mounds and discovered that conditions inside remained remarkably constant and, under a hot African sun, remarkably comfortable. In essence, termite mounds seemed to be air-conditioned – but how? They gleaned much information about the internal structure of termite mounds from these impressive excavations but, in the end, a JCB was just too crude an instrument – even for an engineer. They needed to see much finer details and, in order to do so, they had to exchange the digger for a few tons of gypsum.

The plan involved mixing up a runny solution of gypsum and feeding it into the mound until it filled all the chambers and tunnels, then waiting until it set hard. Now it was time to have fun with a high-pressure hose, washing away the mixture of soil, saliva and faeces that the termites use as a construction material to reveal a sculpture of exquisite beauty. The

(previous page) Species of *Hospitalitermes* are often called 'marching termites', from their extensive foraging columns.

Drywood termites, *Kalotermes* spp., hollow out the remains of trees – and often the timbers that support our buildings.

tracery of tunnels running through the tower remained behind as a cast of white gypsum – and a termite work of art. In fact, these endocasts have been transported for exhibition around quite a few South African museums.

As impressive as they are, it's still hard to use these solid sculptures to analyse how the internal structure provides air-conditioning, so one final step was needed. This entailed building a heavy scaffold frame around a gypsum-filled termite mound and suspending what to all intents and purposes was a giant bacon slicer above it. First, a few millimetres were sliced off from the top of the mound, and the cut surface photographed, revealing a pattern of brown mud and white gypsum. Then, a few more millimetres were removed and the surface, this time showing a slightly different marbling of brown and white, was photographed again. This process was repeated for several days and nights, until the mound had been sliced all the way through. Then, using software similar to that used for turning a series of two-dimensional MRI scans into a three-dimensional model, all of the photographs were combined into a 3D computer model that could be examined in detail from any position and any angle.

Even before such computer models were available, the large-scale structure of termite mounds revealed by excavations suggested one way in which they might work. Hundreds of thousands of termites crammed into the living quarters at the base of the nest generate heat, causing the warmed air to rise up along the central chimney. This air movement, it was thought, would draw in cooler air through openings around the base. Or, as the rising air cooled towards the top of the mound, it might sink back down through the network of tunnels running close to the outer surface, where stale air could be exchanged for fresh, like a giant lung.[4]

Whatever its exact mode of functioning, Zimbabwean architect Mick Pierce was so impressed by the energy-efficient ventilation system of termite mounds that he designed a building to work on similar principles. The result was the Eastgate Centre in Harare, Zimbabwe, the world's first termite-inspired building. Air warmed by activity during the day rises up through a chimney-like structure to draw in cooler air from below in the evenings. It proved a remarkable success. Termites saved the developers around $3.5 million during construction, since they didn't have to buy and install expensive air-conditioning. This passive cooling system also keeps running costs low, which translates into lower rents for office and retail space. All in all, the Eastgate Centre worked like a dream, but the termite nests that had so inspired its architect turned out

Termite mounds cover vast areas of dry forest in Queensland, Australia.

to work in an entirely different way. Ever more precise observations and measurements have revealed layers of further sophistication in termite architecture. We'll get to that shortly, but first we're going to meet a man who recognized the sophistication of termite society nearly a century before Scott Turner, Rupert Soar or Mick Pierce.

Eugène Marais was born in Pretoria in 1871 and became something of an Afrikaaner hero following the Second Boer War of 1899–1902. He was the epitome of the late nineteenth-century polymath, a poet and writer, and a prescient observer of nature. He made one of the first long-term studies of a baboon troop, pre-dating the work of primatologists like Jane Goodall and Diane Fossey by many decades – and he made meticulous studies of termites. After two Boer defeats at the hands of the British, Marais switched to writing only in Afrikaans, so his work on termites, published initially in a series of articles under the title *Die siel van die mier* ('The Soul of the Ant') weren't widely read.

Prone to dark moods, Marais committed suicide in 1936. When, just a year later, his work on termites was published in English, as *The Soul of the White Ant*, his insights became more widely appreciated. Described by writer and scientist Robert Ardrey as a 'worker in a science unborn',

he was among the first to see a colony of social insects as more akin to a single organism than a collection of individuals.

The idea of a superorganism as a level of organization above that of the individual organism grew increasingly popular in the early part of the twentieth century. As Marais made his observations on termites, so the grandfather of ant studies, the Harvard professor William Morton Wheeler, expanded the idea with his observations on ants. In his 1911 essay, 'The Ant Colony as an Organism', he expressed the view that an ant colony really is an organism in its own right and not just the analogue of one.[5] The concept of a superorganism has waxed and waned in popularity, but remains important, because much of the natural history of ants and termites can only be understood by considering their lives at the colony level.

As superorganisms, both ants and termites exert a massive influence on their environments, far beyond their diminutive individual size, with effects comparable to all the much larger vertebrates that live alongside them. Many large tomes have been written on the biology of these extraordinary little creatures, but in this chapter I'll focus on how colony life has allowed them to become such overachievers, and on how social living is the key to their ecological domination. Ants have even been able to create networks of linked colonies – supercolonies – and termites construct such sophisticated architectural edifices that we're only now beginning to understand how they work.

Spending a lot of time around termites has inspired Scott Turner to think even further outside the box, beyond the concept of a colony of individuals as a single organism. Termite nests are so finely tuned to the needs of their builders that Scott sees these castles of clay not as inert lumps of mud, but as an integral part of the superorganism, an extension of the termites' own bodies, complete with their own physiology. It's an idea that Scott now applies to many animal-built structures that extend a creature's physiology beyond the conventional boundaries of its body – boundaries that Scott sees as fuzzy rather than sharply demarcated.[6]

Standing at the base of a mound built by industrious hordes of *Macrotermes michaelseni*, it's hard not to be impressed by the sheer scale of their collective achievements. The chimney towers well over twice my height while the base spreads to nearly 2 metres across. It's just one of dozens of such mounds poking out above scrubby acacia trees. It's an awe-inspiring – if slightly strange – sight, because it looks like a city of leaning towers of Pisa. Each chimney is tilted from the vertical, but at exactly the same angle and in exactly the same direction. Scott Turner worked out that in this part of Namibia, the tilt is nineteen degrees to the north, just the right amount so that, at this latitude, the chimney presents itself end on to the average position of the noonday sun to minimize overheating.[7]

The walls of the mound may just be built from mud, spit and shit, but this is applied with great care to create a porous surface that speeds up the exchange of gases. The termites create tiny holes of two sizes in the wall, the larger ones allowing rapid exchange of carbon dioxide.[8] In some species at least, pore size differs in the upper and lower regions of the mound, being smaller higher up the tower.[9] Together with air circulating through the labyrinthine network of tunnels, these sophisticated porous walls keep carbon dioxide in the busy nest down to tolerable levels. But how does that air circulate?

Ever more detailed examinations of the mounds of *M. michaelseni* suggested that air movement was generated not by warm termites but by wind around the chimney. Wind blowing against the chimney is slowed down and creates a higher pressure on the upwind side, forcing air into the channels that run just below the mound's surface. In the lee of the tower, faster moving wind creates a negative pressure, which sucks air out of the channels on the downwind side. Combined with warm air from the nest rising through the central chimney, these forces set up a circulation of air around the mound that maintains temperature and carbon dioxide at constant levels.[10]

However, it turned out that termites weren't quite ready to give up all their secrets just yet. As more sensitive equipment and more sophisticated computer graphics became available, Scott and his team realized that wind-driven circulation, at least in the nests of *M. michaelseni*, may only be of secondary importance. It now seems that it is the passage of the sun across the sky, heating up different aspects of the tower as the day progresses, that creates the primary driving force. Air is warmed in the tunnels on the side facing the sun and rises up through the network, to sink down again on the cooler face. The circulation pattern changes throughout the day, making sure all parts of the mound are flushed with fresh air.[11]

Nor does Scott consider it strictly correct to see the tower as the air-conditioning unit for the nest at its base. The living quarters are connected to the tower tunnel network by fairly small openings that aren't big enough to allow a mass circulation of air through the occupied nest. Instead, the mound is probably more concerned with gas exchange, maintaining low levels of carbon dioxide in the tunnel network. This gas, which builds up in the nest due to termite activity, can then diffuse out through the small openings into the tunnel network to keep levels within the living quarters tolerable.

This is all just in one species, building its cities on the open, hot and arid semi-deserts of southern Africa. In the forests of the Ivory Coast, *Cubitermes* spp. build low, domed mounds with strange, mushroom-like caps, while in northern Australia, *Amitermes* spp. build great flat slabs, like gravestones. In fact, there are more types of mound than there are mound-building termites because even the same species constructs radically different structures in different conditions. *Macrotermes bellicosus* builds low, domed mounds in forests, but resorts to tall mounds that look like Gothic cathedrals on the nearby open plains. The domed forest mounds are more sheltered from direct sunlight and appear to work in a different way. Air circulation is driven primarily by warm air rising from the central nest chamber, not unlike the original idea that inspired the design of the Eastgate Centre. However, this seems less efficient than the solar-powered ventilation on the sunlit savannah. Carbon dioxide levels in the cathedral mounds on open ground remain pretty close to those in the surrounding air, but they're much higher in the domed forest mounds.[12]

While one single species builds radically different mounds in different environments, sometimes the mounds of unrelated species living continents apart show striking similarities. Gigantic cathedral mounds also dot the outback of northern Australia, in this case made by *Nasutitermes triodiae*. Some of these mounds far outstrip those of *Macrotermes* in Africa, reaching a massive 8 metres in height, although there is an unverified record of a mound of *M. bellicosus* in Africa towering an astonishing 12 metres above ground level.

Drive the Plenty Highway, which runs east–west across Queensland and the Northern Territory, and you'll see dozens of these giant cathedral mounds. The largest of them stands right next to the highway, but in this part of the world you need to re-imagine your mental image of a highway. The Plenty Highway is little more than a dirt track through the vast wilderness of northern Australia. Even so, this massive cathedral mound is a frequent roadside pull-off for outback trekkers – and its scale really does impress. Imagine a termite scaled up to human size, and what they have constructed here is the equivalent of four of Dubai's Burj Khalifa towers (the tallest building in the world at 829.8 metres) stacked on top of each other.

The Plenty Highway joins the Stuart Highway north of Alice Springs, and driving this stretch of the outback brings even more surprises. For reasons I can't fathom, hundreds of termite mounds here have been dressed in clothes of all kinds, from overalls to evening dresses. There's clearly not enough to do in this remote corner of the world. I also wonder what impact smothering these vital lungs of the colony will have on the little creatures that built them.

An even stranger – yet wholly natural – sight greets visitors to Lakefield National Park in the far north of Queensland. Here, *Amitermes meridionalis* builds flat-sided mounds, 4 metres high and 3 metres long, every one of them aligned along a north–south axis. For this reason, they're usually called magnetic termites. So many of them dot the Nifold Plain that the place looks like a giant's graveyard.

It's still not entirely clear why these termites create mounds that look like massive headstones. It's usually said that it's an adaptation to create stable temperatures within the nest. The broad, flat east-facing side warms up quickly in the morning sun after a cold night and the termites move to the eastern face to take advantage. As the sun arches higher in the northern sky (we're in the southern hemisphere, don't forget), only the narrow profile of the nest faces the searing noonday sun, so the nest doesn't overheat. As the cooler evening approaches, the broad western face of the mound can absorb the last rays of sunshine.

There's some merit in this idea. Experimentally rotating mounds off their north–south axis (no mean feat) shows that north–south orientated mounds really do have greater thermal stability. Rounded termite mounds, though, maintain a stable internal temperature even more effectively, so it's unlikely that this is the whole story. *A. meridionalis* only ever builds these 'magnetic' mounds, but there are other species of related termites, such as *A. laurensis* and *A. vitiosus*, which build smaller domed mounds across great tracts of northern Australia. However, colonies of these species living in low-lying or poorly drained areas that flood during the wet season switch to building tombstone mounds. So perhaps these strange mounds are an adaptation to northern Australia's 'Big Wet' – the local rainy season – with large, flat surfaces that allow the nests to dry out more quickly.[13]

Unlike human architects, termites don't have plans or blueprints to work from or even a foreman to manage construction gangs. These extraordinary structures are built by a workforce of hundreds of thousands or even millions, all following simple rules. By manipulating a range of conditions within the mound, Scott Turner has shown how the complex structure of mounds could emerge from such basic behaviour. Termites are genetically pre-programmed to pick up bits of dirt and then put them down again. How and where they do this depends on varying stimuli within the nest. For example, those cleverly tilted spires of *M. michaelseni* mounds could simply be an inevitable consequence of the fact that termite building activity speeds up as they get warmer. The north face of the tower faces the sun and is hotter, so mud deposition is greater on this side and therefore the tower is built out more in this direction.[14]

Similar responses to changes in humidity and carbon dioxide drive the termites to create the network of tunnels within the mound, with smaller tunnels running just below its surface. The mound emerges from the 'mind' of the superorganism – thousands of individual minds each following its own rules that create a kind of communal psyche that Eugène Marais also appreciated back in the early twentieth century. We're a little closer to understanding the soul of these white ants than they were in Marais's day, although a great many mysteries remain. However, these termite societies are not the largest superorganisms. To come face to face with the most impressive ones, we must examine the soul of the black ant.

ANTS – FROM HUNTER-GATHERERS TO FARMERS

Ants are not related to termites in any way. Termites are social cockroaches (Blattodea), while ants belong to the same order as bees and wasps (Hymenoptera). The two groups are separated by one of the most fundamental gulfs to cut across the insect world. Ants are holometabolous – that is, they have distinct larval, pupal and adult stages, while termites are hemimetabolous. Their young stages are more or less miniature versions of the adults. That imposes some major differences in how their colonies and societies work. Yet, despite such a deep distinction, there are also very considerable parallels between the two great insect societies.

Most ants, like many termites, build nests, though not such obviously impressive structures. However, looked at in the right way they are architectural feats of equal beauty – and Walter Tschinkel knows how to look at ant nests. He is Professor Emeritus at Florida State University, Tallahassee, and has perfected a way to appreciate the nests of the many ants that excavate their nests below ground. He was kind enough to demonstrate this for a recent series of films I made on insects.

He heads into the woods of Apalachicola, a place he refers to as 'ant heaven', with a portable furnace that he uses to melt chunks of

The nests of the wood ant, *Formica* spp., are covered in a thatch of vegetation that helps the ants to regulate the temperature in the brood chambers below.

aluminium, usually the chopped-up remains of scuba tanks that have failed pressure tests. Once they have melted, Walter pours the liquefied aluminium into the entrance to an ant nest. It sounds simple – but there is a knack to pouring the molten metal, not least in doing so without setting your trousers on fire.

All that's visible of the ants' home on the surface is a simple hole in the ground, but once the aluminium solidifies and Walter has dug out the resulting metal cast, the true beauty of an ant nest is revealed. Walter became fascinated by invasive fire ants, *Solenopsis invicta*, early in his career and admits that digging out his first aluminium cast of a fire-ant nest was an epiphany. The nest consisted of vertical tunnels linking horizontal chambers in a pattern that, inverted, looks like a stylized image of a bonsai tree.

Fired up by these early observations, Walter has gone on to make casts of close to fifty species of ants, and each one is a stunning work of art.[15] Just as endocasts of termite nests proved popular as museum attractions, so these aluminium ant nests have become centrepieces of exhibitions that draw people to admire them as much for artistic reasons as for scientific ones.

The nests are not just created by the random scrabblings of hordes of ants. Although most conform to a basic pattern of chambers linked by a vertical shaft, they differ between species, for example in how far apart the chambers are. Chambers in the nests of *Monomorium viridum* are stacked close together, like pancakes piled on top of each other, while those of *Dorymyrmex bossutus* are separated along a deep shaft that penetrates several metres underground. Most species build vertical connecting tunnels, but *Pogonomyrmex badius* constructs its central shaft in a loose helix, and the desert species *Veromessor pergandei* builds its central shaft at a shallow angle, perhaps in response to shallow desert soils.[16]

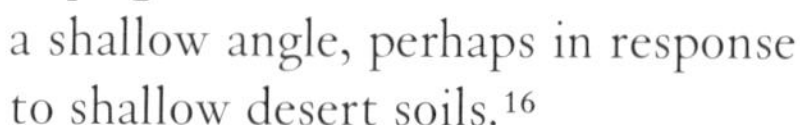

The scale of these constructions beneath a patch of forest floor that you'd never give a second glance is truly amazing, but the workings of one group of ants are on a scale to rival even the works of termites. Leafcutter ants live in the New World tropics and subtropics, with a few species reaching into the south-western US. Their colonies number in the millions, although a census of one nest of *Atta sexdens* recorded an astonishing 8 million inhabitants. To house these enormous populations, their nests are correspondingly vast.

The bulk of the nest is underground, although its scale can be estimated by the size of the area covered by what look like miniature volcanoes – the entrances to this city beneath the soil. Below ground, the central portion can cover 50 square metres and reach down over 8 metres. Contained in this megalopolis are 1,000–2,000 chambers ranging in size from a tennis ball to a football (soccer that is – the round one). The record is 7,864 chambers, painstakingly counted by scientists who had filled a nest with concrete, then excavated its massive bulk.[17]

This principal city may have numerous satellite nests linked by several miles of underground corridors. The central nest is also surrounded by a series of steep channels leading to chimneys that provide ventilation; the whole complex can cover an area the size of several football fields (soccer

Leafcutter ants, *Atta* sp., have cut a broad highway through this lawn in Mexico's Yucatan Peninsula. Such highways allow the rapid transport of cut leaves back to the nest.

against). One group of leafcutter ants, *Acromyrmex* spp., builds curious thatched extensions to the nest's entrance holes, which seem to be visual cues to help the ants find their way back into the sprawling nest, just as entrances to the network of an underground railway system must be marked with conspicuous signs to direct travellers.[18]

At their highest development both ants and termites, by working in close cooperation, construct cities with complex infrastructures. They've achieved this by passing a threshold of social organization such that natural selection works on their society as a whole rather than on individual insects. This threshold has also been crossed by some bees and wasps, such as honeybees and stingless bees, whose nest architecture is as beautiful and complex as that of ants or termites. The ants, bees and wasps evolved their eusocial colonies entirely independently from those of the termites, and at different times in the past. But in both cases, the development of eusociality is seen as one of the great transitions in evolution, a qualitative leap in how life works. So it's natural to ask how the societies of termites and ants arose. How did evolution come up with such a phenomenal trick?

In the previous chapter we explored the world of bees and wasps, which conveniently contains both social and solitary species, allowing us to chart some kind of path from social to solitary, even if that path still hasn't been completely figured out. Unfortunately, there are no living solitary ants so must rely on the fossil record, molecular analyses and good old-fashioned guesswork to work out exactly where ants came from.

DNA evidence points to an origin of ants from wasp ancestors early in the Cretaceous period, perhaps 140 million years ago, although other estimates suggest an origin in the Jurassic period around 180 million years ago.[19] However, up until the middle of the twentieth century, no fossil ants had been found from either the Cretaceous or Jurassic periods. Then, in 1966, in the cliffs of Cliffwood, New Jersey, a chunk of Cretaceous amber* was found that contained an entombed ant, *Sphecomyrmex freyi*. This is generally assumed to resemble the ancestor of all ants and to be a link between wasps and ants since, although it's very clearly an ant, it exhibits a mosaic of wasp and ant features.[20] However, this little creature from a past world is frozen in time and it can't tell us anything about how it lived its life around the feet of dinosaurs.

* Amber is fossilized resin made by a variety of plants as a defensive chemical. When fresh, it oozes from wounds with the consistency of treacle and entraps small insects and other creatures (even small lizards). Once solidified, the entombed insect remains perfectly preserved, as if it died only yesterday.

And so the story moves from New Jersey to Australia, that veritable ark of living fossils, from birds and mammals to the dinosaur ant. The dinosaur ant, or dawn ant, *Nothomyrmecia macrops*, was described from just two workers discovered in 1934 in Western Australia. They bore tantalizing similarities to the extinct *Sphecomyrmex*, so caused quite a stir in the myrmecological world. The dinosaur ant was seen as the Holy Grail for students of ant evolution, but despite numerous expeditions to find a colony of these creatures, no others turned up. Decades passed and it looked like this Holy Grail had – as grails are wont to do – vanished. Then, in 1977, members of yet another expedition, on their way to look for these ants, happened to stop over near the town of Poochera in South Australia. On a walk in the bush they discovered, to their utter astonishment, workers and a foraging queen dinosaur ant, some 1,300 kilometres from the original discovery site.

Subsequent searches in this area turned up whole colonies, no easy task since the entrance to the nest is only a few millimetres across and hidden under leaf litter, and the ants themselves are nocturnal. *Nothomyrmecia* has now become one of the most studied ants on the planet, and the tiny town of Poochera has become the ant capital of Australia. Today, ants are painted on streets around the town and there's a large statue of a dinosaur ant at the Poochera Roadhouse. However, this little speck on the map on the Eyre Peninsula – population thirty-four – has been in decline for many years and even the roadhouse has now gone bust, so the remaining locals are pinning their hopes on being the first place to prosper from ant-based tourism.

Oblivious to these human dramas, the dinosaur ant continues its life as it has for millions of years. Its colonies are tiny, consisting of perhaps just 50–100 workers, a far cry from the teeming millions of leafcutter ants. They scour the forest floor for tiny insects or spiders to feed their larvae – a lifestyle that is presumed to resemble that of the first ants to evolve. This lifestyle includes many of the features we saw in solitary hunting wasps – for example, building a nest and being able to navigate to and from the nest, together with progressive provisioning of the larvae. Alongside genetic evidence, the dinoasaur ant's lifestyle also suggests that ants arose from a group of solitary wasps back in the Jurassic period, and perhaps followed a parallel path to sociality as the wasps that we met in the previous chapter. However, modern molecular analyses are not infallible, and there is still a healthy discussion as to which group of wasps spawned the great dynasty of the ants.[21]

The first ants were undoubtedly hunters on the forest floor, a

Wood ants, *Formica rufa*, exist in such numbers and have such major impacts on the temperate forests where they live that they're sometimes regarded as eco-system engineers.

niche still occupied by one of the more 'primitive' ant groups today. Like dinosaur ants, ponerine ants also live in small colonies in which the queen doesn't look much different from her workers. Nor are these queens particularly fecund, seldom producing more than five eggs a day. However, being socially primitive hasn't stopped the ponerines from becoming ecologically successful.[22] The group also counts among its members some of the most impressive ants you'll ever see. The giant ant, *Dinoponera gigantea*, is a fearsome beast at 4 centimetres long and the bullet ant, *Paraponera clavata*, is not much smaller and packs the most painful sting of any insect.

The redoubtable Justin Schmidt – he of the pain index whom we met in the previous chapter – has, of course, sampled what the bullet ant has on offer, describing it with characteristic panache as 'pure, intense, brilliant pain; like walking over flaming charcoal with a three-inch nail embedded in your heel'. It was, he reckoned, the worst pain he had ever experienced. Indeed, bullet ants get their name because other victims have described the sting as like being shot. All the more bizarre then that the Sateré-Mawé people in the Amazon use bullet ants in a men's initiation rite. To become a man in this part of the world, you must stick your hand in a bag of bullet ants for half an hour, then repeat the process maybe twenty-five times![23]

The ponerines, together with some related groups, pretty much excluded later ants from making a living by hunting on the forest floor in the same way. Instead, they had to come up with some new ways of being an ant, strategies that in the end catapulted them to even greater ecological dominance. The ants that evolved later began to explore the tree canopy as a new living space, although initially they were hampered by a lack of prey – until they hit on the idea of livestock ranching.

Forests abound with hemipteran bugs that feed by plugging their stiletto mouthparts into plant stems to suck sugary sap. They are forced to process so much of this dilute sap just to obtain the nutrients they need that they're faced with an enormous excess of water and sugar, which they excrete as honeydew, much to the delight of ants. From casually sipping this manna from heaven, many ants now manage herds of aphids, mealy bugs or scale insects, and some have become entirely dependent on their livestock.

While exploring the lush rainforests on the slopes of Mount Kinabalu on the island of Borneo, I came across a startling blue orchid, *Cleisocentron gokusingii*. On closer examination, I realized it was swarming with ants. Closer still and I saw that they were picking up mealy bugs and moving

them around the plant. These bugs had unusually long legs and often embraced an ant around its head, clasping on like an animated mask, rather like the creatures in Ridley Scott's *Alien* films. In this way, a mealy bug could easily be moved by the ant to a new feeding area or away from danger. On the island of Java, another kind of mealy bug is an even more adept ant-rider. *Hippeococcus* bugs also have long legs, tipped with sucker-like feet. At the first sign of danger some jump onto the backs of their attending ants, while others are picked up in the ant's jaws, and all are carried swiftly to safety.[24]

Just as we humans have, over millennia, altered the animals we keep, so too ants have domesticated some bugs. The mealy bug *Malaicoccus khooi* can't survive without its ant attendants, in this case *Dolichoderus cuspidatus*. The bugs give birth to live young only in the safety of the ant nest and are dependent on the ants to transport them to young, tender plants that offer the best feeding, sometimes up to 25 metres from the nest. The ants maintain a herd of perhaps 5,000 mealy bugs and when

Ants swarm over an orchid, *Cleisocentron gokusingii*, in the rainforests of Sabah. They're tending herds of tiny mealy bugs that provide them with honeydew.

food around the nest runs out, the whole colony moves, carrying its herd with it to pastures new – like migratory herdsmen.[25]

So important are the milking herds that *Acropyga* queens make sure they carry pregnant mealy bugs on their maiden flights to found new nests, so that they can build up their own herd for their new colony. The ants are as dependent on the mealy bugs as the bugs are on the ants. A *Dolichoderus* ant colony deprived of its herds will soon collapse and die out. Both species are now locked into this symbiosis. Dependence on honeydew is widespread among ants across the planet, although not always in such an extreme way as this. Often it's just a valuable dietary supplement.

Lasius niger is a very common black ant in gardens and cities. They often explore inside houses as they search for anything edible – but they also tend herds of black bean aphids, *Aphis fabae*. If you see a black mass of aphids covering your prize beans, don't be too quick to reach for the bug spray. Settle down and watch closely. You'll almost certainly see black garden ants drinking droplets of honeydew produced by the aphids. The ants, like attentive shepherds, fuss around the aphids, attacking predators such as lacewing or ladybird larvae, the wolves that constantly circle these miniature flocks. The ants also control the behaviour of their aphids. They secrete chemicals that interfere with the normal life cycle of the aphids and stunt their wing growth, so preventing them from flying away from the ants' territory. Any aphid that still has the temerity to grow a set of wings will soon find them chewed off by the ants.

Just occasionally, aphids turn the tables on the ants. *Paracletus cimiciformis* is an aphid with a distinctly dual personality. It is tended by pavement ants, *Tetramorium caespitum*, and over the course of the year produces two distinct forms. One form, like most aphids, feeds on plant sap, while the other is a cunning vampire. This second form is adapted to survive the winter and it does so safe inside the ants' nest. It produces a scent that makes it smell like one of the ants' own larvae, so it's promptly picked up and carried to the nursery area by a diligent worker. Here it repays the kindness by sucking the blood of ant larvae. Fortunately for the ants, it doesn't kill the larvae and, in the spring, as the sap begins to flow again, it switches to a more conventional aphid diet and provides the ants with a supply of honeydew.[26]

This symbiosis with honeydew-producing bugs was one key partnership that fostered ever-increasing ecological success among ants. However, it was a different kind of partnership that allowed ants to develop their most complex societies. They invented agriculture. When humans came

Aphids produce copious quantities of honeydew and are often milked by ants, who protect their valuable flocks like tiny shepherds.

up with this idea around 10,000 years ago, it transformed our culture beyond recognition. And when one group of ants, the attine ants, did the same, around 50 million years earlier, it had much the same effect on their societies.

The more basal lineages of attine ants farm a small range of fungi that they culture on compost made from bits of decomposing plants. The higher attines, the forty-seven species of *Atta* and *Acromyrmex*, are generally known as leafcutter ants and have developed a more intensive system based on the complex processing of fresh leaves chopped from living plants. In addition, they've switched to a monoculture, growing just one single strain of fungus.[27] As we've seen, these leafcutter ant colonies are vast, so their requirements for compost for their fungal gardens are correspondingly large. Big colonies contain so many workers that they can strip an entire citrus tree in twenty-four hours.

Leafcutter ant workers come in a range of different sizes, broadly divisible into majors and minors, with the majors responsible for harvesting leaves. They cut neat discs from a leaf, then, holding these above their heads, make their way in procession back to the nest. Their main trails are busy highways, with bits of leaves or petals streaming back into the nest. It's one of the iconic sights of the insect world, filmed many times over the years – although that didn't stop me from wanting to film it again, albeit from a different perspective.

I contacted a local naturalist and learnt that leafcutters were common around the Mayan remains of Palenque, in the rainforests of Mexico's Yucatán peninsula. The temple site is planted with numerous orange trees, which the ants love, so I felt that shots of trails of ants marching to the nest over archaeological remains with the magnificent stepped pyramid of Palenque in the background would be a fresh, new view of leafcutter ants.

Permits in place, we landed at Mexico City and made our way down from the highlands to the lush rainforests at sea level. After a brief overnight stay in a hotel that looked almost as Mayan as the ruins, we headed off to Palenque – only to find that there wasn't a single leafcutter ant in sight. The authorities responsible for looking after the temple had grown tired of seeing all the leaves and flowers from their lovely orange trees disappearing in a continuous stream into leafcutter ant nests and had sprayed the whole area with insecticide the day before.

Such things are not uncommon when trying to film wildlife, and my usual recourse is to head to the nearest bar for some inspiration on how to solve the problem. After a couple of welcome cold beers, the cameraman and I fell silent. No realistic ideas on how to proceed had emerged. Then, as I sat looking out at the hotel garden, I noticed movement on the steps that led from the garden to the bar. A closer examination revealed a trail of leafcutter ants, helping themselves to leaves and flowers from the plants around the bar, then marching off down the steps to their nest somewhere in the undergrowth on the edge of the jungle.

Lying on the steps with my eyes at leafcutter level, I looked like someone who had definitely drunk way more than a couple of beers but, from that perspective, the bar steps looked exactly like the steps that mount the pyramid at Palenque, and the ancient hotel behind could easily pass for some of the Mayan buildings that surround the pyramid. As usual, a few beers had solved the problem.

Back in the UK, there are colonies of leafcutter ants maintained for research, and these ants are also now common in displays at zoos and butterfly houses. So, it wasn't too complicated a matter to set up a colony in our studio where we could film activity inside the nest, normally hidden deep underground, and follow the final stages of the journey of the stream of leaves. The leaf discs are destined for the fungus gardens, sponge-like structures called fungus combs that fill many of the chambers scattered around the nest, but the leaves are handed on to specialist gardeners for further processing.

These ants are tiny compared to the foragers because they need to be able to access the smallest crevices in the honeycomb structures on which the fungus grows, in order to feed their crops and keep the gardens clean. The gardening ants break down the leaves into smaller fragments and insert them into the fungus combs, where the fungal hyphae* get to work digesting the tough cellulose that makes up the leaves. Many animals have trouble digesting plant matter without help from other organisms, and leafcutter ants have outsourced their digestion to fungi. The fungi, in return for their board and lodging, produce structures called gongylidia, the swollen tips of their hyphae, packed with nutrients that the ants eat. Although some of the lower attine ants farm fungi that are also found in the forest at large, the fungi cultivated by leafcutters are more like domesticated plants – they can't survive out on their own.[28] Nor can the ants live without the fungi, so when a mated queen sets out to establish a new nest, she must take a piece of the fungus with her to establish a new garden.

* *Hyphae* (singular: *hypha*) are long, branching structures that form a large network called a mycelium, the main body of a fungus. The more familiar mushrooms are reproductive structures that grow from the mycelium.

Like farmers everywhere, leafcutter ants have to control pests and parasites. In particular, they fear an invasion by another fungus called *Escovopsis*. If it gains a foothold, this parasite can quickly overrun their gardens and destroy their crops, spelling doom for the whole colony. Not surprisingly, leafcutter ants are pretty hot on biosecurity. Ants that forage in the outside world are barred from entering the fungus chambers in case they are carrying foreign spores. The nest also has dedicated areas for waste disposal, well away from the precious gardens. The poor ants that handle the refuse collections are shunned by every other ant in the colony.

Leafcutter ants have also mastered the use of chemicals to control pests. They cultivate bacteria on specialized parts of their bodies that produce anti-fungal chemicals which specifically target invading *Escovopsis* fungi.[29] The ants themselves are also little chemical factories; one species of leafcutter ant, *Acromyrmex octospinosus*, has been found to produce twenty different chemicals in its metapleural glands.* Many of these are organic acids, such as carboxylic acid, which reduce the pH of the substrate on which they grow their fungi. The cultivated fungi prefer this more acid growing medium, but it has the added advantage of inhibiting the growth of parasites.[30] *A. octospinosus* also secretes indole-acetic acid, a common plant hormone that promotes cell division and growth, and which may therefore encourage the growth of the cultivated fungus.

Agricultural ants such as these are often considered the apex of social evolution in the insect world, a view perhaps biased by the parallel trajectory of our own society. Nevertheless, their colonies are remarkable superorganisms, composite beings whose constituent organs are made up of several very different life forms – ants, fungi and bacteria – all locked in a tight symbiosis that allows these colonies to have as much impact on their world as we human farmers do on our own. However, ants are not the only insects to have climbed this evolutionary peak.

Deep inside the base of those enormous mud towers of *Macrotermes* termites there are also fungus gardens, carefully cultivated by hordes of hard-working termites. Farming insects have divided the world between them. Leafcutter ants are only found in the New World, while fungus-growing termites are confined to parts of Africa and Asia. Yet, despite both great geographical and phylogenetic gulfs, there are many parallels between these two societies.

* *Hyphae* (singular: *hypha*) are long, branching structures that form a large network called a mycelium, the main body of a fungus. The more familiar mushrooms are reproductive structures that grow from the mycelium.

Termites, like leafcutter ants, grow a monocultural strain of one particular fungus – in the termites' case, species of *Termitomyces*. Termites have also partnered with symbiotic bacteria that produce anti-fungal agents, which work against invading parasitic fungi.[31] In fact, termites seem to have the better techniques when it comes to sterile culture. Unlike the combs of leafcutter ants, termite combs are largely parasite-free, because they incorporate their symbiotic bacteria into the comb structure itself and, in addition, *Termitomyces* can produce its own anti-fungal compounds. Compared to the gardens of leafcutter ants, those of termites are low maintenance.[32] Take the ants away from a leafcutter ant comb and it's overrun by pests and parasites in just a few days. Like the ants, termites are also careful not to bring infections into their gardens. Foraging termites collect plant material and even dung as a substrate for their fungi, but take great care to avoid anything contaminated with pathogens that might infect their fungi. They take less good care of their own health and seem oblivious to anything infected with potentially lethal insect diseases.[33]

Despite such parallels, these two great insect societies evolved in very different ways. We've seen that ants evolved from a group of wasps whose parasitoid lifestyle already contained elements that could be put to good use in a social context. The final impetus for the switch to social living probably came from ecological circumstances that favoured cooperation. But what of the termites? Termites used to reside in their own order, the Isoptera, but new genetic studies have shown that they are unequivocally cockroaches.[34] Termites are just highly social roaches or, to put it in another way, roaches are solitary termites, which means we'll need to look for clues to the origins of termite society among the cockroaches.

TERMITES – FROM HUMBLE BEGINNINGS

For most people, the word cockroach conjures up an unloved pest scurrying to find cover when a light is turned on. However, there's much more to the world of roaches than that, although I should probably declare an abiding interest in this group of insects, so my views might be slightly biased. I keep breeding colonies of quite a few species from around the world: giant tiger roaches, *Princisia vanwaerebeki*, from Madagascar; question-mark roaches, *Therea olegrandjeani*, from India, beautifully marked in black and white; and my personal favourites, emerald roaches, *Pseudoglomeris magnifica*, from the rainforests of south-east Asia. The latter are bright iridescent green and as beautiful as any tropical beetle.

Cockroaches also exhibit intriguing behaviour, which includes differing levels of parental care. While some lay eggs, packed into a hardened purse called an ootheca, others give birth to live young. Some cockroach mothers, like those of the Pacific beetle roach, *Diploptera punctata*, even nourish their unborn young in a way that parallels us mammals. They secrete a cockroach version of milk that is swallowed by the growing embryos.[35] It's pretty wholesome stuff. It contains three times the calories of most mammals' milk, so baby beetle roaches grow at a phenomenal rate.[36]

For many roaches, parental duties don't end with birth. The newly born nymphs of *Perisphaerus* roaches cling to the underside of their mother, where she feeds them on secretions from four openings between her legs, conjuring up images of lactating mammals.[37] Quite a few roaches carry their young until they are well grown. *Thorax porcellana* looks more like a beetle than a roach, in part owing to its domed wing-cases. Between these raised wing-cases and the body of the roach, there's a cosy space for a brood of nymphs. They remain like this, carried safely by their mother, for seven weeks, until they simply can't squeeze into their mobile home any longer. During this time, their mother also feeds them. They have sharp, elongated mouthparts with which they pierce their mother's body to drink her blood.

The scrubby gum forests of Queensland are home to the most impressive roaches of all – at least to my mind. There are several species of very large burrowing roach here, but the granddaddy of them all is the giant burrowing roach, *Macropanesthia rhinoceros*. Up to 8 centimetres in length and weighing in at more than 30 grams, this is the world's heaviest roach. It's a bulldozer of an insect, with large and powerful front legs with which it digs deep burrows in the hard soil. These burrows go down a metre in a broad, sweeping helix, and end with a nest chamber well insulated from the harsh world of the outback.

These chambers are also nurseries, housing the roaches' young until they're large enough to head off into the big wide world on their own. They only forage at night and spend the rest of their time a metre beneath the ground, which makes them difficult animals to observe, but it seems that, at first, both males and females remain in the nest with their newly born young. Later, males wander off, leaving their partners to continue to care for the young for most of the summer, until they disperse in the early autumn.[38]

With the help of Alan Henderson, who knows as much about Queensland insects as anyone I've met, we recently set up some observation burrows of this species at his home near Kuranda. The mother leaves her burrow by night to gather bits of dried leaves, not the most appetizing looking diet, but she diligently drags these back to the bottom of her burrow. In our observation burrows, we were able to watch her young eagerly scurrying around her, feeding on small leaf fragments broken off as the female chewed up the leaves. It's hard to say whether she deliberately brings food home for the family since, even without young, she drags leaves into her burrow where she can munch on them in safety. Even so, it's a thoroughly heart-warming sight, although, as I mentioned, I may be a little biased.

A long association of the female with her offspring in a secure nest – these features could be the prerequisites for taking the step from roach to termite. However, to find a creature with an even closer resemblance to that ancestral termite necessitates a visit to the forests of the Appalachian Mountains in North Carolina – equipped with a crowbar and in the company of Christine Nalepa, a professor at North Carolina State University with a passion for roaches that far outstrips my own. In the early 1990s we went looking for wood roaches, to film them for the BBC/PBS series *Alien Empire*.

The Eastern Wood Roach, *Cryptocercus punctulatus*, is found throughout the forests of eastern North America. It dwells in galleries excavated in rotting wood. Here, both parents live with thirty or forty young, a family life that may last three years or more. Wood roaches feed on rotting wood as well as making their homes in it, but wood is even less appetizing than dried *Eucalyptus* leaves. To digest it, these roaches depend on a diverse gut flora, including bacteria and single-celled flagellates. The young are born without these vital microbes and protozoans, so must acquire them by nibbling on their parents' droppings.

Christine knows what kind of logs these roaches prefer and, breaking them open, we soon find a few families. Eastern wood roaches shun light, which makes filming them hard, but they can't see infrared light. Using an infrared camera we could watch the tiny youngsters behaving naturally, constantly following their parents, licking their bodies and chewing on specially produced faecal pellets. In their early stages, young wood roaches are entirely dependent on their parents for food. It's not until around the third instar that their own gut floras become fully established, allowing them to digest their own food – a very good reason to remain in the family home.[39]

In Asia, an unrelated group of roaches, *Salganea* spp., has adopted a very similar lifestyle, living inside rotting logs where the roaches spend years rearing their families. After mating, the pair begin to dig out galleries in their chosen log but, before they move in, they get rid of their wings either by rubbing them off or by each chewing off their partner's wings.[40] The queens of both ants and termites also de-alate like this once they've made their maiden flight and have no other use for their wings.

Salganea roaches differ in one other important respect from wood roaches. Wood roaches produce all their young at once, then raise a single large family. *Salganea* roaches give birth to fewer young but do so several times, so that many nests are occupied by young at varying stages of growth. I've already alluded to a very significant difference between the termites and roaches compared to the ants, bees and wasps The larvae of bees, wasps and ants remain as helpless grubs until they metamorphose, via the pupal stage, into adults, but roach and termite nymphs are small replicas of their parents and, once they grow to a certain size, they can become useful helpers in the nest.

All it took was for older nymphs to begin helping to rear their younger siblings and the termite dominion began.[41] Like wood roaches,

the first termites to evolve lived by burrowing in wood and needed the help of symbiotic protozoans to help digest it. This dependence forced young proto-termites to stay close to their parents and provided the opportunity for such helping behaviour to evolve.

Wood roaches are close to the point that roaches become termites, but to find a creature on the other side of that divide we need to return to the continent of living fossils – Australia. Not only are there dinosaur ants down under, but there are also 'dinosaur' termites. Darwin's termite, *Mastotermes darwiniensis*, is the world's most primitive termite. It's also the largest and the most roach-like of all the termites. The termites live beneath the soil and inside trees, often in such numbers that they hollow out the entire trunk and kill the tree. They're also partial to the timber that we've used in our own dwellings, and they wreak havoc in homes across northern Australia. Darwin's termites are perhaps the most destructive of all termites. They feed on wood, but they'll have a go at just about anything, including plastic, lead piping, bags of salt – and even dung.

The workers are massive, the soldiers reaching well over a centimetre in length, and the galleries they carve through wood and soil can stretch

(*above*) The vast numbers of both termites, such as these *Hospitalitermes* sp., and ants have massive impacts on their environments.

(*left*) The guts of drywood termites are packed with micro-organisms that digest their dry wood diet.

for a hundred metres. Unlike the more 'advanced' termites, the queen is not much larger than her retinue and looks very much like a cockroach. She also lays eggs in a very cockroach-like way, packed inside an ootheca (or egg-case). In other termites the eggs are laid singly, each one picked up and cared for by the workers.

Once both ants and termites adopted a eusocial lifestyle, there was no looking back. They soon rose to dominate ecosystems around the planet. In Africa, mound-building termites have a bigger effect on the vegetation than all the herds of big grazing mammals,[42] and globally the total biomass of termites may even equal that of all terrestrial vertebrates.

THE EMPIRES OF TERMITES AND ANTS

A good place to gain a dramatic sense of the scale of the termite dominion is on the dry Cerrado savannahs of Brazil. Here 2–4-metre-high mounds, known locally as *murundus*, occur in densities of thirty-five mounds per hectare. They stretch like this over nearly a quarter of a million square kilometres. That adds up to around 200 million mounds.[43] The mounds are the spoil heaps created by subterranean termites as they excavate an extensive network of tunnels and seem to be the work of several different kinds. As far as anyone can tell, they were started by *Syntermes dirus*, which builds mounds only a little over a metre in height, but were then occupied by other species that expanded the mounds further. Construction of these mounds began as long as 4,000 years ago. Although they are sometimes said to be long deserted, some of them at least still have termites in residence.[44] They are also now occupied by ants. The giant Amazonian ant *Dinoponera quadriceps* is far more abundant in these mound fields than elsewhere, not in the mounds themselves, but in the labyrinth of tunnels excavated by the termites around the mounds.[45]

Similar mounds occur in South Africa, where they are called *heuweltjies*. These mounds are only a couple of metres high but reach 25 metres in diameter and cover vast areas of the Clanwilliam district of the Western Cape. They are active colonies of a species of harvester termite, *Microhodotermes viator*, and excavations have revealed trace fossils of the same termites in a hardened layer of soil at the very bottom of the mounds. This layer has also been dated to around 4,000 years old and has led some to believe that these mounds were originally made by termites and that they have been continuously occupied for at least four millennia.[46] There are, however, other theories as to their origin. They may simply be ancient accumulations of wind-blown soil that has become stabilized around the bases of long-vanished shrubs and trees.[47] Termites may have colonized these mounds later. There are other areas of similar, neatly spaced mounds in different parts of the world and many of these are also attributed to termites. Even if only some of these mound fields owe their origins to termites, it's clear that these tiny insects can certainly have landscape-scale effects.

Compared to many groups of insects, there aren't that many termites in the world – only around 2,500 species – but their ecological impact goes way beyond that of most other insects. They aerate soils, remove huge amounts of dung and bring up minerals from deep beneath the surface, and there is even one kind that pollinates a rare orchid. The underground orchid, *Rhizanthella gardneri*, is a very strange plant, blooming beneath leaf litter in Western Australia. Not surprisingly, it's hardly ever seen as it's almost impossible to find. However, Australian harvester termites, *Drepanotermes* spp., have no such trouble. They're quick to find newly opened flowers and carry their pollen masses to other orchids.[48]

Ants have just as much impact on global ecology as termites. There are many more ant species than termite species – 14,000 so far known to science[49] – but although there are doubtless many more out there living their secret lives, this figure is still only a paltry 2 per cent of all insect species. However, if there's any doubt that, as it did for termites, sociality gave ants a huge evolutionary edge, then consider that ants make up about a third of the biomass of all insects.[50] It's been worked out that the total weight of ants in the Amazon is some four times the combined weight of all the vertebrates there.[51]

We've already seen that leafcutter ants in the New World are major grazers, having as much impact on local vegetation as do termites elsewhere, although, unlike termites, many other ants are predators and have equally enormous impacts on animal populations. Just what it must be like for a small, tasty insect in an ant-dominated world became abundantly clear to me a few years back. Exploring the rainforests along the Daintree river in northern Queensland, I came across peculiar clumps of leaves that looked as though they'd been glued together. When I got closer, I realized that they had been glued together – it was part of the nest of a colony of green tree ants, *Oecophylla smaragdina*, often called weaver ants. Once I got my eye in, I could see dozens of these leafy nests, all part of the same large colony that claimed this stretch of the forest canopy as its own.

There are only two species of weaver ants, the green tree ant of south-east Asia and Australia and *O. longinoda*, which lives in a broad band across

equatorial Africa, although in the future taxonomists may split these up into many more species since they show considerable variations across their large range.[52] However, both species live in much the same way – as the scourge of the forest – ranging through tree tops and shrubs, and even over the ground in search of anything edible. Nowhere is beyond their reach. As I watched, a band of foragers reached the end of a twig, but rather than turn back, several braced themselves at the tip. Leaning out, they allowed others to crawl over their backs. These too took a firm hold on their sisters, and before long they had formed a living bridge that reached to the next tree.

No other insect was safe in these trees. Large cicadas and katydids were quickly overwhelmed. I watched a dragonfly that hadn't been quick enough to take to the air being hauled along a branch by a large group of ants, all tugging together. They were heading back to a nearby leaf nest where their prize would be butchered and its vital juices imbibed, to be fed to the growing larvae. Green tree ant colonies often reach half a million inhabitants, but unlike ground-nesting mega-colonies of leafcutters, which can build a massive central nest to house their whole workforce, tree-living weaver ants don't have that option. Instead, they build multiple small shelters from leaves, dotted all around their territory. This suits their lifestyle well. They aggressively defend their territory from neighbouring colonies and nests placed close to their boundaries serve as guard posts and a rapid-response first line of defence. These peripheral nests are occupied by the oldest and therefore most

Green tree ants, *Oecophylla smaragdina*, build living bridges to cross from twig to twig as they hunt through the branches in search of prey.

expendable workers.[53] It's been said that while humans send their young men to war, ants send their old ladies.

The nests themselves must be replaced at regular intervals since the leaves soon wither and die. Having found a suitable new patch of leaves, workers line up along the edge of one leaf, taking a firm grip with their legs. Then they reach up to another leaf and clamp on with their powerful jaws. They all pull together to draw the second leaf close to the first. If the two leaves are some distance apart, workers climb over each other to form another living chain until they can reach the second leaf. Now it's time for the weavers to move in.

We've seen that one difference between termites and ants is that termite nymphs, being miniature versions of the adults, can help around the colony once they're old enough, whereas ant larvae are just helpless grubs, dependent on their adult sisters until they too become adult. Not so among weaver ants. Weaver ant larvae play a vital role in the life of their colony. Weavers arrive at the construction site each with a large, last instar larva clamped in its jaws. The larvae secrete copious quantities of silk, and the worker ants use them a bit like glue guns, passing them from leaf edge to leaf edge until the folded leaves are secured with strong silk.

There is one very large problem with building nests like this. New queens have nowhere to hide. If you watch what's going on during a 'flying ant day' in late summer, you'll see lots of queens racing around over the ground. They've recently mated and now urgently need to find somewhere to start their new nest. They shed their wings and, without these cumbersome objects, they can dive into tiny crevices, safe from the hordes of predators that eagerly await flying ant days. Here they can dig out a small but secure nest and lay their first batch of eggs.

A queen green tree ant can't build a nest without larvae, and being arboreal nor can she retreat to a crack in the ground. She must find whatever shelter she can. I've seen queen green tree ants sitting exposed on leaves, brooding their eggs like mother hens, doing their best to look intimidating and defending the clutch against all comers. Only last instar larvae produce silk, so the new queen must wait for around two weeks until the larvae are large enough, then she pulls a nearby leaf down and glues it in place with silk to make a simple shelter. Needless to say, this is an extremely dangerous time for the queen and very few survive long enough to begin a new colony.

Many social insects pass through such a solitary phase. Most ants, along with queen bumblebees as well as queen hornets and yellowjackets, are also faced with starting new colonies on their own. However, once large eusocial colonies evolved, another possibility presented itself. We've already met swarm-founding wasps and swarm-founding bees, such as honeybees, none of which face setting up new nests on their own. In Central America, on several occasions, I've worked with some very impressive ants that do the same.

An excited chirping and twittering in the dense rainforests of Panama alerts us to a kind of ant that has an even greater impact on life in the forest than green tree ants. An army ant column is on the march. Army ants are found in both the Old and New Worlds and are thought to have evolved from a common ancestor, sometime in the mid-Cretaceous period, which took up a nomadic existence.[54]

Colonies of army ants can be immense. One colony of an Old World species, *Dorylus wilverthi*, was estimated to contain 20 million individuals. They spend part of their lives on the move, swarming over the forest floor in vast marching columns, overwhelming even large invertebrates and devouring them. Anything that can flee does – which is why we were able to locate a column of the widespread New World species, *Eciton burchellii*, by the sound of forest birds. All manner of birds flock to these columns to feast on small creatures flushed out by the advancing army. At least eighteen species of antbird (Thamnophilidae) are entirely dependent on army ants for their food, and are almost never seen foraging away from the marching columns.[55] There are also ant butterflies. Several species of skipper (Hesperiidae) obtain vital minerals from bird droppings and

For a lone queen, starting a new colony is a dangerous time. Even before she has finished mating, this winged queen has come to grief in a spider's web.

often home in on army ant columns, where they know bird droppings from lots of well-fed birds will be abundant.[56]

Some 300 other forest species are dependent on army ants in varying degrees for food and shelter. Colonies of *E. burchellii* and their hangers-on are thought to be the largest association of any animals on the planet. These ants shape the ecology of forests where they occur in many ways. Even though they kill as many as 30,000 invertebrates a day, many other invertebrates find unlikely sanctuary among army ants. Perhaps the most improbable is a species of snail. *Allopeas myrmekophilos* is only found within colonies of an Asian army ant, *Leptogenys distinguenda*, where it crawls around among the ants unmolested. When the ants move, they even pick up the little snails and carry them to their new hunting grounds. The snail's mucus seems to mimic the smell that the ants use to identify each other as colony members, so fooling the ants into thinking it's one of their own larvae.[57]

Army ants carry their larvae with them in their nomadic phase. Their colonies are so large and contain so many hungry larvae that they'd soon strip any single patch of forest bare. So they keep moving, making temporary camps each night. After a few weeks of daily marching,

the larvae all pupate at the same time and now the colony enters its stationary phase. Workers make a bivouac by linking their legs together, perhaps half a million of them, and, deep inside this living nest, the pupae develop into a new generation of soldiers. At the same time the queen takes advantage of the pause to lay more eggs.

She'd been deliberately slimmed down for the nomadic phase so she could keep up with the moving columns. Now she swells up and goes into egg-laying mode. After around twenty days, the eggs have hatched, and new adults have emerged from their pupae in tight synchrony. With more hungry mouths to feed, the colony goes on the march again – and so the cycle continues for maybe three years as the colony grows ever larger. Eventually, the time comes to found a new colony, but army ants don't expose their new queens to all the risks of starting a colony on their own.

The workers fatten up a few larvae as new queens, then choose among themselves which one is going to reign over a new colony. It's a democratic vote for a new monarch, since about half of the workers will head off with the newly elected queen. The colony only ever splits into two, so just one queen, the one with the most worker support, will be elected as head of a new army. In this way, army ants never pass through a solitary phase. They always have a retinue of workers, as do the honeybees and swarm-founding wasps that we met in the previous chapter. Among ants, this

strategy gives some species the opportunity for world domination, to transform from a superorganism to a supercolony.

Ants are great colonists, especially with human help. It's all too easy to transport queens or small colonies around the world and, once in a new place, often free from predators and parasites, the ants go on to exploit these new opportunities on an unimaginable scale. The tropical fire ant, *Solenopsis geminata*, is native to the New World tropics but probably began its global travels with the first conquistadors, who ferried looted treasure (and fire ants) from Mexico to Manila in the Philippines. From there they spread to the rest of the world. The conquistadors came to conquer but, in the end, it was the ants that went on to achieve a truly global dominion. Fire ants are tiny and easily overlooked so, not surprisingly, other related species have begun their own conquests. The red imported fire ant, *S. invicta*, is a more recent hitchhiker, but has rapidly colonized the southern United States. These ants do well in open grassland like golf courses, which means that golfers are at particular risk from their excruciating stings. In the year 2000 alone, 30,000 golfers were hospitalized by fire ants and over 100 died from anaphylactic shock, so ants have huge impacts on the human world as well as on the natural world.

Because most of these invasions often began with just one queen, all the resulting colonies are genetically very similar, which has helped their rapid spread. Argentine ants, *Linepithema humile*, have followed a similar path to the fire ants and are now found around the world. Analyses of their DNA reveal that colonies from Australia, Europe and North America show much less genetic diversity than colonies across their native range in South America. Critically, this is reflected in a lower diversity of the cuticular chemicals that the ants use to recognize nest mates. In other words, most invasive Argentine ants smell the same.[58]

An ant from a colony in Australia could march into a colony in Europe and be recognized as a nest mate. This has allowed Argentine Ants to establish colonies that reach for thousands of kilometres. One such Argentine ant hegemony stretches around the western Mediterranean and along the Atlantic coast of Portugal. All the ants over 6,000 kilometres of coastline treat each other as nest mates. It's one enormous supercolony spread across millions of nests and containing billions of individuals. A second, smaller supercolony has become established in Catalonia, presumably from a second introduction. It is distinctive enough that the ants don't recognize each other, but that still leaves most of Europe in the hands of just two mega-families of ants. Instead of wasting time competing with neighbouring colonies, as Argentine ants

Ants are just as happy foraging on city streets as they are in the countryside.

do in South America, these ants all work together and consequently have a devastating effect on local native ant populations.

However, invasive ants don't have it all their own way. Just when North America's red imported fire ants thought they had their new world to themselves, they faced another invasion. Tawny crazy ants, *Nylanderia fulva*, hail from the same part of the world as the fire ants – and they're old adversaries. They were first seen in the US in 2002 in Houston, Texas, but have now spread around the Gulf states – and they don't like fire ants. The two species fight aggressively but the tawny crazy ant usually gains the upper hand because, back in the old country, it developed a secret weapon.

When attacking other ants, fire ants spray a toxic cocktail of chemicals onto their victims, which usually incapacitates or kills them. So far, that has allowed fire ants to wipe out large populations of native North American ants – but tawny crazy ants can detoxify the venom spray. If targeted by a fire ant, crazy ants secrete a droplet of formic acid and smear it over their bodies. This chemical neutralizes the toxic effects of fire-ant venom and the crazy ants are soon back on their feet and ready to carry on the fight.

The evolution of the superorganism among ants, bees and wasps along with termites was a huge evolutionary advance, allowing ants and termites in particular to have impacts on local ecology far beyond their diminutive size. Added to that, the recent rise, with human help, of supercolonies of superorganisms is having an unprecedented global impact. Nor is this story finished yet. Doubtless new species will accidentally find their way to places far from their homes and create further ecological havoc. Even so, the sheer scale of these invasions only reinforces the view that the evolution of social behaviour is one of the most successful strategies that nature has ever fashioned. It is one big reason why insects are the most successful animals ever to have lived.

★

I hope, after this breakneck ride through a world of a million tiny lives, that you share something of the same joy I feel when I meet any of these extraordinary creatures face to face. If so, you will also share my dismay at what we are doing to their world, especially in the last few decades. Over this short period of time, well within my own experience as an entomologist, insect populations around the world have crashed in an alarming way. In some places there were ten times as many insects

Elephant hawk-moth, *Deilephila elpenor*.

(overleaf) Azure damselflies, *Coenagrion puella*, egg-laying. Favoured sites often attract large numbers of pairs.

crawling, hopping and flying about when I began my studies as there are now. This makes it feel like a very personal loss. It's utterly crazy. As we've seen throughout this book, so much of the ecology of the world – our world – depends on insects.

At the end of 2021, we lost one of the greatest ambassadors for insects when Edward Wilson died. It was abundantly clear to him that our destruction of the insect world, both deliberate and accidental, was, in the end, also self-destruction. 'If all mankind were to disappear, the world would regenerate back to the rich state of equilibrium that existed ten thousand years ago. If insects were to vanish, the environment would collapse into chaos.'[59]

Despite their importance, insects remain largely unappreciated by most people, who don't think twice about squishing any bug they find. Never mind that each has its place in the complex web of life on our planet. Recent research points to the fact that insect brains, as tiny as they are, possess enough complexity to generate a basic level of consciousness. So don't just see a pesky bug in need of swatting – meet a fellow sentient being – one whose kin have been living here for at least two hundred times longer than we have.

Acknowledgements

In almost four decades of travelling the world to film insects (and a few other creatures, too), I've worked with a great many scientists and naturalists who have been unfailingly generous in sharing their knowledge and expertise. We've come across a few of these good people in the preceding pages but for those for whom space has denied a mention, I offer my heartfelt thanks.

On many of these trips, I've been accompanied by Kevin Flay, a wildlife cameraman who seems to have an almost psychic connection with the animals that we were trying to film. He has the knack of persuading even the most reticent of creatures to perform for the camera. We've had a lot of adventures over the last forty years, from climbing volcanoes or trekking over ice fields to battling mosquitoes and dripping with buckets of sweat in remote rainforests – all with consistently good humour. Quite a few of our exploits appear in the preceding chapters and I thank him for his many years of companionship. I also thank him for allowing me to use some of his excellent photographs in the book.

In similar vein, I'm also grateful to Nathan Small for generously allowing me to include some of his photographs. I've only worked with Nathan for the last few years but in that time, I've come to appreciate his enormous talents as a cameraman. He has a deep understanding of all that modern technology can offer and combines this with an uncommon eye for lighting and atmosphere with both moving and still images. The results are invariably stunning.

Thanks also to Rupert Barrington, a fellow producer and director as well as a fellow enthusiast for insect-kind, for the use of some of his photographs.

Despite these generous contributions and my own extensive library of insect photographs, there are some things that I haven't had the chance to capture on film. So I'm enormously grateful to my wife, Vicky Coules, for stepping in. She's a very talented artist and has contributed a number of original artworks to the book which both broaden the scope of the images and add to the visual impact.

I'm also grateful to Vicky for reading and commenting on several early drafts of the manuscript for no reward other than pizza and good red wine. As a non-entomologist her guidance on what was working for a more general audience was invaluable.

For other illustrations, I'm delighted to have been able to work with a long-standing friend and colleague, Alfred Vendl, a professor at the University of Applied Arts in Vienna. Alfred is also a documentary filmmaker and we've worked together on many projects over the years.

During this time, I discovered that he has a visionary approach to science visualization. He has developed ways of combining data from micro-CT scans and those from a scanning electron microscope to produce images of the tiniest objects in which both surface and internal details are visible, and which take macro photography to a whole new level. Thanks also to Stephan Handschuh at the Vetcore Facility of the University of Veterinary Medicine in Vienna for providing the micro-CT scans and to Martina R. Fröschl in the Science Visualization Lab Angewandte for producing the final images. For the book, Alfred and the science visualization team have provided extraordinary images of a developing insect egg and the universe of microorganisms that inhabits the guts of termites.

To photograph insects with any of this technology also requires the insects themselves. Thanks to Graham and Janice Smith of Metamorphosis who have a mind-boggling collection of insects and other invertebrates from around the world. They provided many of the insects for photography in my studio, as well as many pleasurable days and wine-fuelled evenings of swapping insect stories. I was extremely grateful for the opportunity to discuss with them the many aspects of insect life as I developed the ideas for the book.

I'd also like to thank Stephen Buchmann, an entomologist based at the University of Arizona and a talented author himself. I've known Steve for most of my career and we've worked together on several occasions in the Sonoran Desert. As an expert in bees, I'm grateful for his time in looking through the chapter dealing with these insects as well as his guiding hand and expertise in exploring the Sonoran Desert.

Heartfelt thanks are also due to Richard Milbank at Head of Zeus, for his sensitive and insightful editing of the manuscript. His talents have greatly improved my deliberations on the world of insects. This is the second of my books that he has edited, and both have been extremely pleasurable and creative experiences.

Thanks also to the rest of the team at Head of Zeus. It's a rare experience to work with such accomplished people who also share the same love of books as myself.

Thanks to my agent, Kate Hordern of KHLA. I'm hugely grateful for her enthusiasm and guidance in the early stages of developing this project and for finding a great home for it, then tackling all the boring contractual bits with a reassuring eye for detail. I feel safe in her experienced hands.

Notes

INTRODUCTION

1 C. Darwin, Letter to William Fox, 12 June 1828.

2 I. Flatow, Interview with E. O. Wilson, in *Talk of the Nation*, PBS (US) (2013).

3 E. A. Hartop, B. V. Brown and R. H. L. Disney, 'Opportunity in our ignorance: urban biodiversity study reveals 30 new species and one new Nearctic record for *Megaselia* (Diptera: Phoridae) in Los Angeles (California, USA)', *Zootaxa*, 2015, 3941(4), pp. 451–484.

4 D. Grimaldi and M. S. Engel, *Evolution of the Insects*, Cambridge University Press (2005).

5 Z.-Q. Zhang, 'Phylum Arthropoda', in Z.-Q. Zhang (ed.), 'Animal biodiversity: an outline of higher-level classification and survey of taxonomic richness (Addenda 2013)', *Zootaxa*, 2013, 3703(1), pp. 17–26.

6 P. Van Roy, A. C. Daley and D. E. Briggs, 'Anomalocaridid trunk limb homology revealed by a giant filter-feeder with paired flaps', *Nature*, 2015, 522(7554), pp. 77–80.

7 J. L. Rainford et al., 'Phylogenetic distribution of extant richness suggests metamorphosis is a key innovation driving diversification in insects', *PLOS One*, 2014, 9(10), p. e109085.

8 E. O. Wilson, 'The little things that run the world (the importance and conservation of invertebrates)', *Conservation Biology*, 1987, 1(4), pp. 344–346.

1 TEEMING HORDES

1 E. M. May, 'Biological diversity: how many species are there?', *Nature*, 1986, 324(6097), pp. 514–515.

2 N. E. Stork, 'How many species of insects and other terrestrial arthropods are there on Earth?' *Annual Review of Entomology*, 2018, 63, pp. 31–45.

3 T. Contador et al., 'Assessing distribution shifts and ecophysiological characteristics of the only Antarctic winged midge under climate change scenarios', *Scientific Reports*, 2020, 10(1), pp. 1–12.

4 S. A. Hayward et al., 'Slow dehydration promotes desiccation and freeze tolerance in the Antarctic midge *Belgica antarctica*', *Journal of Experimental Biology*, 2007, 210(5), pp. 836–844.

5 G. Mahadik et al., 'Superhydrophobicity and size reduction enabled Halobates (Insecta: Heteroptera, Gerridae) to colonize the open ocean', *Scientific Reports*, 2020, 10(1), pp. 1–12.

6 S. Maddrell, 'Why are there no insects in the open sea?', *Journal of Experimental Biology*, 1998, 201(17), pp. 2461–2464.

7 G. Mahadik et al., 2020.

8 N. M. Andersen and L. Cheng, 'The marine insect Halobates (Heteroptera: Gerridae): biology, adaptations, distribution, and phylogeny', *Oceanography and Marine Biology: An Annual Review*, 2004 (42), pp. 119–180.

9 L. Cheng and J. Collins, 'Observations on behavior, emergence and reproduction of the marine midges *Pontomyia* (Diptera: Chironomidae)', *Marine Biology*, 1980, 58(1), pp. 1–5.

10 C. Bessey and A. Cresswell, 'Masses of the marine insect *Pontomyia oceana* at Ningaloo Reef, Western Australia', *Coral Reefs*, 2016, 35(4), pp. 1225–1225.

11 M. S. Leonardi et al., 'How did seal lice turn into the only truly marine insects?', *Insects*, 2021, 13(1), p. 46.

12 H. Hinton, 'Respiratory adaptations of marine insects', *Marine Insects*, 1976, 43, p. 79.

13 M. Price, 'For these intrepid crickets, lava is home sweet home', *Science*, 2019 (363), p. 1262.

14 J. L. Heinen-Kay et al., 'Lava crickets (*Caconemobius* spp.) on Hawai'i Island: first colonisers or persisters in extreme habitats?', *Ecological Entomology*, 2021, 46(3), pp. 505–513.

15 J. Aubert, '*Andiperla willinki* n. sp., Plécoptère nouveau des Andes de Patagonie', *Mitteilungen der Schweizerischen Entomologischen Gesellschaft*, 1956, 29(2), pp. 229–232.

16 H. Hinton, 'A fly larva that tolerates dehydration and temperatures of −270° to +102° C', *Nature*, 1960, 188(4747), pp. 336–337.

17 P. S. Cranston, 'A new putatively cryptobiotic midge, *Polypedilum ovahimba* sp. nov. (Diptera: Chironomidae), from southern Africa', *Austral Entomology*, 2014, 53(4), pp. 373–379.

18 R. Cornette et al., 'A new anhydrobiotic midge from Malawi, *Polypedilum pembai* sp. n. (Diptera: Chironomidae), closely related to the desiccation tolerant midge, *Polypedilum vanderplanki Hinton*', *Systematic Entomology*, 2017, 42(4), pp. 814–825.

19 R. Cornette, S. N. Motitsoe and M. C. Mlambo, 'A new desiccation-resistant midge from ephemeral rock pools in South Africa, *Polypedilum (Pentapedilum) cranstoni* sp. nov. (Diptera: Chironomidae)', *Zootaxa*, 2022, 5128(3), pp. 397–410.

20 G. F. Barrowclough et al., 'How many kinds of birds are there and why does it matter?', *PLOS One*, 2016, 11(11), p. e0166307.

21 J. Ray, *The Wisdom of God Manifested in the Works of the Creation... The Sixth Edition, Corrected. [With a Portrait.]*, 1714, William Innys.

22 W. Kirby and W. Spence, *An Introduction to Entomology: [or, Elements of the Natural History of Insects]*, London, 1826, Longman, Hurst, Rees, Orme and Brown.

23 T. L. Erwin, 'Tropical forests: their richness in Coleoptera and other arthropod species', *The Coleopterists' Bulletin*, 1982, 36(1), pp.74–75.

24 K. J. Gaston, 'The magnitude of global insect species richness', *Conservation Biology*, 1991, 5(3), pp. 283–296.

25 A. J. Hamilton et al., 'Quantifying uncertainty in estimation of tropical arthropod species richness', *The American Naturalist*, 2010, 176(1), pp. 90–95.

26 N. E. Stork, 2018.

27 D. Yeo et al., 'Mangroves are an overlooked hotspot of insect diversity despite low plant diversity', *BMC Biology*, 2021, 19(1), pp. 1–17.

28 V. Dincă et al., 'Unexpected layers of cryptic diversity in wood white Leptidea butterflies', *Nature Communications*, 2011, 2(1), pp. 1–8.

29 J. Waters et al., 'Niche differentiation of a cryptic bumblebee complex in the Western Isles of Scotland', *Insect Conservation and Diversity*, 2011, 4(1), pp. 46–52.

30 N. E. Stork, 2018.

31 B. B. Larsen et al., 'Inordinate fondness multiplied and redistributed: the number of species on earth and the new pie of life', *The Quarterly Review of Biology*, 2017, 92(3), pp. 229–265.

32 G. Lamas, 'The butterflies of Cosñipata. An altitudinal transect study of a megadiverse fauna in southeast Peru', presented at *Entomologentagung* 2017, Freising, Germany.

33 J. T. Huber, 'Biodiversity of Hymenoptera', in *Insect Biodiversity: Science and Society*, 2017, John Wiley and Sons Ltd, pp. 419–461.

34 R. S. Peters et al., 'Evolutionary history of the Hymenoptera', *Current Biology*, 2017, 27(7), pp. 1013–1018.

35 N. Fatouros et al., 'How to escape from insect egg parasitoids: a review of potential factors explaining parasitoid absence across the Insecta', *Proceedings of the Royal Society B*, 2020, 287(1931), p. 20200344.

36 C. García-Robledo et al., 'The Erwin equation of biodiversity: from little steps to quantum leaps in the discovery of tropical insect diversity', *Biotropica*, 2020, 52(4), pp. 590–597.

37 V. Dincă et al., 2011.

38 P. Bouchard et al., 'Biodiversity of Coleoptera', in *Insect Biodiversity: Science and Society*, 2017, pp. 337–417.

39 K. D. Klass et al., 'Mantophasmatodea: a new insect order with extant members in the Afrotropics', *Science*, 2002, 296(5572), pp. 1456–1459.

40 R. S. Peters et al., 2017.

41 R. S. Peters et al., 2017.

42 A. A. Forbes et al., 'Quantifying the unquantifiable: why Hymenoptera, not Coleoptera, is the most speciose animal order', *BMC Ecology*, 2018 18(1), p. 21.

43 P. D. Hebert et al., 'Counting animal species with DNA barcodes: Canadian insects', Philosophical Transactions of the Royal Society B, 2016, 371(1702), p. 20150333.

44 E. A. Hartop, B. V. Brown and R. H. L. Disney, 'Opportunity in our ignorance: urban biodiversity study reveals 30 new species and one new Nearctic record for *Megaselia* (Diptera: Phoridae) in Los Angeles (California, USA)', *Zootaxa*, 2015, 3941(4), pp. 451–484.

45 E. Tihelka et al., 'The evolution of insect biodiversity', *Current Biology*, 2021, 31(19), pp. R1299–R1311.

46 P. J. Mayhew, 'Why are there so many insect species? Perspectives from fossils and phylogenies', *Biological Reviews of the Cambridge Philosophical Society*, 2007, 82(3), pp. 425–454.

47 A. Minelli and G. Fusco, 'No limits: breaking constraints in insect miniaturization', *Arthropod Structure & Development*, 2019, 48, pp. 4–11.

48 A. A. Polilov and R. G. Beutel, 'Miniaturisation effects in larvae and adults of *Mikado* sp. (Coleoptera: Ptiliidae), one of the smallest free-living insects', *Arthropod Structure & Development*, 2009, 38(3), pp. 247–270.

49 S. Fischer, V. B. Meyer-Rochow and C. H. Müller, 'Compound eye miniaturization in Lepidoptera: a comparative morphological analysis', *Acta Zoologica*, 2014, 95(4), pp. 438–464.

50 A. Kaiser et al., 'Increase in tracheal investment with beetle size supports hypothesis of oxygen limitation on insect gigantism', *Proceedings of the National Academy of Sciences*, 2007, 104(32), pp. 13198–13203.

51 F. H. Hennemann and O. V. Conle, 'Revision of Oriental Phasmatodea: the tribe Pharnaciini Günther, 1953, including the description of the world's longest insect, and a survey of the family Phasmatidae Gray, 1835 with keys to the subfamilies and tribes (Phasmatodea: "Anareolatae": Phasmatidae)', *Zootaxa*, 2008, 1906(1), pp. 1–316.

52 T. P. Atkinson, 'Arthropod body fossils from the Union Chapel Mine, in Pennsylvanian Footprints in the Black Warrior Basin of Alabama', *Alabama Palaeontological Society Mongraph*, 2005, pp. 169–176.

53 A. Nel et al., 'Palaeozoic giant dragonflies were hawker predators', *Scientific Reports*, 2018, 8(1), pp. 1–5.

54 N. S. Davies et al., 'The largest arthropod in Earth's history: insights from newly discovered Arthropleura remains (Serpukhovian Stainmore Formation, Northumberland, England)', *Journal of the Geological Society*, vol. 179, 2021.

55 R. Dudley, 'Atmospheric oxygen, giant Paleozoic insects and the evolution of aerial locomotor performance', *Journal of Experimental Biology*, 1998, 201(8), pp. 1043–1050.

56 G. Chapelle and L. S. Peck, 'Polar gigantism dictated by oxygen availability', *Nature*, 1999, 399(6732), pp. 114–115.

57 A. E. Cannell, 'The engineering of the giant dragonflies of the Permian: revised body mass, power, air supply, thermoregulation and the role of air density', *Journal of Experimental Biology*, 2018, 221(19) p.jeb185405.

58 M. Wang et al., 'A new Jurassic scansoriopterygid and the loss of membranous wings in theropod dinosaurs', *Nature*, 2019, 569(7755), pp. 256–259.

59 M. McIntyre, 'The ecology of some large weta species', in 'The Biology of Wetas, King Crickets and their Allies', 2001, CABI Publishing UK, pp. 225–242.

60 C. A. Hallmann et al., 'More than 75 percent decline over 27 years in total flying insect biomass in protected areas', *PLOS One*, 2017, 12(10), p. e0185809.

61 S. R. Leather, '"Ecological Armageddon" – more evidence for the drastic decline in insect numbers', *Annals of Applied Biology*, 2018, 172(1), pp. 1–3.

62 R. Van Klink et al., 'Meta-analysis reveals declines in terrestrial but increases in freshwater insect abundances', *Science*, 2020, 368(6489), pp. 417–420.

63 B. C. Lister and A. Garcia, 'Climate-driven declines in arthropod abundance restructure a rainforest food web', *Proceedings of the National Academy of Sciences*, 2018, 115(44), pp. E10397–E10406.

64 T. D. Schowalter et al., 'Arthropods are not declining but are responsive to disturbance in the Luquillo Experimental Forest, Puerto Rico', *Proceedings of the National Academy of Sciences*, 2021, 118(2), p. e2002556117.

65 D. L. Wagner et al., 'Insect decline in the Anthropocene: death by a thousand cuts', *Proceedings of the National Academy of Sciences*, 2021, 118(2), p. e2023989118.

66 D. Goulson, 'The insect apocalypse, and why it matters', *Current Biology*, 2019, 29(19), pp. R967–R971.

2 ORIGINS

1 M. S. Engel, 'Insect evolution', *Current Biology*, 2015, 25(19), pp. R868–R872.

2 D. M. Rudkin, G. A. Young and G. S. Nowlan, 'The oldest horseshoe crab: a new xiphosurid from Late Ordovician Konservat-Lagerstätten deposits, Manitoba, Canada', *Palaeontology*, 2008, 51(1), pp. 1–9.

3 J. Lozano-Fernandez et al., 'A molecular palaeobiological exploration of arthropod terrestrialization', *Philosophical Transactions of the Royal Society of London B: Biological Sciences*, 2016, 371(1699), p. 20150133.

4 D. E. Briggs, 'Paleontology: a new Burgess Shale fauna', *Current Biology*, 2014, 24(10), pp. R398–R400.

5 O. Rota-Stabelli, A. C. Daley and D. Pisani, 'Molecular timetrees reveal a Cambrian colonization of land and a new scenario for ecdysozoan evolution', *Current Biology*, 2013, 23(5), pp. 392–398.

6 D. Fox, 'What sparked the Cambrian explosion?' *Nature*, 2016, 530(7590), p. 268.

7 A. C. Daley et al., 'Early fossil record of Euarthropoda and the Cambrian Explosion', *Proceedings of the National Academy of Sciences*, 2018, 115(21), pp. 5323–5331.

8 T. Negus, 'An account of the discovery of Charnia', 2007, [10/11/21], available from: www.charnia.org.uk/newsletter/2007/discovery_charnia_2007.htm.

9 G. D. Edgecombe, 'Arthropod origins: integrating paleontological and molecular evidence', *Annual Review of Ecology, Evolution, and Systematics*, 2020, 51, pp. 1–25.

10 O. Rota-Stabelli, A. C. Daley and D. Pisani, 2013.

11 G. Giribet and G. D. Edgecombe, 'The phylogeny and evolutionary history of arthropods', *Current Biology*, 2019, 29(12), pp. R592–R602.

12 D. A. Grimaldi, '400 million years on six legs: on the origin and early evolution of Hexapoda', *Arthropod Structure & Development*, 2010, 39(2–3), pp. 191–203.

13 C. Haug and J. T. Haug, 'The presumed oldest flying insect: more likely a myriapod?' *PeerJ*, 2017, 5, p. e3402.

14 J. Lozano-Fernandez et al., 'A Cambrian–Ordovician terrestrialization of arachnids', *Frontiers in Genetics*, 2020, 11, p. 182.

15 B. Misof et al., 'Phylogenomics resolves the timing and pattern of insect evolution', *Science*, 2014, 346(6210), pp. 763–767.

16 J. Yager, 'Remipedia, a new class of Crustacea from a marine cave in the Bahamas', *Journal of Crustacean Biology*, 1981, pp. 328–333.

17 B. M. von Reumont et al., 'The first venomous crustacean revealed by transcriptomics and functional morphology: remipede venom glands express a unique toxin cocktail dominated by enzymes and a neurotoxin', *Molecular Biology and Evolution*, 2014, 31(1), pp. 48–58.

18 M. Schwentner et al., 'A phylogenomic solution to the origin of insects by resolving crustacean–hexapod relationships', *Current Biology*, 2017, 27(12), pp. 1818–1824.e5.

19 H. Enghoff, 'The ground-plan of chilognathan millipedes', in *Proceedings of the 7th International Congress of Myriapodology*, 1990, Brill.

20 P. E. Marek et al., 'The first true millipede – 1306 legs long', *Scientific Reports*, 2021, 11(1), pp. 1–8.

21 R. G. Beutel et al., 'The phylogeny of Hexapoda (Arthropoda) and the evolution of megadiversity', *Proc. Arthropod. Embryol. Soc. Jpn.*, 2017, pp.1–15.

22 B. Wipfler et al., 'Evolutionary history of Polyneoptera and its implications for our understanding of early winged insects', *Proceedings of the National Academy of Sciences*, 2019, 116(8), pp. 3024–3029.

23 R. Garrouste et al., 'A complete insect from the Late Devonian period', *Nature*, 2012, 488(7409), pp. 82–85.

24 P. Ward et al., 'Confirmation of Romer's Gap as a low oxygen interval constraining the timing of initial arthropod and vertebrate terrestrialization', *Proceedings of the National Academy of Sciences*, 2006, 103(45), pp. 16818–16822.

25 M. S. Engel and D. A. Grimaldi, 'New light shed on the oldest insect', *Nature*, 2004, 427(6975), pp. 627–630.

26 P. E. Marek et al., 2021.

27 S. R. Schachat et al., 'Phanerozoic p O2 and the early evolution of terrestrial animals', *Proceedings of the Royal Society B: Biological Sciences*, 2018, 285(1871), p. 20172631.

28 M. S. Engel, 2015.

29 A. Nel et al., 'The earliest holometabolous insect from the Carboniferous: a "crucial" innovation with delayed success (Insecta Protomeropina Protomeropidae)', *Annales de la Société entomologique de France*, 2007, 43(3), pp.349–355.

30 C. Darwin, *Journal of Researches into the Geology and Natural History of the various countries visited by H.M.S. Beagle*, 1839, London, Henry Colburn.

31 J. Rolff, P. R. Johnston and S. Reynolds, 'Complete metamorphosis of insects', 2019, Philosophical Transcations if the Royal Society B 374.1783 (2019):10190063.

32 S. P. Nicholls, 'Metamorphosis of the Malpighian tubules of *Libellula quadrimaculata L.*: structure and physiology (Anisoptera: Libellulidae)', *Odonatologica*, 1984, 13(2), pp. 249–258.

33 C. R. Darwin, *The Origin of Species by Means of Natural Selection*, 1866, London, UK, John Murray.

34 S. H. P. Maddrell, 'How the simple shape and soft body of the larvae might explain the success of endopterygote insects', *Journal of Experimental Biology*, 2018, 221(11), p. jeb177535.

3 Hexapods

1 D. Huang et al., 'Diverse transitional giant fleas from the Mesozoic era of China', *Nature*, 2012, 483(7388), pp. 201–204.

2 M. Rothschild et al., 'The flying leap of the flea', *Scientific American*, 1973, 229(5), pp. 92–101.

3 J.-B. Gouyon, 'BBC Wildlife Documentaries in the Age of Attenborough', 2019, Springer Nature.

4 G. Giribet and G. D. Edgecombe, 'The phylogeny and evolutionary history of arthropods', *Current Biology*, 2019, 29(12), pp. R592–R602.

5 G. A. Boxshall, 'The evolution of arthropod limbs', *Biological Reviews*, 2004, 79(2), pp. 253–300.

6 R. J. Full and M. S. Tu, 'Mechanics of a rapid running insect: two-, four- and six-legged locomotion', *Journal of Experimental Biology*, 1991, 156(1), pp. 215–231.

7 P. Miller, 'A possible sensory function for the stop–go patterns of running in phorid flies', *Physiological Entomology*, 1979, 4(4), pp. 361–370.

8 R. Wehner, 'Blick ins Cockpit von Cataglyphis – Hirnforschung en miniature', *Naturwiss. Rdsch.*, 2003, 657, pp. 134–140.

9 R. Wehner, A. Marsh and S. Wehner, 'Desert ants on a thermal tightrope', *Nature*, 1992, 357(6379), pp. 586–587.

10 R. Wehner, 'The architecture of the desert ant's navigational toolkit (Hymenoptera: Formicidae)', *Myrmecol News*, 2009, 12 (September), pp. 85–96.

11 V. Wahl, S. E. Pfeffer and M. Wittlinger, 'Walking and running in the desert ant *Cataglyphis fortis*', *Journal of Comparative Physiology, Neuroethology, Sensory, Neural, and Behavioral Physiology*, 2015, 201(6), pp. 645–656.

12 C. Zollikofer, 'Stepping patterns in ants–influence of speed and curvature', *Journal of Experimental Biology*, 1994, 192(1), pp. 95–106.

13 S. Kamoun and S. A. Hogenhout, 'Flightlessness and rapid terrestrial locomotion in tiger beetles of the *Cicindela L.* subgenus *Rivacindela* van Nidek from saline habitats of Australia (Coleoptera: Cicindelidae)', *The Coleopterists' Bulletin*, 1996, 50(3), pp. 221–230.

14 S. Kamoun and S. A. Hogenhout, 1996.

15 C. Gilbert, 'Visual control of cursorial prey pursuit by tiger beetles (Cicindelidae)', *Journal of Comparative Physiology A: Sensory, Neural, and Behavioral Physiology*, 1997, 181(3), pp. 217–230.

16 D. B. Zurek and C. Gilbert, 'Static antennae act as locomotory guides that compensate for visual motion blur in a diurnal, keen-eyed predator', *Proceedings of the Royal Socikety B*. 281.1779 (2014): 20133072.

17 S. Nicolson, G. Bartholomew and M. Seely, 'Ecological correlates of locomotion speed, morphometries and body temperature in three Namib Desert tenebrionid beetles', *African Zoology*, 1984, 19(3), pp. 131–134.

18 G. A. Bartholomew, J. R. B. Lighton and G. N. Louw, 'Energetics of locomotion and patterns of respiration in tenebrionid beetles from the Namib Desert', *Journal of Comparative Physiology B*, 1985, 155(2), pp. 155–162.

19 P. P. Goodwyn, 'Water striders: the biomechanics of water locomotion and functional morphology of the hydrophobic surface (Insecta: Hemiptera-Heteroptera)', *Journal of Bionic Engineering*, 2008, 5(2), pp. 121–126.

20 N. M. Andersen, 'A comparative study of locomotion on the water surface in semiaquatic bugs (Insecta, Hemiptera, Gerromorpha)', *Videnskabelige meddelelser fra Dansk Naturhistorisk Forening i København*, 1976, 139, pp. 337–396.

21 D. L. Hu and J. W. M. Bush, 'The hydrodynamics of water-walking arthropods', *Journal of Fluid Mechanics*, 2010, 644, pp. 5–33.

22 P. P. Goodwyn, 2008.

23 F. Moreira and J. F. Barbosa, 'A new Rhagovelia (Hemiptera: Heteroptera: Veliidae) from the Brazilian Amazon, with a key to species of the robusta group known from the country', *Zootaxa*, 2014, 3790, pp. 595–600.

24 N. M. Andersen, 1976.

25 W. Nachtigall, 'Mechanics of swimming in water-beetles', in H. Y. Elder and E. R. Trueman (eds), *Aspects of Animal Movement*. Seminar Series 5. Cambridge University Press, Cambridge, pp 107–124.

26 W. Nachtigall, 'Funktionelle morphologie, kinematik und hydromechanik des Ruderapparates von *Gyrinus*', *Zeitschrift für vergleichende physiologie*, 1961, 45(2), pp. 193–226.

27 W. Nachtigall, 'Locomotion: mechanics and hydrodynamics of swimming in aquatic insects', in *The Physiology of Insecta*, 1974, Elsevier, pp. 381–432.

28 J. Voise and J. Casas, 'The management of fluid and wave resistances by whirligig beetles', *Journal of the Royal Society Interface*, 2010, 7(43), pp. 343–352.

29 Z. Xu et al., 'Experimental studies and dynamics modeling analysis of the swimming and diving of whirligig beetles (Coleoptera: Gyrinidae)', *PLOS Computational Biology*, 2012, 8(11), p. e1002792.

30 L. Hendrich, M. Manuel and M. Balke, 'The return of the Duke – locality data for *Megadytes ducalis* Sharp, 1882, the world's largest diving beetle, with notes on related species (Coleoptera: Dytiscidae)', *Zootaxa*, 2019, 4586(3), pp. 517–535.

31 K. K. Jones et al., 'Gas exchange and dive characteristics of the free-swimming backswimmer *Anisops deanei*', Journal of Experimental Biology, 2015, 218(Pt 21), pp. 3478–3486.

32 M. Picker, J. F. Colville and M. Burrows, 'A cockroach that jumps', *Biology Letters*, 2012, 8(3), pp. 390–392.

33 Š. Schrader, 'The function of the cercal sensory system in escape behavior of the cave cricket *Troglophilus neglectus* Krauss', *Pflügers Archiv – European Journal of Physiology*, 2000, 439(7), pp. R187–R189.

34 G. P. Sutton and M. Burrows, 'Insect jumping springs', *Current Biology*, 2018, 28(4), pp. R142–R143.

35 T. Weis-Fogh, 'A rubber-like protein in insect cuticle', *Journal of Experimental Biology*, 1960, 37(4), pp. 889–907.

36 M. Burrows and G. P. Sutton, 'Locusts use a composite of resilin and hard cuticle as an energy store for jumping and kicking', *Journal of Experimental Biology*, 2012, 215(Pt 19), pp. 3501–3512.

37 M. Burrows and M. Dorosenko, 'Jumping performance of flea hoppers and other mirid bugs (Hemiptera, Miridae)', *Journal of Experimental Biology*, 2017, 220(Pt 9), pp. 1606–1617.

38 H. Bennet-Clark and E. Lucey, 'The jump of the flea: a study of the energetics and a model of the mechanism', *Journal of Experimental Biology*, 1967, 47(1), pp. 59–76.

39 M. L. Rothschild et al., 'The jumping mechanism of *Xenopsylla cheopis* III. Execution of the jump and activity', *Philosophical Transactions of the Royal Society of London B: Biological Sciences*, 1975, 271(914), pp. 499–515.

40 G. P. Sutton and M. Burrows, 'Biomechanics of jumping in the flea', *Journal of Experimental Biology*, 2011, 214(Pt 5), pp. 836–847.

41 G. P. Sutton and M. Burrows, 2011.

42 M. Burrows, 'Froghopper insects leap to new heights', *Nature*, 2003, 424(6948), p. 509.

43 M. Burrows, 'Jumping performance of planthoppers (Hemiptera, Issidae)', *Journal of Experimental Biology*, 2009, 212(17), pp. 2844–2855.

44 M. Burrows and G. Sutton, 'Interacting gears synchronize propulsive leg movements in a jumping insect', *Science*, 2013, 341(6151), pp. 1254–1256.

45 T. M. Ali, C. B. Urbani and J. Billen, 'Multiple jumping behaviors in the ant *Harpegnathos saltator*', *Naturwissenschaften*, 1992, 79(8), pp. 374–376.

46 D. Ye, J. C. Gibson and A. V. Suarez, 'Effects of abdominal rotation on jump performance in the ant *Gigantiops destructor* (Hymenoptera, Formicidae)', *Integrative Organismal Biology*, 2020, 2(1), p. obz033.

47 S. N. Patek et al., 'Multifunctionality and mechanical origins: ballistic jaw propulsion in trap-jaw ants', *Proceedings of the National Academy of Sciences*, 2006, 103(34), pp. 12787–12792.

48 D. M. Sorger, 'Snap! Trap-jaw ants in Borneo also jump using their legs', *Frontiers in Ecology and the Environment*, 2015, 13(10), pp. 574–575.

49 F. J. Larabee, A. A. Smith and A. V. Suarez, 'Snap-jaw morphology is specialized for high-speed power amplification in the Dracula ant, *Mystrium camillae*', *Royal Society Open Science*, 2018, 5(12), p. 181447.

50 F. Wieland and G. J. Svenson, 'Biodiversity of Mantodea', *Insect Biodiversity: Science and Society*, 2018, 2, pp. 389–416.

51 Wieland, F. (2013). The phylogenetic system of Mantodea (Insecta: Dictyoptera). Species, Phylogeny and Evolution. https://doi.org/10.17875/gup2013-711', 2013, *Universitätsverlag Göttingen Göttingen*.

52 M. Nyffeler, M. R. Maxwell and J. Remsen Jr, 'Bird predation by praying mantises: a global perspective', *The Wilson Journal of Ornithology*, 2017, 129(2), pp. 331–344.

53 R. Battiston, R. Puttaswamaiah and N. Manjunath, 'The fishing mantid: predation on fish as a new adaptive strategy for praying mantids (Insecta: Mantodea)', *Journal of Orthoptera Research*, 2018, 27(2), pp. 155–158.

54 F. R. Prete, P. H. Wells and L. E. Hurd, *The Praying Mantids*, 1999, JHU Press.

55 J. O'Hanlon, G. Holwell and M. Herberstein, 'Predatory pollinator deception: does the orchid mantis resemble a model species?', *Current Zoology*, 2014, 60(1), pp. 90–103.

56 T. Mizuno et al., '"Double-trick" visual and chemical mimicry by the juvenile orchid mantis *Hymenopus coronatus* used in predation of the Oriental Honeybee *Apis cerana*', *Zoological Science*, 2014, 31(12), pp. 795–801.

57 J. Rivera and Y. Callohuari, 'A new species of praying mantis from Peru reveals impaling as a novel hunting strategy in Mantodea (Thespidae: Thespini)', *Neotropical Entomology*, 2020, 49(2), pp. 234–249.

58 C. Weirauch, D. Forero and D. H. Jacobs, 'On the evolution of raptorial legs – an insect example (Hemiptera: Reduviidae: Phymatinae)', *Cladistics*, 2011, 27(2), pp. 138–149.

59 E. A. McMahan, 'Adaptations, feeding preferences, and biometrics of a termite-baiting assassin bug (Hemiptera: Reduviidae)', *Annals of the Entomological Society of America*, 1983, 76(3), pp. 483–486.

60 D. Forero, D. H. Choe and C. Weirauch, 'Resin gathering in neotropical resin bugs (Insecta: Hemiptera: Reduviidae): functional and comparative morphology', *Journal of Morphology*, 2011, 272(2), pp. 204–229.

61 F. G. Soley, 'Fine-scale analysis of an assassin bug's behaviour: predatory strategies to bypass the sensory systems of prey', Royal Society Open Science, 2016, 3(10), p. 160573.

62 C. T. Brues, 'Peculiar tracheal dilatations in *Bittacomorpha clavipes* Fabr', *The Biological Bulletin*, 1900, 1(3), pp. 155–160.

4 First in Flight

1 R. Dudley and G. Pass, 'Wings and powered flight: core novelties in insect evolution', *Arthropod Structure & Development*, 2018, 47(4), pp. 319–448.

2 R. Pei et al., 'Potential for powered flight neared by most close avialan relatives, but few crossed its thresholds', *Current Biology*, 2020, 30(20), pp. 4033–4046. e8.

3 S. C. Anderson and G. D. Ruxton, 'The evolution of flight in bats: a novel hypothesis', *Mammal Review*, 2020, 50(4), pp. 426–439.

4 D. E. Alexander, 'A century and a half of research on the evolution of insect flight', *Arthropod Structure & Development*, 2018, 47(4), pp. 322–327.

5 C. Gegenbaur, *Grundzüge der vergleichenden Anatomie*, 1870, W. Engelmann.

6 J. H. Marden and M. G. Kramer, 'Surface-skimming stoneflies: a possible intermediate stage in insect flight evolution', *Science*, 1994, 266(5184), pp. 427–430.

7 L. Ruffieux, J. Elouard and M. Sartori, 'Flightlessness in mayflies and its relevance to hypotheses on the origin of insect flight', *Proceedings of the Royal Society of London. Series B: Biological Sciences*, 1998, 265(1410), pp. 2135–2140.

8 H. Mukundarajan et al., 'Surface tension dominates insect flight on fluid interfaces', *Journal of Experimental Biology*, 2016, 219(5), pp. 752–766.

9 K. W. Will, 'Plecopteran surface-skimming and insect flight evolution', *Science*, 1995, 270, p. 8.

10 J. H. Marden and M. A. Thomas, 'Rowing locomotion by a stonefly that possesses the ancestral pterygote condition of co-occurring wings and abdominal gills', *Biological Journal of the Linnean Society*, 2003, 79(2), pp. 341–349.

11 J. Kukalova-Peck, 'Fossil history and the evolution of Hexapod structures', in *The Insects of Australia* (I.D. Naumann, ed.), Vol. 1, 1991, Melbourne University Press, Victoria.

12 S. B. Carroll, S. D. Weatherbee and J. A. Langeland, 'Homeotic genes and the regulation and evolution of insect wing number', *Nature*, 1995, 375(6526), pp. 58–61.

13 B. Wipfler et al., 'Evolutionary history of Polyneoptera and its implications for our understanding of early winged insects', *Proceedings of the National Academy of Sciences*, 2019, 116(8), pp. 3024–3029.

14 S. P. Yanoviak, R. Dudley and M. Kaspari, 'Directed aerial descent in canopy ants', *Nature*, 2005, 433(7026), pp. 624–626.

15 S. Yanoviak, B. Fisher and A. Alonso, 'Directed aerial descent behavior in African canopy ants (Hymenoptera: Formicidae)', *Journal of Insect Behavior*, 2008, 21(3), pp. 164–171.

16 S. P. Yanoviak, M. Kaspari and R. Dudley, 'Gliding hexapods and the origins of insect aerial behaviour', *Biology Letters*, 2009, 5(4), pp. 510–512.

17 S. P. Yanoviak, M. Kaspari and R. Dudley, 2009.

18 R. Wootton and C. Ellington, 'Biomechanics and the origin of insect flight', *Biomechanics in Evolution*, 1991, pp. 99–112.

19 J. Prokop and M. S. Engel, 'Palaeodictyopterida', *Current Biology*, 2019, 29(9), pp. R306–R309.

20 N. Niwa et al., 'Evolutionary origin of the insect wing via integration of two developmental modules', *Evolution & Development*, 2010, 12(2), pp. 168–176.

21 J. Prokop et al., 'Paleozoic nymphal wing pads support dual model of insect wing origins', *Current Biology*, 2017, 27(2), pp. 263–269.

22 A. Yanagawa, A. Guigue and F. Marion-Poll, 'Hygienic grooming is induced by contact chemicals in *Drosophila melanogaster*', *Frontiers in Behavioral Neuroscience*, 2014, 8, p. 254.

23 G. Pass, 'Beyond aerodynamics: the critical roles of the circulatory and tracheal systems in maintaining insect wing functionality', *Arthropod Structure & Development*, 2018, 47(4), pp. 391–407.

24 R. J. Bomphrey et al., 'Flight of the dragonflies and damselflies', *Philosophical Transactions of the Royal Society of London B: Biological Sciences* 371(1704), p.20150389.

25 A. Magnan, *Le Vol des insectes*, 1934, Paris, Éditions Hermann.

26 T. Weis-Fogh, 'Quick estimates of flight fitness in hovering animals, including novel mechanisms for lift production', *Journal of Experimental Biology*, 1973, 59(1), pp. 169–230.

27 L. Johansson and P. Henningsson, 'Butterflies fly using efficient propulsive clap mechanism owing to flexible wings', *Journal of the Royal Society Interface*, 2021, 18(174), p. 20200854.

28 M. R. Nabawy and W. J. Crowther, 'The role of the leading edge vortex in lift augmentation of steadily revolving wings: a change in perspective', *Journal of the Royal Society Interface*, 2017, 14(132), p. 20170159.

29 M. H. Dickinson, F.-O. Lehmann and S. P. Sane, 'Wing rotation and the aerodynamic basis of insect flight', *Science*, 1999, 284(5422), pp. 1954–1960.

30 R. J. Bomphrey et al., 'Smart wing rotation and trailing-edge vortices enable high frequency mosquito flight', *Nature*, 2017, 544(7648), pp. 92–95.

31 S. E. Farisenkov et al., 'Novel flight style and light wings boost flight performance of tiny beetles', *Nature*, 2022, 602(7895), pp. 96–100.

32 G. Pass, 2018.

33 A. L. Thomas et al., 'Dragonfly flight: free-flight and tethered flow visualizations reveal a diverse array of unsteady lift-generating mechanisms, controlled primarily via angle of attack', *Journal of Experimental Biology*, 2004, 207(24), pp. 4299–4323.

34 S. N. Gorb, 'Evolution of the dragonfly head-arresting system', *Proceedings of the Royal Society B: Biological Sciences*, 1999, 266(1418), pp. 525–535.

35 A. Mizutani, J. S. Chahl and M. V. Srinivasan, 'Motion camouflage in dragonflies', *Nature*, 2003, 423(6940), p. 604.

36 S. Agrawal, D. Grimaldi and J. L. Fox, 'Haltere morphology and campaniform sensilla arrangement across Diptera', *Arthropod Structure & Development*, 2017, 46(2), pp. 215–229.

37 R. Goulard et al., 'Behavioural evidence for a visual and proprioceptive control of head roll in hoverflies (*Episyrphus balteatus*)', *Journal of Experimental Biology*, 2015, 218(23), pp. 3777–3787.

38 S. T. Fabian et al., 'Avoiding obstacles while intercepting a moving target: a miniature fly's solution', *Journal of Experimental Biology*, 2022, 225(4), p. jeb243568.

39 T. Geisler, 'Analysis of the structure and mechanism of wing folding and flexion in *Xylotrupes gideon* beetle (L. 1767)(Coloptera, Scarabaeidae)', *Acta Mechanica et Automatica*, 2012, 6(3), pp. 37–44.

40 K. Saito et al., 'Asymmetric hindwing foldings in rove beetles', *Proceedings of the National Academy of Sciences*, 2014, 111(46), pp. 16349–16352.

41 F. Haas, S. Gorb and R. Blickhan, 'The function of resilin in beetle wings', *Proceedings of the Royal Society B: Biological Sciences*, 2000, 267(1451), pp. 1375–1381.

42 J. Sun et al., 'The hydraulic mechanism of the unfolding of hind wings in *Dorcus titanus platymelus* (order: Coleoptera)', *International Journal of Molecular Sciences*, 2014, 15(4), pp. 6009–6018.

43 F. Haas and R. J. Wootton, 'Two basic mechanisms in insect wing folding', *Proceedings of the Royal Society B: Biological Sciences*, 1996, 263(1377), pp. 1651–1658.

44 J. Deiters, W. Kowalczyk and T. Seidl, 'Simultaneous optimisation of earwig hindwings for flight and folding', *Biology Open*, 2016, 5(5), pp. 638–644.

45 R. Dudley, *The Biomechanics of Insect Flight: Form, Function, Evolution*, 2002, Princeton University Press.

46 T. J. Bradley et al., 'Episodes in insect evolution', *Integrative and Comparative Biology*, 2009, 49(5), pp. 590–606.

47 T. J. Bradley et al., 2009.

5 Wings over the World

1 C. Williams, *Insect Migration*, The New Naturalist Series, Vol. 36, 1958 Collins.

2 D. T. Flockhart et al., 'Tracking multi-generational colonization of the breeding grounds by monarch butterflies in eastern North America', *Proceedings of the Royal Society B: Biological Sciences*, 2013, 280(1768), p. 20131087.

3 F. Urquhart, 'Monarch butterflies found at last. The Monarch's winter home', *National Geographic* magazine, 1976, 150, pp.161–173.

4 W. Beebe, 'Migration of Danaidae, Ithomiidae, Acraeidae and Heliconidae (butterflies) at Rancho Grande, north-central Venezuela', *Zoologica* (New York), 1950, 35, pp. 57–68.

5 C. Williams, 1958.

6 R. A. Holland, M. Wikelski and D. S. Wilcove, 'How and why do insects migrate?', *Science*, 2006, 313(5788), pp. 794–796.

7 L. P. Brower, L. S. Fink and P. Walford, 'Fueling the fall migration of the monarch butterfly', *Integrative and Comparative Biology*, 2006, 46(6), pp. 1123–1142.

8 V. Bhaumik and K. Kunte, 'Female butterflies modulate investment in reproduction and flight in response to monsoon-driven migrations', *Oikos*, 2018, 127(2), pp. 285–296.

9 S. Santhosh and S. Basavarajappa, 'Migratory behaviour of two butterfly species (Lepidoptera: Nymphalidae) amidst agriculture ecosystems of South-Western Karnataka, India', *Journal of Entomology and Zoological Studies*, 2017, 5(1), pp. 758–765.

10 B. Senthilmurugan, 'Mukurthi National Park: a migratory route for the butterflies', *Journal of the Bombay Natural History Society*, 2005, 102(2), p. 241.

11 T. van der Heyden, 'Notes on recent migrations of *Urania fulgens* (Walker, 1854) in Costa Rica (Lepidoptera: Uraniidae)', *Arquivos Entomolóxicos*, 2015(13), pp. 277–279.

12 N. G. Smith, 'Host plant toxicity and migration in the dayflying moth *Urania*', *The Florida Entomologist*, 1983, 66(1), pp. 76–85.

13 E. Warrant et al., 'The Australian bogong moth *Agrotis infusa*: a long-distance nocturnal navigator', *Frontiers in Behavioral Neuroscience*, 2016, 10, p. 77.

14 J. Flood, *The Moth Hunters: Aboriginal Prehistory of the Australian Alps*, 1980, Australian Institute of Aboriginal and Torres Strait Island.

15 A. Smith and L. Broome, 'The effects of season, sex and habitat on the diet of the mountain pygmy-possum (*Burramys parvus*)', *Wildlife Research*, 1992, 19(6), pp. 755–767.

16 A. R. Clarke and M. P. Zalucki, 'Monarchs in Australia: on the winds of a storm?', *Biological Invasions*, 2004, 6(1), pp. 123–127.

17 J. Fernández-Haeger, D. Jordano and M. P. Zalucki, 'Monarchs across the Atlantic Ocean. Monarchs in a changing world', *Biology and Conservation of an Iconic Butterfly*, 2015, p. 247.

18 R. I. Vane-Wright,'The Columbus hypothesis: an explanation for the dramatic 19th century range expansion of the monarch butterfly', S.B. Malcolm, M. Zalucki (eds). *Biology and Conservation of the Monarch Butterfly*, published by Natural History Museum of Los Angeles County, 1993, p. 179.

19 M. P. Zalucki and A. R. Clarke; editors: S. B. Malcolm and M. Zalucki. Published by the Natural History Museum of Los Angeles County, Los Angeles 'Monarchs across the Pacific: the Columbus hypothesis revisited', *Biological Journal of the Linnean Society*, 2004, 82(1), pp. 111–121.

20 S. M. Reppert and J. C. de Roode, 'Demystifying Monarch butterfly migration', *Current Biology*, 2018, 28(17), pp. R1009–R1022.

21 A. Vervloet, 'Long-distance migration in butterflies and moths: a cross-species review', 2012, Master's Thesis, University of Utrecht.

22 M. Devaud and M. Lebouvier, 'First record of *Pantala flavescens* (Anisoptera: Libellulidae) from the remote Amsterdam Island, southern Indian Ocean', *Polar Biology*, 2019, 42(5), pp. 1041–1046.

23 C. Stefanescu et al., 'Back to Africa: autumn migration of the painted lady butterfly *Vanessa carduiis* timed to coincide with an increase in resource availability', Ecological Entomology, 2017, 42(6), pp. 737–747.

24 M. Menchetti, M. Guéguen and G. Talavera, 'Spatio-temporal ecological niche modelling of multigenerational insect migrations', *Proceedings of the Royal Society B: Biological Sciences*, 2019, 286(1910), p. 20191583.

25 C. Stefanescu et al., 'Long-distance autumn migration across the Sahara by painted lady butterflies: exploiting resource pulses in the tropical savannah', *Biology Letters*, 2016, 12(10), p. 20160561.

26 K. S. Oberhauser et al., 'Lincoln Brower, champion for Monarchs', *Frontiers in Ecology and Evolution*, 2019, 7 p.149.

27 D. T. Flockhart et al., 2013.

28 E. Howard and A. K. Davis, 'The fall migration flyways of monarch butterflies in eastern North America revealed by citizen scientists', *Journal of Insect Conservation*, 2009, 13(3), pp. 279–286.

29 L. J. Brindza et al., 'Comparative success of monarch butterfly migration to overwintering sites in Mexico from inland and coastal sites in Virginia', *Journal of the Lepidopterists' Society*, 2008, 62(4), pp. 189–200.

30 H. B. Vander Zanden et al., 'Alternate migration strategies of eastern monarch butterflies revealed by stable isotopes', *Animal Migration*, 2018, 5(1), pp. 74–83.

31 R. M. Pyle, 'New Perspectives on Monarch Distribution in the Pacific Northwest', in *Monarchs in a Changing World: Biology and Conservation of an Iconic Butterfly*, K. Oberhauser, S. Altizer and K. Nail (eds), 2015, Cornell University Press, Ithaca.

32 J. W. Chapman, D. R. Reynolds and K. Wilson, 'Long-range seasonal migration in insects: mechanisms, evolutionary drivers and ecological consequences', *Ecology Letters*, 2015, 18(3), pp. 287–302.

33 S. M. Knight et al., 'Radio-tracking reveals how wind and temperature influence the pace of daytime insect migration', *Biology Letters*, 2019, 15(7), p. 20190327.

34 E. Howard and A. K. Davis, 2009.

35 D. T. T. Flockhart et al., 'Migration distance as a selective episode for wing morphology in a migratory insect', *Movement Ecology*, 2017, 5, p. 7.

36 S. M. Reppert, P. A. Guerra and C. Merlin, 'Neurobiology of monarch butterfly migration', *Annual Review of Entomology*, 2016, 61, pp. 25–42.

37 H. Mouritsen et al., 'An experimental displacement and over 50 years of tag-recoveries show that monarch butterflies are not true navigators', *Proceedings of the National Academy of Sciences*, 2013, 110(18), pp. 7348–7353.

38 D. Dreyer et al., 'The Earth's magnetic field and visual landmarks steer migratory flight behavior in the nocturnal Australian Bogong moth', *Current Biology*, 2018, 28(13), pp. 2160–2166.e5.

39 S. Heinze and E. Warrant, 'Bogong moths', *Current Biology*, 2016, 26(7), pp. R263-265.

40 W. S. Robinson, 'Migrating giant honey bees (*Apis dorsata*) congregate annually at stopover site in Thailand', *PLOS One* e44976, 2012.

41 M. Wikelski et al., 'Simple rules guide dragonfly migration', *Biology Letters*, 2006, 2(3), pp. 325–329.

42 M. L. May, 'A critical overview of progress in studies of migration of dragonflies (Odonata: Anisoptera), with emphasis on North America', *Journal of Insect Conservation*, 2013, 17(1), pp. 1–15.

43 M. T. Hallworth et al., 'Tracking dragons: stable isotopes reveal the annual cycle of a long-distance migratory insect', *Biology Letters*, 14.12 (2018):20180741.

44 S. Heinze and E. Warrant, 2016.

45 A. Vervloet, 2012.

46 P. S. Corbet, '*Pantala flavescens* (Fabricius) in New Zealand (Anisoptera: Libellulidae)', *Odonatologica*, 1979, 8(2), pp. 115–121.

47 J. Hedlund et al., 'New records of the Paleotropical migrant *Hemianax ephippiger* in the Caribbean and a review of its status in the Neotropics', *International Journal of Odonatology*, 2020, 23(4), pp. 315–325.

48 S. Chowdhury et al., 'Migration in butterflies: a global overview', *Biological Reviews*, 2021, 96(4), 1462–1483.

49 I. A. Dublon and D. J. Sumpter, 'Flying insect swarms', *Current Biology*, 2014, 24(18), pp. R828–R830.

50 G. Hu, et al., 'Mass seasonal bioflows of high-flying insect migrants', *Science*, 2016, 354(6319), pp. 1584–1587.

51 K. R. Wotton et al., 'Mass seasonal migrations of hoverflies provide extensive pollination and crop protection services', *Current Biology*, 2019, 29(13), pp. 2167–2173 e5.

52 L. P. Brower et al., 'Effect of the 2010–2011 drought on the lipid content of monarchs migrating through Texas to overwintering sites in Mexico', in *Monarchs in a Changing World: Biology and Conservation of an Iconic Butterfly*, K. S. Oberhauser, K. R. Nail and S. Altizer (eds), 2015, pp. 117–129.

53 E. Rendon-Salinas, 'Area of forest occupied by the colonies of monarch butterflies in Mexico during the 2021–2022 overwintering period', 2022. Available from: monarchjointventure.org/images/uploads/documents/Area_of_Forest_Occupied_by_Monarch_Butterfly_Survey_(2021-2022)_24may22.pdf (accessed 14 July 2022).

54 'New Year's count of Western Monarchs tracks population decline during overwintering season', 2022. Available from: https://xerces.org/blog/new-years-count-of-western-monarchs-tracks-population-decline-during-overwintering-season#:~:text=Between%20December%2025%2C%202021%20and,monarchs%20at%20209%20overwintering%20sites (accessed 14 July 2022).

55 D. D. McKenna et al., 'Mortality of Lepidoptera along roadways in central Illinois', *Journal of the Lepidopterists' Society*, 2001, 55(2), pp. 63–68.

56 T. Kantola et al., 'Spatial risk assessment of eastern monarch butterfly road mortality during autumn migration within the southern corridor', *Biological Conservation*, 2019, 231, pp. 150–160.

57 L. P. Brower and S. B. Malcolm, 'Animal migrations: endangered phenomena', *American Zoologist*, 1991, 31(1), pp. 265–276.

58 C. Sáenz-Romero et al., '*Abies religiosa* habitat prediction in climatic change scenarios and implications for monarch butterfly conservation in Mexico', *Forest Ecology and Management*, 2012, 275, pp. 98–106.

6 FLOWER POWER

1 J. L. Bronstein, R. Alarcon and M. Geber, 'The evolution of plant–insect mutualisms', *New Phytologist*, 2006, 172(3), pp. 412–428.

2 F. Burkhardt, *The Correspondence of Charles Darwin*, Vol. 18, 2010, Cambridge University Press.

3 W. E. Friedman, 'The meaning of Darwin's "abominable mystery"', *American Journal of Botany*, 2009, 96(1), pp. 5–21.

4 Y. Asar, S. Y. Ho and H. Sauquet, 'Early diversifications of angiosperms and their insect pollinators: were they unlinked?', *Trends in Plant Science*, 2022 27(9), pp. 858–869.

5 D. H. Janzen, 'When is it coevolution?', *Evolution*, 1980, 34, pp. 611–612.

6 Y. Asar, S. Y. Ho and H. Sauquet, 2022.

7 S. R. Schachat et al., 'A Cretaceous peak in family-level insect diversity estimated with mark–recapture methodology', *Proceedings of the Royal Society B: Biological Sciences*, 2019, 286(1917), p. 20192054.

8 S. Ramírez-Barahona, H. Sauquet and S. Magallón, 'The delayed and geographically heterogeneous diversification of flowering plant families', *Nature Ecology & Evolution*, 2020, 4(9), pp. 1232–1238.

9 T. Suchan and N. Alvarez, 'Fifty years after Ehrlich and Raven, is there support for plant–insect coevolution as a major driver of species diversification?', *Entomologia Experimentalis et Applicata*, 2015, 157(1), pp. 98–112.

10 P. Bouchard et al., 'Biodiversity of Coleoptera', R. G. Foottit, P. H. Adler (eds), in *Insect Biodiversity: Science and Society*, 2017, John Wiley and Sons Ltd, pp. 337–417.

11 D. E. Dussourd and R. F. Denno, 'Deactivation of plant defense: correspondence between insect behavior and secretory canal architecture', *Ecology*, 1991, 72(4), pp. 1383–1396.

12 P. R. Ehrlich and P. H. Raven, 'Butterflies and plants: a study in coevolution', *Evolution: International Journal of Organic Evolution*, vol. 18, 1964, pp. 586–608.

13 M. W. Chase et al., 'Murderous plants: Victorian Gothic, Darwin and modern insights into vegetable carnivory', *Botanical Journal of the Linnean Society*, 2009, 161(4), pp. 329–356.

14 D. Voigt, E. Gorb and S. Gorb, 'Hierarchical organisation of the trap in the protocarnivorous plant *Roridula gorgonias* (Roridulaceae)', *Journal of Experimental Biology*, 2009, 212(19), pp. 3184–3191.

15 D. Voigt and S. Gorb, 'An insect trap as habitat: cohesion-failure mechanism prevents adhesion of *Pameridea roridulae* bugs to the sticky surface of the plant *Roridula gorgonias*', *Journal of Experimental Biology*, 2008, 211(16), pp. 2647–2657.

16 J. A. Moran et al., 'Nutritional mutualisms of *Nepenthes* and *Roridula*', in *Carnivorous Plants: Physiology, Ecology, and Evolution*, A. M. Ellison and L. Adamec (eds), 2018, Oxford University Press.

17 U. Bauer, H. F. Bohn and W. Federle, 'Harmless nectar source or deadly trap: *Nepenthes* pitchers are activated by rain, condensation and nectar', *Proceedings of the Royal Society B: Biological Sciences*, 2008, 275(1632), pp. 259–265.

18 D. Voigt and S. Gorb, 2008.

19 C. Clarke and R. Kitching, 'Swimming ants and pitcher plants: a unique ant–plant interaction from Borneo', *Journal of Tropical Ecology*, 1995, 11(4), pp. 589–602.

20 D. G. Thornham et al., 'Setting the trap: cleaning behaviour of *Camponotus schmitzi* ants increases long-term capture efficiency of their pitcher plant host, *Nepenthes bicalcarata*', *Functional Ecology*, 2012, 26(1), pp. 11–19.

21 M. A. Merbach et al., 'Mass march of termites into the deadly trap', *Nature*, 2002, 415(6867), pp. 36–37.

22 J. A. Moran, C. M. Clarke and B. J. Hawkins, 'From carnivore to detritivore? Isotopic evidence for leaf litter utilization by the tropical pitcher plant *Nepenthes ampullaria*', *International Journal of Plant Sciences*, 2003, 164(4), pp. 635–639.

23 W. N. Lam and H. T. Tan, 'The crab spider–pitcher plant relationship is a nutritional mutualism that is dependent on prey-resource quality', *Journal of Animal Ecology*, 2019, 88(1), pp. 102–113.

24 W. Adlassnig, M. Peroutka and T. Lendl, 'Traps of carnivorous pitcher plants as a habitat: composition of the fluid, biodiversity and mutualistic activities', *Annals of Botany*, 2011, 107(2), pp. 181–194.

25 W. N. Lam et al., 'Inquiline predator increases nutrient-cycling efficiency of *Nepenthes rafflesiana* pitchers', *Biology Letters*, 2019, 15(12), p. 20190691.

26 J. A. Moran, C. M. Clarke and B. J. Hawkins, 2003.

27 L. S. Bittleston et al., 'Convergence between the microcosms of Southeast Asian and North American pitcher plants', *Elife*, 2018, 7, p. e36741.

28 D. Peris et al., 'False blister beetles and the expansion of gymnosperm-insect pollination modes before angiosperm dominance', *Current Biology*, 2017, 27(6), pp. 897–904.

29 S. Hu et al., 'Early steps of angiosperm pollinator coevolution', *Proceedings of the National Academy of Sciences*, 2008, 105(1), pp. 240–245.

30 S. Hu et al., 2008.

31 V. A. Albert et al., 'The *Amborella* genome and the evolution of flowering plants', *Science*, 2013, 342(6165), p. 1241089.

32 C. Darwin, *On the Various Contrivances by Which British and Foreign Orchids are Fertilised by Insects*, 1862, John Murray.

33 C. Darwin, *On the Origin of Species by Means of Natural Selection, or the Preservation of Favoured Races in the Struggle for Life*, 1859, London, John Murray.

34 A. R. Wallace, 'Creation by law', *QJ Sci*, 1867, 4(16), pp. 470–488.

35 L. Wasserthal, 'The pollinators of the Malagasy star orchids *Angraecum sesquipedale*, *A. sororium* and *A. compactum* and the evolution of extremely long spurs by pollinator shift', *Botanica Acta*, 1997, 110(5), pp. 343–359.

36 A. Pauw, J. Stofberg and R. J. Waterman, 'Flies and flowers in Darwin's race', *Evolution: International Journal of Organic Evolution*, 2009, 63(1), pp. 268–279.

37 K. E. Steiner and V. Whitehead, 'Oil flowers and oil bees: further evidence for pollinator adaptation', *Evolution: International Journal of Organic Evolution*, 1991, 45(6), pp. 1493–1501.

38 A. Pauw et al., 'Long-legged bees make adaptive leaps: linking adaptation to coevolution in a plant–pollinator network', *Proceedings of the Royal Society B: Biological Sciences*, 2017, 284(1862), p. 20171707.

39 B. Kahnt et al., 'Should I stay or should I go? Pollinator shifts rather than cospeciation dominate the evolutionary history of South African *Rediviva* bees and their *Diascia* host plants', *Molecular Ecology*, 2019, 28(17), pp. 4118–4133.

40 J. M. Cook and J.-Y. Rasplus, 'Mutualists with attitude: coevolving fig wasps and figs', *Trends in Ecology & Evolution*, 2003, 18(5), pp. 241–248.

41 O. Pellmyr, 'Yuccas, yucca moths, and coevolution: a review', *Annals of the Missouri Botanical Garden*, 2003, pp. 35–55.

42 M. Shrestha et al., 'Rewardlessness in orchids: how frequent and how rewardless?', *Plant Biology*, 2020, 22(4), pp. 555–561.

43 A. G. Ellis and S. D. Johnson, 'Floral mimicry enhances pollen export: the evolution of pollination by sexual deceit outside of the Orchidaceae', *The American Naturalist*, 2010, 176(5), pp. E143–E151.

44 A. Gaskett, 'Orchid pollination by sexual deception: pollinator perspectives', *Biological Reviews*, 2011, 86(1), pp. 33–75.

45 D. Rakosy et al., 'Looks matter: changes in flower form affect pollination effectiveness in a sexually deceptive orchid', *Journal of Evolutionary Biology*, 2017, 30(11), pp. 1978–1993.

46 F. P. Schiestl et al., 'Sex pheromone mimicry in the early spider orchid (*Ophrys sphegodes*): patterns of hydrocarbons as the key mechanism for pollination by sexual deception', *Journal of Comparative Physiology A*, 2000, 186, pp. 567–574.

47 R. D. Phillips et al., 'Caught in the act: pollination of sexually deceptive trap-flowers by fungus gnats in *Pterostylis* (Orchidaceae)', *Annals of Botany*, 2014, 113(4), pp. 629–641.

48 C. Martel, W. Francke and M. Ayasse, 'The chemical and visual bases of the pollination of the Neotropical sexually deceptive orchid *Telipogon peruvianus*', *New Phytologist*, 2019, 223(4), pp. 1989–2001.

49 C. Cohen et al., 'Sexual deception of a beetle pollinator through floral mimicry', *Current Biology*, 2021, 31(9), pp. 1962–1969.e6.

50 W. R. Liltved and S. Johnson, *The Cape Orchids: A Regional Monograph of the Orchids of the Cape Floristic Region*, 2012, Sandstone Editions.

51 D. Rakosy et al., 2017.

52 T. van der Niet, D. M. Hansen and S. D. Johnson, 'Carrion mimicry in a South African orchid: flowers attract a narrow subset of the fly assemblage on animal carcasses', *Annals of Botany*, 2011, 107(6), pp. 981–992.

53 A. Jürgens and A. Shuttleworth, Carrion and Dung Mimicry in Plants, chapter in *Carrion Ecology, Evolution, and their Applications*, CRC Press, ed. M. E. Benbow, J. K. Tomberlin, A. M. Tarone, 2015, pp. 361–386.

54 D. H. Janzen, 'Coevolution of mutualism between ants and acacias in Central America', *Evolution: International Journal of Organic Evolution*, 1966, 20(3), pp. 249–275.

55 M. E. Frederickson, M. J. Greene and D. M. Gordon, '"Devil's gardens" bedevilled by ants', *Nature*, 2005, 437(7058), pp. 495–496.

56 A. Salas-Lopez, S. Talaga and H. Lalagüe, 'The discovery of devil's gardens: an ant–plant mutualism in the cloud forests of the Eastern Amazon', *Journal of Tropical Ecology*, 2016, 32(3), pp. 264–268.

57 M. J. Benton, P. Wilf and H. Sauquet, 'The Angiosperm Terrestrial Revolution and the origins of modern biodiversity', *New Phytologist*, 2022, 233(5), pp. 2017–2035.

58 G. J. Vermeij and R. K. Grosberg, 'The great divergence: when did diversity on land exceed that in the sea?' *Integrative and Comparative Biology*, 2010, 50(4), pp. 675–682.

7 The Mating Game

1 M. Twain, *Roughing It*, 1872, Hartford, CT, American Publishing Company.

2 P. Vorster, *A Water Balance Forecast Model for Mono Lake, California*, 1985, California State University, Hayward.

3 M. Twain, 1872.

4 F. van Breugel and M. H. Dickinson, 'Superhydrophobic diving flies (*Ephydra hians*) and the hypersaline waters of Mono Lake', *Proceedings of the National Academy of Sciences*, 2017, 114(51), pp. 13483–13488.

5 J. R. Jehl, *Biology of the Eared Grebe and Wilson's Phalarope in the Nonbreeding Season: A Study of Adaptations to Saline Lakes*, 1988, Cooper Ornithological Society, Los Angeles, CA.

6 H. Caspers, 'On the ecology of hypersaline lagoons on Laysan Atoll and Kauai Island, Hawaii, with special reference to the Laysan duck, *Anas laysanensis Rothschild*', in *Salt Lakes*, 1981, Springer, pp. 261–270.

7 C. Heling, 'Spatial distribution of benthic invertebrates in Lake Winnebago, Wisconsin', 2016, MS thesis, University of Wisconsin, Oshkosh.

8 R. S. Stelzer et al., 'Carbon sources for lake sturgeon in Lake Winnebago, Wisconsin', *Transactions of the American Fisheries Society*, 2008, 137(4), pp. 1018–1028.

9 D. Szaz et al., 'Lamp-lit bridges as dual light-traps for the night-swarming mayfly, *Ephoron virgo*: interaction of polarized and unpolarized light pollution', *PLOS One*, 2015, 10(3), p. e0121194.

10 P. M. Stepanian et al., 'Declines in an abundant aquatic insect, the burrowing mayfly, across major North American waterways', *Proceedings of the National Academy of Sciences*, 2020, 117(6), pp. 2987–2992.

11 Á. Egri et al., 'Method to improve the survival of night-swarming mayflies near bridges in areas of distracting light pollution', *Royal Society Open Science*, 2017, 4(11), p. 171166.

12 Thornhill, R., 'Sexual selection within mating swarms of the lovebug, *Plecia nearctica* (Diptera: Bibionidae)', *Animal Behaviour*, 1980, 28(2), pp. 405–412.

13 N. C. Leppla, 'Living with lovebugs', *EDIS*, 2007, 2007(7), pp. 1–7.

14 K.-G. Heller and C. Hemp, 'Fiddler on the tree: a bush-cricket species with unusual stridulatory organs and song', *PLOS One*, 2014, 9(3), p. e92366.

15 B. Chivers et al., 'Ultrasonic reverse stridulation in the spider-like katydid *Arachnoscelis* (Orthoptera: Listrosceledinae)', *Bioacoustics*, 2014, 23(1), pp. 67–77.

16 M. B. Fenton and G. P. Bell, 'Recognition of species of insectivorous bats by their echolocation calls', *Journal of Mammalogy*, 1981, 62(2), pp. 233–243.

17 F. A. Sarria-S et al., 'Shrinking wings for ultrasonic pitch production: hyperintense ultra-short-wavelength calls in a new genus of neotropical katydids (Orthoptera: Tettigoniidae)', *PLOS One*, 2014, 9(6), p. e98708.

18 M. Van Staaden and H. Römer, 'Sexual signalling in bladder grasshoppers: tactical design for maximizing calling range', *Journal of Experimental Biology*, 1997, 200(20), pp. 2597–2608.

19 H. Bennet-Clark, 'The mechanism and efficiency of sound production in mole crickets', *Journal of Experimental Biology*, 1970, 52(3), pp. 619–652.

20 L. Prozesky-Schulze et al., 'Use of a self-made sound baffle by a tree cricket', *Nature*, 1975, 255(5504), pp. 142–143.

21 N. Mhatre, 'Tree cricket baffles are manufactured tools', *Ethology*, 2018, 124(9), pp. 691–693.

22 J. M. Petti, 'Loudest', in *University of Florida Book of Insect Records*, Chapter 24: 1997.

23 R. Miles, D. Robert and R. Hoy, 'Mechanically coupled ears for directional hearing in the parasitoid fly *Ormia ochracea*', *Journal of the Acoustical Society of America*, 1995, 98(6), pp. 3059–3070.

24 M. Zuk, J. T. Rotenberry and R. M. Tinghitella, 'Silent night: adaptive disappearance of a sexual signal in a parasitized population of field crickets', *Biology Letters*, 2006, 2(4), pp. 521–524.

25 S. Pascoal et al., 'Rapid convergent evolution in wild crickets', *Current Biology*, 2014, 24(12), pp. 1369–1374.

26 A. Eriksson et al., 'Inter-plant vibrational communication in a leafhopper insect', *PLOS One*, 2011, 6(5), p. e19692.

27 R. T. Cardé, 'Moth navigation along pheromone plumes', in *Pheromone Communication in Moths*. Pub. by University of California Press/R. T. Carde and Jeremy D. Allison (eds), 2016, pp. 173–189.

28 F. E. Regnier and J. H. Law, 'Insect pheromones', *Journal of Lipid Research*, 1968, 9(5), pp. 541–551.

29 J. Rennie, *Insect Miscellanies*, Vol. 23, 1831, C. Knight.

30 J.-H. Fabre, *The Life of the Caterpillar*, 1916, Dodd, Mead.

31 R. F. Chapman, *The Insects: Structure and Function*, 1998, Cambridge University Press, Cambridge.

32 A. Eriksson et al., 2011.

33 M. A. Willis and M. C. Birch, 'Male lek formation and female calling in a population of the arctiid moth *Estigmene acrea*', *Science*, 1982, 218(4568), pp. 168–170.

34 H. Wunderer et al., 'Sex pheromones of two Asian moths (*Creatonotos transiens*, *C. gangis*; Lepidoptera: Arctiidae), behavior, morphology, chemistry and electrophysiology', *Experimental Biology*, 1986, 46(1), pp. 11–27.

35 D. Schneider et al., 'Scent organ development in *Creatonotos* moths: regulation by pyrrolizidine alkaloids', *Science*, 1982, 215(4537), pp. 1264–1265.

36 R. L. Dressler, 'Biology of the orchid bees (*Euglossini*)', *Annual Review of Ecology and Systematics*, 1982, 13(1), pp. 373–394.

37 T. Eltz, A. Sager and K. Lunau, 'Juggling with volatiles: exposure of perfumes by displaying male orchid bees', *Journal of Comparative Physiology A*, 2005, 191(7), pp. 575–581.

38 J. E. Lloyd, 'Firefly mating ecology, selection and evolution', in *The Evolution of Mating Systems in Insects and Arachnids*, B. J. Crespi and J. C. Choe (eds), 1997, Cambridge University Press: Cambridge, pp. 184–192.

39 S. Lewis, *Silent Sparks: The Wondrous World of Fireflies*, 2016, Princeton University Press.

40 S. M. Lewis et al., 'Firefly tourism: advancing a global phenomenon toward a brighter future', *Conservation Science and Practice*, 2021, 3(5), p. e391.

41 D. T. Gwynne and D. C. Rentz, 'Beetles on the bottle: male buprestids mistake stubbies for females (Coleoptera)', *Australian Journal of Entomology*, 1983, 22(1), pp. 79–80.

42 T. Eltz, A. Sager and K. Lunau, 2005.

43 M. A. Branham and M. D. Greenfield, 'Flashing males win mate success', *Nature*, 1996, 381(6585), pp. 745–746.

44 F. V. Vencl and A. D. Carlson, 'Proximate mechanisms of sexual selection in the firefly *Photinus pyralis* (Coleoptera: Lampyridae)', *Journal of Insect Behavior*, 1998, 11(2), pp. 191–207.

45 C. K. Cratsley and S. M. Lewis, 'Female preference for male courtship flashes in *Photinus ignitus* fireflies', *Behavioral Ecology*, 2003, 14(1), pp. 135–140.

46 S. K. Sakaluk et al., 'The troublesome gift: the spermatophylax as a purveyor of sexual conflict and coercion in crickets', *Advances in the Study of Behavior*, 2019, 51, pp. 1–30.

47 R. Thornhill, 'Fighting and assessment in Harpobittacus scorpionflies', *Evolution; International Journal of Organic Evolution*, 1984, pp. 204–214.

48 C. Bull, 'The function of complexity in the courtship of the grasshopper *Myrmeleotettix maculatus*', *Behaviour*, 1979, 69(3–4), pp. 201–216.

49 K. Riede, 'Influence of the courtship song of the acridid grasshopper *Gomphocerus rufus* L. on the female', *Behavioral Ecology and Sociobiology*, 1983, 14(1), pp. 21–27.

50 M. Fea, *The Evolution of Orthopteran Mating Systems and the Reproductive Ecology of New Zealand Cave Wētā*, 2018, ResearchSpace, University of Auckland.

51 T. C. Mendelson and K. L. Shaw, 'Rapid speciation in an arthropod', *Nature*, 2005, 433(7024), pp. 375–376.

52 D. M. O'Brien, M. Katsuki and D. J. Emlen, 'Selection on an extreme weapon in the frog-legged leaf beetle (*Sagra femorata*', *Evolution; International Journal of Organic Evolution*, 2017, 71(11), pp. 2584–2598.

53 M. Fea and G. Holwell, 'Combat in a cave-dwelling wētā (Orthoptera: Rhaphidophoridae) with exaggerated weaponry', *Animal Behaviour*, 2018, 138, pp. 85–92.

54 E. L. Haley and D. A. Gray, 'Mating behavior and dual-purpose armaments in a camel cricket', *Ethology*, 2012. 118(1), pp. 49–56.

55 K. Vahed et al., 'Functional equivalence of grasping cerci and nuptial food gifts in promoting ejaculate transfer in katydids', *Evolution; International Journal of Organic Evolution*, 2014, 68(7), pp. 2052–2065.

56 A. Khila, E. Abouheif and L. Rowe, 'Function, developmental genetics, and fitness consequences of a sexually antagonistic trait', *Science*, 2012, 336(6081), pp. 585–589.

57 S. K. Sakaluk et al., 'The gin trap as a device facilitating coercive mating in sagebrush crickets', *Proceedings of the Royal Society B: Biological Sciences*, 1995, 261(1360), pp. 65–71.

8 THE NEXT GENERATION

1 G. V. Amdam et al., 'Complex social behaviour derived from maternal reproductive traits', *Nature*, 2006, 439(7072), pp. 76–78.

2 S. T. Trumbo, 'Patterns of parental care in invertebrates', 2012, pp. 81–100.

3 A. Córdoba-Aguilar, D. González-Tokman and I. González-Santoyo, *Insect Behavior: From Mechanisms to Ecological and Evolutionary Consequences*, 2018, Oxford University Press.

4 M. P. Scott, 'The ecology and behavior of burying beetles', *Annual Review of Entomology*, 1998, 43(1), pp. 595–618.

5 A. Capodeanu-Nägler et al., 'Offspring dependence on parental care and the role of parental transfer of oral fluids in burying beetles', *Frontiers in Zoology*, 2018, 15(1), pp. 1–12.

6 M. P. Scott, 1998.

7 E. C. de Castro et al., 'The arms race between heliconiine butterflies and *Passiflora* plants – new insights on an ancient subject', *Biological Reviews*, 2018, 93(1), pp. 555–573.

8 C. Wiklund, 'Egg-laying patterns in butterflies in relation to their phenology and the visual apparency and abundance of their host plants', *Oecologia*, 1984, 63(1), pp. 23–29.

9 H. E. Hinton, *Biology of Insect Eggs, Vols I–III*, 1981, Pergammon Press.

10 J. Zhang et al., 'Molecular phylogeny of Harpactorinae and Bactrodinae uncovers complex evolution of sticky trap predation in assassin bugs (Heteroptera: Reduviidae)', *Cladistics*, 2016, 32(5), pp. 538–554.

11 D. Forero, D.-H. Choe and C. Weirauch, 'Resin gathering in neotropical resin bugs (Insecta: Hemiptera: Reduviidae): functional and comparative morphology', *Journal of Morphology*, 2011, 272(2), pp. 204–229.

12 D. H. Choe and M. K. Rust, 'Use of plant resin by a bee assassin bug, *Apiomerus flaviventris* (Hemiptera: Reduviidae), *Annals of the Entomological Society of America*, 2007, 100(2), pp. 320–326.

13 L. Hughes and M. Westoby, 'Capitula on stick insect eggs and elaiosomes on seeds: convergent adaptations for burial by ants', *Functional Ecology*, 1992, 6(6), pp. 642–648.

14 N. E. Pierce and M. A. Elgar, 'The influence of ants on host plant selection by *Jalmenus evagoras*, a myrmecophilous lycaenid butterfly', *Behavioral Ecology and Sociobiology*, 1985, 16(3), pp. 209–222.

15 R. L. Kitching, E. Scheermeyer and R. Jones, *Biology of Australian Butterflies*, Vol. 6, 1999, CSIRO Publishing.

16 J. Thomas, 'Why did the large blue become extinct in Britain?' *Oryx*, 1980, 15(3), pp. 243–247.

17 I. Wynhoff, 'The recent distribution of the European *Maculinea* species', *Journal of Insect Conservation*, 1998, 2(1), pp. 15–27.

18 F. Barbero et al., '*Myrmica* ants and their butterfly parasites with special focus on the acoustic communication', *Psyche: A Journal of Entomology*, 2012, 2012(1), pp. 1–11.

19 J. A. Thomas et al., 'Corruption of ant acoustical signals by mimetic social parasites: *Maculinea* butterflies achieve elevated status in host societies by mimicking the acoustics of queen ants', *Communicative & Integrative Biology*, 2010, 3(2), pp. 169–171.

20 J. Thomas, G. Elmes and J. Wardlaw, 'Polymorphic growth in larvae of the butterfly *Maculinea rebeli*, a social parasite of *Myrmica* ant colonies', *Proceedings of the Royal Society B: Biological Sciences*, 1998, 265(1408), pp. 1895–1901.

21 P. Reguera and M. Gomendio, 'Flexible oviposition behavior in the golden egg bug (*Phyllomorpha laciniata*) and its implications for offspring survival', *Behavioral Ecology*, 2002, 13(1), pp. 70–74.

22 A. Kaitala, G. Gamberale-Stille and S. Swartling, 'Egg carrying attracts enemies in a cryptic coreid bug (*Phyllomorpha laciniata*), *Journal of Insect Behavior*, 2003, 16(3), pp. 319–328.

23 S.-Y. Ohba and A. Maeda, 'Paternal care behaviour of the giant water bug *Kirkaldyia deyrolli* (Heteroptera: Belostomatidae) against ants', *Ecological Entomology*, 2017, 42(4), pp. 402–410.

24 A. Kaitala, G. Gamberale-Stille and S. Swartling, 2003.

25 J. S. Ralston, 'Egg guarding by male assassin bugs of the genus *Zelus* (Hemiptera: Reduviidae)', *Psyche*, 1977, 84(1), pp. 103–107.

26 A. Manica and R. A. Johnstone, 'The evolution of paternal care with overlapping broods', *The American Naturalist*, 2004, 164(4), pp. 517–530.

27 S. T. Trumbo, 2012.

28 P. Thrasher, E. Reyes and H. Klug, 'Parental care and mate choice in the giant water bug *Belostoma lutarium*', *Ethology*, 2015, 121(10), pp. 1018–1029.

29 S. Roth, W. Adaschkiewitz and C. Fischer, 'Notes on the bionomics of *Elasmucha grisea* (LINNAEUS 1758) (*Heteroptera, Acanthosomatidae*) with special regard to joint brood guarding', Rabitsch W, editor. Hug the bug – For love of true bugs. Festschrift zum 70. Geburtstag von Ernst Heiss. Denisia. 19th ed. 2006. p. 1153–67.

30 J. T. Costa, *The Other Insect Societies*, 2006, Harvard University Press.

31 J. T. Costa, 2006.

32 T. Wood, 'Alarm behavior of brooding female *Umbonia crassicornis* (Homoptera: Membracidae)', *Annals of the Entomological Society of America*, 1976, 69(2), pp. 340–344.

33 M. Staerkle and M. Kölliker, 'Maternal food regurgitation to nymphs in earwigs (*Forficula auricularia*)', *Ethology*, 2008, 114(9), pp. 844–850.

34 A. F. Agrawal, J. M. Brown and E. D. Brodie, 'On the social structure of offspring rearing in the burrower bug, *Sehirus cinctus* (Hemiptera: Cydnidae)', *Behavioral Ecology and Sociobiology*, 2004, 57(2), pp. 139–148.

35 S. Suzuki, M. Kitamura and K. Matsubayashi, 'Matriphagy in the hump earwig, *Anechura harmandi* (Dermaptera: Forficulidae), increases the survival rates of the offspring', *Journal of Ethology*, 2005, 23(2), pp. 211–213.

36 F. Vegliante and A. Zilli, 'Larval morphology of *Heterogynis* (Lepidoptera: Heterogynidae)', *European Journal of Entomology*, 2004, 101(1), pp. 165–184.

37 L. Tsukamoto and S. Tojo, 'A report of progressive provisioning in a stink bug, *Parastrachia japonensis* (Hemiptera: Cydnidae)', *Journal of Ethology*, 1992, 10(1), pp. 21–29.

38 M. Hironaka et al., 'Trophic egg production in a subsocial shield bug, *Parastrachia japonensis Scott* (Heteroptera: Parastrachiidae), and its functional value', *Ethology*, 2005, 111(12), pp. 1089–1102.

39 S. T. Trumbo, 2012.

40 G. Halffter and W. D. Edmonds, *The Nesting Behavior of Dung Beetles (Scarabaeinae). An Ecological and Evolutive Approach*. Man and Biosphere Program, UNESCO, 1982, Instituto de Ecologica, Mexico.

41 H. Klemperer, 'Parental behaviour in *Copris lunaris* (Coleoptera, Scarabaeidae): care and defence of brood balls and nest', *Ecological Entomology*, 1982, 7(2), pp. 155–167.

42 E. Matthews, 'La biogeografía ecológica de los escarabajos del estiércol', *Acta Politécnica Mexicana* (México), 1975, 16(72), pp. 89–98.

43 J. J. Midgley et al., 'Faecal mimicry by seeds ensures dispersal by dung beetles', *Nature Plants*, 2015, 1(10), pp. 1–3.

44 J. J. Midgley and J. D. White, 'Two dung beetle species that disperse mimetic seeds both feed on eland dung', *South African Journal of Science*, 2016, 112(7–8), pp. 1–3.

45 H. Howden and P. Ritcher, 'Biology of *Deltochilum gibbosum* (Fab.) with a description of the larva', *The Coleopterists' Bulletin*, 1952, 6(4), pp. 53–57.

46 L. C. Forti et al., 'Predatory behavior of *Canthon virens* (Coleoptera: Scarabaeidae): a predator of leafcutter ants', *Psyche* 2012, vol. 2012, pp. 921465.

47 T. H. Larsen et al., 'From coprophagy to predation: a dung beetle that kills millipedes', *Biology Letters*, 2009, 5(2), pp. 152–155.

48 S. A. Forgie, *Phylogeny of the Scarabaeini (Coleptera: Scarabaeidae)*, 2003, University of Pretoria.

49 P. B. Edwards and H. Aschenborn, 'Maternal care of a single offspring in the dung beetle *Kheper nigroaeneus*: the consequences of extreme parental investment', *Journal of Natural History*, 1989, 23(1), pp. 17–27.

50 M. P. Scott, 'Competition with flies promotes communal breeding in the burying beetle, *Nicrophorus tomentosus*', *Behavioral Ecology and Sociobiology*, 1994, 34(5), pp. 367–373.

51 D. Inward, G. Beccaloni and P. Eggleton, 'Death of an order: a comprehensive molecular phylogenetic study confirms that termites are eusocial cockroaches', *Biology Letters*, 2007, 3(3), pp. 331–335.

52 J. T. Costa, 2006.

9 Living Together

1 E. O. Wilson, *The Insect Societies*, published by Harvard University Press, 1971.

2 P. Brunton, *The Endeavour Journal of Joseph Banks: The Australian Journey*, Angus and Robertson, Sydney, 1998.

3 C. J. Burwell and T. D. Edwards, 'The identity of Sir Joseph Banks' "wrathfull militia": the larvae of "*Doratifera stenora*" Turner (Lepidoptera: Limacodidae)', *The Australian Entomologist*, 2003, 30(1), pp. 39–43.

4 W. H. Hudson, *The Naturalist in La Plata*, 1892, J.M. & Company.

5 J. Alcock, 'Observations on the behaviour of the grasshopper *Taeniopoda eques* (Burmeister), Orthoptera, Acrididae', *Animal Behaviour*, 1972, 20(2), pp. 237–242.

6 J. M. Mathieu, 'Biological studies on *Chromacris colorata* (Orthoptera: Romaleinae)', *Journal of the Kansas Entomological Society*, 1970, 43(3), pp. 262–269.

7 E. Despland, 'Ontogenetic shift from aposematism and gregariousness to crypsis in a *Romaleid* grasshopper', *PLOS One*, 2020, 15(8), P. e0237594.

8 M. J. Wade, 'The biology of the imported willow leaf beetle, *Plagiodera versicolora (Laicharting)*', in *Novel Aspects of the Biology of Chrysomelidae*, pub. by Kluwer Academic, Dordrecht/P. H. Jolivet, M. L. Cox & E. Petitpierre (eds), 1994, pp. 541–547.

9 R. Chapman, W. Page and A. McCaffery, 'Bionomics of the variegated grasshopper (*Zonocerus variegatus*) in West and Central Africa', *Annual Review of Entomology*, 1986, 31(1), pp. 479–505.

10 S. Bazazi et al., 'Collective motion and cannibalism in locust migratory bands', *Current Biology*, 2008, 18(10), pp. 735–739.

11 R. M. Peck, 'Locust plagues: insect invasions on a biblical scale', *Natural History*, 2020, 2020(7–8), pp. 16–21.

12 J. Rosenberg and P. J. Burt, 'Windborne displacements of desert locusts from Africa to the Caribbean and South America', *Aerobiologia*, 1999, 15(3), pp. 167–175.

13 R. M. Peck, 2020.

14 N. Lovejoy et al., 'Ancient trans-Atlantic flight explains locust biogeography: molecular phylogenetics of *Schistocerca*', *Proceedings of the Royal Society B: Biological Sciences*, 2006, 273(1588), pp. 767–774.

15 J. A. Lockwood, *Locust: The Devastating Rise and Mysterious Disappearance of the Insect that Shaped the American Frontier*, 2009, Basic Books.

16 J. A. Lockwood, 2009.

17 N. Waloff and G. Popov, 'Sir Boris Uvarov (1889–1970): the father of acridology', *Annual Review of Entomology*, 1990, 35(1), pp. 1–26.

18 S. Fuzeau-Braesch et al., 'Composition and role of volatile substances in atmosphere surrounding two gregarious locusts, *Locusta migratoria* and *Schistocerca gregaria*', *Journal of Chemical Ecology*, 1988, 14(3), pp. 1023–1033.

19 S. Simpson et al., 'Gregarious behavior in desert locusts is evoked by touching their back legs', *Proceedings of the National Academy of Sciences*, 2001, 98(7), pp. 3895–3897.

20 J. Lockwood, 'The fate of the Rocky Mountain locust, *Melanoplus spretus* Walsh: implications for conservation biology', *Terrestrial Arthropod Reviews*, 2010, 3(2), pp. 129–160.

21 W. Chapco and G. Litzenberger, 'A DNA investigation into the mysterious disappearance of the Rocky Mountain grasshopper, mega-pest of the 1800s', *Molecular Phylogenetics and Evolution*, 2004, 30(3), pp. 810–814.

22 A. F. Pinto et al., '*Lonomia obliqua* venom: in vivo effects and molecular aspects associated with the hemorrhagic syndrome', *Toxicon*, 2010, 56(7), pp. 1103–1112.

23 J. A. Fordyce and A. A. Agrawal, 'The role of plant trichomes and caterpillar group size on growth and defence of the pipevine swallowtail *Battus philenor*', *Journal of Animal Ecology*, 2001, 70(6), pp. 997–1005.

24 W. H. Edwards, *The Butterflies of North America*, Vol. 3, 1897, Houghton, Mifflin and Company.

25 J.-H. Fabre, *The Life of the Caterpillar*, Vol. 6, 1916, Dodd, Mead.

26 T. Edwards, 'On the procession and pupation of the larva of *Cnethocampa pinivora*', *Proc. Cambridge Phil. Soc.*, 1910, Issue 15, pp. 431–436.

27 J.-H. Fabre, 1916.

28 T. Fitzgerald, 'Role of trail pheromone in foraging and processionary behavior of pine processionary caterpillars *Thaumetopoea pityocampa*', *Journal of Chemical Ecology*, 2003, 29(3), pp. 513–532.

29 S. C. Peterson, 'Chemical trail marking and following by caterpillars of *Malacosoma neustria*', *Journal of Chemical Ecology*, 1988, 14(3), pp. 815–824.

30 T. D. Fitzgerald, *The Tent Caterpillars*, 1995, Ithaca NY, Cornell University Press.

31 B. Joos et al., 'Roles of the tent in behavioral thermoregulation of eastern tent caterpillars', *Ecology*, 1988, 69(6), pp. 2004–2011.

32 N. E. Stamp and M. D. Bowers, 'Body temperature, behavior, and growth of early-spring caterpillars (*Hemileuca lucina: Saturniidae*)', *Journal of the Lepidopterists' Society*, 1990, 44(3), pp. 143–155.

33 J. Thomas, *The Butterflies of Britain and Ireland*, 2020, Bloomsbury Publishing.

34 R. Reed, 'Gregarious oviposition in butterflies', *Journal of the Lepidopterists' Society*, 2005, 59(1), pp. 40–43.

35 A. H. Porter et al., 'Relatedness and population differentiation in a colonial butterfly, *Eucheira socialis* (Lepidoptera: Pieridae)', *Annals of the Entomological Society of America*, 1997, 90(2), pp. 230–236.

36 T. D. Fitzgerald and D. L. Underwood, 'Winter foraging patterns and voluntary hypothermia in the social caterpillar *Eucheira socialis*', *Ecological Entomology*, 2000, 25(1), pp. 35–44.

37 W. Wheeler and W. Mann, 'A singular habit of sawfly larvae', *Psyche*, 1923, 30(1), pp. 9–12.

38 G. J. Dury, J. C. Bede and D. M. Windsor, 'Preemptive circular defence of immature insects: definition and occurrences of cycloalexy revisited', *Psyche*, 2014, 2014(9–10), pp. 1–13.

39 J. Korb and J. Heinze, 'Major hurdles for the evolution of sociality', *Annual Review of Entomology*, 2016, 61, pp. 297–316.

40 D. L. Underwood and A. M. Shapiro, 'A male-biased primary sex ratio and larval mortality in *Eucheira socialis* (Lepidoptera: Pieridae)', *Evolutionary Ecology Research*, 1999, 1(6), pp. 703–717.

41 T. Fitzgerald, 2003.

42 P. Weinstein and D. Maelzer, 'Leadership behaviour in sawfly larvae *Perga dorsalis* (Hymenoptera: Pergidae)', *Oikos*, 1997, pp. 450–455.

43 L.K. Hodgkin, M. Symonds and M. Elgar, 'Leaders benefit followers in the collective movement of a social sawfly', *Proceedings of the Royal Society B: Biological Sciences*, 2014, 281(1796), p. 20141700.

44 L. K. Hodgkin, M. R. Symonds and M. A. Elgar, 'Leadership through knowledge and experience in a social sawfly', *Animal Behaviour*, 2017, 134, pp. 177–181.

45 J. Ramos-Elorduy et al., 'Edible Lepidoptera in Mexico: geographic distribution, ethnicity, economic and nutritional importance for rural people', *Journal of Ethnobiology and Ethnomedicine*, 2011, 7(1), pp. 1–22.

46 J. C. Schuster and L. B. Schuster, 'The evolution of social behavior in Passalidae (Coleoptera)', in *The Evolution of Social Behaviour in Insects and Arachnids*, B. J. Crespi and J. C. Choe (eds), 1997, Cambridge University Press, Cambridge, pp. 260–269.

47 C. Palestrini et al., 'Differences in distress signals of adult passalid beetles (Coleoptera Passalidae)', *Bollettino della Societa Entomologica Italiana*, 2003, 135, pp. 45–53.

48 J. T. Costa, *The Other Insect Societies*, 2006, Harvard University Press.

49 D. Kent and J. Simpson, 'Eusociality in the beetle *Austroplatypus incompertus* (Coleoptera: Curculionidae)', *Naturwissenschaften*, 1992, 79(2), pp. 86–87.

50 D. L. Stern and W. A. Foster, 'The evolution of sociality in aphids: a clone's-eye view', in *The Evolution of Social Behaviour in Insects and Arachnids*, B. J. Crespi and J. C. Choe (eds), 1997, Cambridge University Press, Cambridge, pp. 150–165.

51 U. Kurosu and S. Aoki, 'Monomorphic first instar larvae of *Colophina clematicola* (Homoptera, Aphidoidea) attack predators', *Kontytu*, 1988, 56(4), pp. 867–871.

52 D. L. Stern and W. A. Foster, 1997.

53 J. T. Costa, 2006.

54 B. Crespi, 'Behavioural ecology of Australian gall thrips (Insecta, Thysanoptera)', *Journal of Natural History*, 1992, 26(4), pp. 769–809.

55 P. Abbot and T. Chapman, 'Sociality in Aphids and Thrips', in *Comparative Social Evolution*, D. R. Rubenstein and P. Abbot (eds), 2017, Cambridge University Press, Cambridge, pp. 154–187.

56 J. S. Edgerly, 'Biodiversity of Embiodea', *Insect Biodiversity: Science and Society*, 2018, 2, pp. 219–244.

57 J. S. Edgerly, 2018.

10 For Queen and Colony

1 C. Darwin, *On the Origin of Species by Means of Natural Selection, or the Preservation of Favoured Races in the Struggle for Life*, 1859, London, John Murray.

2 K. V. Frisch, *Bees: Their Vision, Chemical Senses and Language*, Cornell University Press, 1950.

3 C. M. Bradley and D. Colodner, 'The Sonoran Desert', in *Encyclopedia of the World's Biomes*, M. I. Goldstein and D. A. DellaSala (eds), 2020, Elsevier, Oxford, pp. 110–125.

4 J. O. Schmidt, M. S. Blum and W. L. Overal, 'Hemolytic activities of stinging insect venoms', *Archives of Insect Biochemistry and Physiology*, 1983, 1(2), pp. 155–160.

5 J. O. Schmidt, *The Sting of the Wild*, 2016, JHU Press.

6 J. H. Hunt and A. L. Toth, 'Sociality in Wasps', in *Comparative Social Evolution*, D. R. Rubenstein and P. Abbot (eds), 2017, Cambridge University Press, Cambridge, pp. 84–123.

7 M. Coutts, 'The mechanism of pathogenicity of *Sirex noctilio* on *Pinus radiata* II. Effects of *S. noctilio* mucus', *Australian Journal of Biological Sciences*, 1969, 22(5), pp. 1153–1162.

8 C. Darwin, 'Letter from Charles Darwin to Asa Gray, May 22nd. 1860', Darwin Correspondence Project. DCP LETT 2814.xml.

9 L. M. Henry and B. D. Roiberg, 'Parasitoids', in *Encyclopedia of Animal Behaviour*, 2nd ed., 2019, Elsevier, pp. 407–412.

10 K. V. Yeargan and S. K. Braman, 'Life history of the parasite *Diolcogaster facetosa* (Weed) (Hymenoptera: Braconidae) and its behavioral adaptation to the defensive response of a lepidopteran host', Annals of the Entomological Society of America, 1986, 79(6), pp. 1029–1033.

11 K. V. Yeargan and S. K. Braman, 'Life history of the hyperparasitoid *Mesochorus discitergus* (Hymenoptera: Ichneumonidae) and tactics used to overcome the defensive behavior of the green cloverworm (Lepidoptera: Noctuidae)', Annals of the Entomological Society of America, 1989, 82(3), pp. 393–398.

12 E. R. Eaton, *Wasps: The Astonishing Diversity of a Misunderstood Insect*, 2021, Princeton University Press.

13 L. Masner, 'A revision of ecitophilous diapriid-genus *Mimopria Holmgren* (Hym., Proctotrupoidea), *Insectes Sociaux*, 1959, 6(4), pp. 361–367.

14 M. R. Strand, 'Polyembryony', in *Encyclopedia of Insects*, 2009, Elsevier, pp. 821–825.

15 H. J. Brockmann, 'Tool use in digger wasps (Hymenoptera: Sphecinae)', *Psyche*, 1985, 92(2–3), pp. 309–329.

16 M. Kaltenpoth et al., 'Symbiotic bacteria protect wasp larvae from fungal infestation', *Current Biology*, 2005, 15(5), pp. 475–479.

17 T. S. Collett, A. Philippides and N. H. de Ibarra, 'Insect navigation: how do wasps get home?', *Current Biology*, 2016, 26(4), pp. R166–R168.

18 R. S. Peters et al., 'Evolutionary history of the Hymenoptera', *Current Biology*, 2017, 27(7), pp. 1013–1018.

19 B. N. Danforth et al., *The Solitary Bees: Biology, Evolution, Conservation*, 2019, Princeton University Press.

20 C. Scholtz, J. Scholtz and H. de Klerk, *Pollinators, Predators and Parasites: The Ecological Roles of Insects in Southern Africa*, 2021, Cape Town, Struik Nature.

21 K. Krombein and B. Norden, 'Nesting behavior of *Krombeinictus nordenae* Leclercq, a sphecid wasp with vegetarian larvae (Hymenoptera: Sphecidae: Crabroninae)', *Entomological Society of Washington* (US), 1997, 99(1), pp.42–49.

22 C. D. Michener, *The Bees of the World*, Vol. 1, 2000, JHU Press.

23 B. N. Danforth et al., 2019.

24 J. H. Cane, 'A native ground-nesting bee (*Nomia melanderi*) sustainably managed to pollinate alfalfa across an intensively agricultural landscape', *Apidologie*, 2008, 39(3), pp. 315–323.

25 B. N. Danforth et al., 2019.

26 B. Danforth,'Bees', *Current Biology*, 2007, 17(5), pp. R156–R161.

27 B. N. Danforth et al., 2019.

28 K. Hogendoorn and H. Velthuis, 'The role of young guards in *Xylocopa pubescens*', *Insectes Sociaux*, 1995, 42(4), pp. 427–448.

29 T. Dunn and M. H. Richards, 'When to bee social: interactions among environmental constraints, incentives, guarding, and relatedness in a facultatively social carpenter bee', *Behavioral Ecology*, 2003, 14(3), pp. 417–424.

30 J. Field et al., 'Cryptic plasticity underlies a major evolutionary transition', *Current Biology*, 2010, 20(22), pp. 2028–2031.

31 C. Plateaux-Quénu, 'Biology of *Halictus marginatus* Brullé', *Journal of Apicultural Research*, 1962, 1(1), pp. 41–51.

32 J. Gibbs et al., 'Phylogeny of halictine bees supports a shared origin of eusociality for *Halictus* and *Lasioglossum* (Apoidea: Anthophila: Halictidae)', Molecular Phylogenetics and Evolution, 2012, 65(3), pp. 926–939.

33 B. N. Danforth et al., 2019.

34 B. N. Danforth, L. Conway and S. Ji, 'Phylogeny of eusocial *Lasioglossum* reveals multiple losses of eusociality within a primitively eusocial clade of bees (Hymenoptera: Halictidae)', *Systematic Biology*, 2003, 52(1), pp. 23–36.

35 B. N. Danforth et al., 2019.

36 H. E. Evans and A. W. Hook, 'Communal nesting in the digger wasp *Cerceris australis* (Hymenoptera: Sphecidae)', *Australian Journal of Zoology*, 1982, 30(4), pp. 557–568.

37 J. Alcock, 'Communal nesting in an Australian solitary wasp, *Cerceris antipodes* Smith (Hymenoptera, Sphecidae)', *Australian Journal of Entomology*, 1980, 19(3), pp. 223–228.

38 C. Polidori, 'Interactions between the social digger wasp, *Cerceris rubida*, and its brood parasitic flies at a Mediterranean nest aggregation', *Journal of Insect Behavior*, 2017, 30(1), pp. 86–102.

39 R. W. Matthews, 'Evolution of social behavior in sphecid wasps', in *The Social Biology of Wasps*, 2018, Cornell University Press, pp. 570–602.

40 E. Lucas et al., 'Social and genetic structure in colonies of the social wasp *Microstigmus nigrophthalmus*', *Insectes Sociaux*, 2011, 58(1), pp. 107–114.

41 P. K. Piekarski et al., 'Phylogenomic evidence overturns current conceptions of social evolution in wasps (Vespidae)', *Molecular Biology and Evolution*, 2018, 35(9), pp. 2097–2109.

42 E. A. Sugden and R. L. McAllen, 'Observations on foraging, population and nest biology of the Mexican honey wasp, *Brachygastra mellifica* (Say) in Texas [Vespidae: Polybiinae]', *Journal of the Kansas Entomological Society*, 1994, 67(2), pp. 141–155.

43 J. H. Hunt and A. L. Toth, 2017.

44 G. P. Elliott et al., 'Declines in common, widespread native birds in a mature temperate forest', *Biological Conservation*, 2010, 143(9), pp. 2119–2126.

45 K. Harano, C. Maia-Silva and M. Hrncir, 'Why do stingless bees (*Melipona subnitida*) leave their nest with resin loads?', *Insectes Sociaux*, 2020, 67(1), pp. 195–200.

46 C. Grüter et al., 'Warfare in stingless bees', *Insectes Sociaux*, 2016, 63(2), pp. 223–236.

47 C. Grüter et al., 'Repeated evolution of soldier sub-castes suggests parasitism drives social complexity in stingless bees', *Nature Communications*, 2017, 8(1), pp. 1–8.

48 B. Danforth, 2007.

49 M. A. Nowak, C. E. Tarnita and E. O. Wilson, 'The evolution of eusociality', Nature, 2010, 466(7310), pp. 1057–1062.

11 Superorganism

1 B. Hölldobler and E. O. Wilson, *The Ants*, 1990, Harvard University Press.

2 E. Marais, *The Soul of the White Ant*, 1937. First published as *Die Siel van die Mier* in 1925, Dodd, Mead/ Translated by Winifred de Kok.

3 S. P. Nicholls and A. Vendl, 'Nature Tech', 2003, *ORF*. Episode 3.

4 M. Lüscher, 'Air-conditioned termite nests', *Scientific American*, 1961, 205(1), pp. 138–147.

5 W. M. Wheeler, 'The ant-colony as an organism', *Journal of Morphology*, 1911, 22(2), pp. 307–325.

6 J. S. Turner, *The Extended Organism: The Physiology of Animal-built Structures*, 2000, Harvard University Press.

7 J. S. Turner, 'Architecture and morphogenesis in the mound of *Macrotermes michaelseni* (Sjöstedt) (Isoptera: Termitidae, Macrotermitinae) in northern Namibia', *Cimbebasia*, 2000, 16, pp. 143–175.

8 K. Singh et al., 'The architectural design of smart ventilation and drainage systems in termite nests', *Science Advances*, 2019, 5(3), p. eaat8520.

9 N. Zachariah et al., 'Bi-layered architecture facilitates high strength and ventilation in nest mounds of fungus-farming termites', *Scientific Reports*, 2020, 10(1), p. 13157.

10 J. S. Turner, 'On the mound of *Macrotermes michaelseni* as an organ of respiratory gas exchange', *Physiological and Biochemical Zoology*, 2001, 74(6), pp. 798–822.

11 S. A. Ocko et al., 'Solar-powered ventilation of African termite mounds', *Journal of Experimental Biology*, 2017, 220(Pt 18), pp. 3260–3269.

12 J. Korb, 'Termite mound architecture, from function to construction', in *Biology of Termites: A Modern Synthesis*, published by Springer (Dordrecht), 2010, pp. 349–373.

13 M. Ozeki et al., 'Phylogeography of an Australian termite, *Amitermes laurensis* (Isoptera, Termitidae), with special reference to the variety of mound shapes', *Molecular Phylogenetics and Evolution*, 2007, 42(1), pp. 236–247.

14 J. S. Turner, 2000.

15 W. R. Tschinkel, *Ant Architecture: The Wonder, Beauty, and Science of Underground Nests*, 2021, Princeton University Press.

16 W. R. Tschinkel, 'The architecture of subterranean ant nests: beauty and mystery underfoot', *Journal of Bioeconomics*, 2015, 17(3), pp. 271–291.

17 S. Foitzik and O. Fritsche, *Empire of Ants: The Hidden Worlds and Extraordinary Lives of Earth's Tiny Conquerors*, 2021, The Experiment.

18 I. J. S. Moreira, M. F. Santos and M. S. Madureira, 'Why do *Acromyrmex* nests have thatched entrance structures? Evidence for use as a visual homing cue', *Insectes Sociaux*, 2018, 66(1), pp. 165–170.

19 R. Crozier, L. Jermiin and M. Chiotis, 'Molecular evidence for a Jurassic origin of ants', *Naturwissenschaften*, 1997, 84(1), pp. 22–23.

20 E. O. Wilson, F. M. Carpenter and W. L. Brown, 'The first Mesozoic ants', *Science*, 1967, 157(3792), pp. 1038–1040.

21 P. S. Ward, 'The phylogeny and evolution of ants', *Annual Review of Ecology, Evolution, and Systematics*, 2014, 45, pp. 23–43.

22 E. O. Wilson and B. Hölldobler, 'The rise of the ants: a phylogenetic and ecological explanation', *Proceedings of the National Academy of Sciences*, 2005, 102(21), pp. 7411–7414.

23 S. Foitzik and O. Fritsche, 2021.

24 M. J. Way, 'Mutualism between ants and honeydew-producing Homoptera', *Annual Review of Entomology*, 1963, 8(1), pp. 307–344.

25 U. Maschwitz and H. Hänel, 'The migrating herdsman *Dolichoderus* (*Diabolus*) *cuspidatus*: an ant with a novel mode of life', *Behavioral Ecology and Sociobiology*, 1985, 17(2), pp. 171–184.

26 A. Salazar et al., 'Aggressive mimicry coexists with mutualism in an aphid', *Proceedings of the National Academy of Sciences*, 2015, 112(4), pp. 1101–1106.

27 B. Hölldobler and E. O. Wilson, *The Leafcutter Ants: Civilization by Instinct*, 2010, W.W. Norton & Company.

28 T. R. Schultz and S. G. Brady, 'Major evolutionary transitions in ant agriculture', *Proceedings of the National Academy of Sciences*, 2008, 105(14), pp. 5435–5440.

29 C. R. Currie et al., 'Fungus-growing ants use antibiotic-producing bacteria to control garden parasites', *Nature*, 1999, 398(6729), pp. 701–704.

30 D. Ortius-Lechner et al., 'Metapleural gland secretion of the leaf-cutter ant *Acromyrmex octospinosus*: new compounds and their functional significance', *Journal of Chemical Ecology*, 2000, 26(7), pp. 1667–1683.

31 S. Um et al., 'The fungus-growing termite *Macrotermes natalensis* harbors bacillaene-producing *Bacillus* sp. that inhibit potentially antagonistic fungi', *Scientific Reports*, 2013, 3(1), pp. 1–7.

32 S. Otaniet et al., 'Disease-free monoculture farming by fungus-growing termites', *Scientific Reports*, 2019, 9(1), pp. 1–10.

33 K. H. Bodawatta, M. Poulsen and N. Bos, 'Foraging *Macrotermes natalensis* fungus-growing termites avoid a mycopathogen but not an entomopathogen', *Insects*, 2019, 10(7), p. 185.

34 D. Inward, G. Beccaloni and P. Eggleton, 'Death of an order: a comprehensive molecular phylogenetic study confirms that termites are eusocial cockroaches', *Biology Letters*, 2007, 3(3), pp. 331–335.

35 B. Stay and A. C. Coop, '"Milk" secretion for embryogenesis in a viviparous cockroach', *Tissue and Cell*, 1974, 6(4), pp. 669–693.

36 S. Banerjee et al., 'Structure of a heterogeneous, glycosylated, lipid-bound, in vivo-grown protein crystal at atomic resolution from the viviparous cockroach *Diploptera punctata*', *IUCrJ*, 2016, 3(4), pp. 282–293.

37 C. A. Nalepa and W. J. Bell, 'Postovulation parental investment and parental care in cockroaches', in *The Evolution of Social Behavior in Insects and Arachnids*, Cambridge University Press, ed. J. C. Choe & B. J. Crespi, 1997, pp. 26–51.

38 T. Matsumoto, 'Familial association, nymphal development and population density in the Australian giant burrowing cockroach, *Macropanesthia rhinoceros* (Blattaria: Blaberidae)', *Zoological Science*, 1992, 9(4), pp. 835–842.

39 C. A. Nalepa, 'Altricial Development in Wood-Feeding Cockroaches: The Key Antecedent of Termite Eusociality', in *Biology of Termites: A Modern Synthesis*, ed. D. Bignell & Y. Roisin, 2010, pp. 69–95.

40 K. Maekawa, T. Matsumoto and C. A. Nalepa, 'Social biology of the wood-feeding cockroach genus *Salganea* (Dictyoptera, Blaberidae, Panesthiinae): did ovoviviparity prevent the evolution of eusociality in the lineage?', *Insectes Sociaux*, 2008, 55(2), pp. 107–114.

41 C. A. Nalepa, 2010.

42 P. Okullo and S. R. Moe, 'Termite activity, not grazing, is the main determinant of spatial variation in savanna herbaceous vegetation', *Journal of Ecology*, 2012, 100(1), pp. 232–241.

43 S. J. Martin et al., 'A vast 4,000-year-old spatial pattern of termite mounds', *Current Biology*, 2018, 28(22), pp. R1292–R1293.

44 R. R. Funch, 'Termite mounds as dominant land forms in semiarid northeastern Brazil', *Journal of Arid Environments*, 2015, 122, pp. 27–29.

45 T. Batista et al., 'Association of giant ant *Dinoponera quadriceps* nests with termite mounds and landscape variables in a Caatinga dry forest, Brazil', *Insectes Sociaux*, 2021, 68(1), pp. 41–47.

46 J. Moore and M. Picker, '*Heuweltjies* (earth mounds) in the Clanwilliam district, Cape Province, South Africa: 4,000-year-old termite nests', *Oecologia*, 1991, 86(3), pp. 424–432.

47 M. D. Cramer, S. N. Innes and J. J. Midgley, 'Hard evidence that *heuweltjie* earth mounds are relictual features produced by differential erosion', *Palaeogeography, Palaeoclimatology, Palaeoecology*, 2012, 350, pp. 189–197.

48 N. A. Van Der Cingel, *An Atlas of Orchid Pollination: America, Africa, Asia and Australia*, 2001, A. A. Balkema Publishers.

49 J. Heinze, K. Kellner and J. Seal, 'Sociality in ants', in *Comparative Social Evolution*, D. R. Rubenstein and P. Abbot (eds), 2017, Cambridge University Press, Cambridge, pp. 21–49.

50 E. O. Wilson and B. Hölldobler, 2005.

51 E. Fittkau and H. Klinge, 'On biomass and trophic structure of the central Amazonian rain forest ecosystem', *Biotropica*, 1973, pp. 2–14.

52 R. H. Crozier et al., 'A masterpiece of evolution – *Oecophylla* weaver ants (Hymenoptera: Formicidae)', *Myrmecological News*, 2010, 13(5), pp. 57–71.

53 B. Hölldobler, 'Territorial behavior in the green tree ant (*Oecophylla smaragdina*), *Biotropica*, 1983, pp. 241–250.

54 S. G. Brady, 'Evolution of the army ant syndrome: the origin and long-term evolutionary stasis of a complex of behavioral and reproductive adaptations', *Proceedings of the National Academy of Sciences*, 2003, 100(11), pp. 6575–6579.

55 K. Zimmer and M. Isler, 'Family Thamnophilidae (Typical Antbirds)', in *Handbook of the Birds of the World*, J. del Hoyo, A. Elliot and D. Christie (eds), 2003, Lynx Ediciones, Barcelona, pp. 459–531.

56 C. W. Rettenmeyer et al., 'The largest animal association centered on one species: the army ant *Eciton burchellii* and its more than 300 associates, *Insectes Sociaux*, 2010, 58(3), pp. 281–292.

57 V. Witte et al., '*Allopeas myrmekophilos* (Gastropoda, Pulmonata), the first myrmecophilous mollusc living in colonies of the ponerine army ant *Leptogenys distinguenda* (Formicidae, Ponerinae)', *Insectes Sociaux*, 2002, 49(4), pp. 301–305.

58 M. Brandt, E. Van Wilgenburg and N. Tsutsui, 'Global-scale analyses of chemical ecology and population genetics in the invasive Argentine ant', *Molecular Ecology*, 2009, 18(5), pp. 997–1005.

59 R. Jarski, *Words from the Wise: Over 6,000 of the Smartest Things Ever Said*, 2007, Skyhorse Publishing Inc.

Bibliography

UK INSECTS

Butterflies (British Wildlife Series Vol. 10). Martin Warren. Bloomsbury Publishing 2021

Ants (British Wildlife Series Vol. 11). Richard Jones. Bloomsbury Publishing 2022

The Royal Entomological Society Book of British Insects. Peter C. Barnard. Wiley-Blackwell. 2011

Britain's Hoverflies: An Introduction to the Hoverflies of Britain (WildGuides series). Stuart Ball and Roger Morris. Princeton University Press 2013

A Comprehensive Guide to Insects of Britain and Ireland (2nd Edition). Paul D. Brock. Pisces Publications. 2019

Beetles (New Naturalist Series, Vol. 136). Richard Jones. HarperCollins. 2018

Field Guide to the Dragonflies of Britain and Europe. Klaas-Douwe B. Dijkstra. British Wildlife Publishing. 2006

NORTH AMERICAN INSECTS

Bumble Bees of North America: An Identification Guide. (Princeton Field Guide Series Vol. 87). Paul W. Williams, Robbin W. Thorp, Leif Richardson and Sheila Colla. Princeton University Press. 2014

A Swift Guide to Butterflies of North America. Jeffrey Glassberg. Princeton University Press. 2017

The Bees in Your Backyard: A Guide to North America's Bees. Joseph S. Wilson, Olivia J. Messinger Carril. Princeton University Press. 2015

Insects – Their Natural History and Diversity: With a Photographic Guide to Insects of Eastern North America. Stephen A. Marshall. Firefly Books. 2017

Garden Insects of North America: The Ultimate Guide to Backyard Bugs. Whitney Cranshaw, David J. Shetlar. Princeton University Press. 2017

Kaufman Field Guide to Insects of North America (Kaufman Field Guide Series). Eric R. Easton and Kenn Kaufman. Houghton Mifflin Harcourt. 2007

American Insects: A Handbook of the Insects of America North of Mexico. Ross H. Arnett. CRC Press. 2000

Monarchs in a Changing World: Biology and Conservation of an Iconic Butterfly. Edited by Karen S. Oberhauser, Kelly R. Nail and Sonia M. Altizer. Comstock Publishing Associates. 2015

Locust: The Devastating Rise and Mysterious Disappearance of the Insect That Shaped the American Frontier. Jeffery A. Lockwood. Basic Books. 2004

GENERAL READING

encyclopedia of Insects. Edited by Vincent H. Resh and Ring T. Cardé. Academic Press. 2009

The Solitary Bees: Biology, Evolution, Conservation. Bryan N. Danforth. Princeton University Press. 2019

A World of Insects: The Harvard University Press Reader. Edited by Ring T. Cardé and Vincent H. Resh. Harvard University Press. 2012

Wasps: The Astonishing Diversity of a Misunderstood Insect. Eric R. Eaton. Princeton University Press. 2021

Butterflies of the World. Adrian Hoskins. New Holland Publishers. 2015

The Mind of a Bee. Lars Chittka. Princeton University Press. 2022

The Other Insect Societies. James T. Costa. Harvard University Press. 2006

For Love of Insects. Thomas Eisner. Belknap Harvard. 2003

The Superorganism: The Beauty, Elegance, and Strangeness of Insect Societies. Bert Hölldobler and Edward O. Wilson. W. W. Norton & Co. 2009

The Inside Out of Flies. Erica McAlister. Natural History Museum, London. 2020

Dragonflies of the World. Jill Silsby. Smithsonian Institution Press. 2001

Journey to the Ants: A Story of Scientific Exploration. Bert Hölldobler and Edward O. Wilson. Harvard University Press. 1994

Bugs in the System: Insects and their Impact on Human Affairs. May Berenbaum. Perseus Books. 1995

Life on a Little-known Planet. Howard Ensign Evans. André Deutsch. 1970

Life in the Undergrowth. David Attenborough. BBC Books. 2005

INSECT CONSERVATION

pollinators & Pollination: Nature and Society. Jeff Ollerton. Pelagic Publishing. 2020

Silent Earth: Averting the Insect Apocalypse. Dave Goulson. Vintage. 2021

The Insect Crisis: The Fall of the Tiny Empires that Run the World. Oliver Milman. Atlantic Books. 2022

Pollinators in Crisis: How We Can All Give Them a Helping Hand. Michael Fogden. NatureBureau. 2022

Index

Q

R

S

T

U

V